T0331266

GREEN COMPOSITES *from* NATURAL RESOURCES

GREEN COMPOSITES *from* NATURAL RESOURCES

Edited by Vijay Kumar Thakur

CRC Press
Taylor & Francis Group
Boca Raton London New York

CRC Press is an imprint of the
Taylor & Francis Group, an **informa** business

CRC Press
Taylor & Francis Group
6000 Broken Sound Parkway NW, Suite 300
Boca Raton, FL 33487-2742

First issued in paperback 2017

© 2014 by Taylor & Francis Group, LLC
CRC Press is an imprint of Taylor & Francis Group, an Informa business

No claim to original U.S. Government works
Version Date: 20131021

ISBN 13: 978-1-138-07729-4 (pbk)
ISBN 13: 978-1-4665-7069-6 (hbk)

This book contains information obtained from authentic and highly regarded sources. Reasonable efforts have been made to publish reliable data and information, but the author and publisher cannot assume responsibility for the validity of all materials or the consequences of their use. The authors and publishers have attempted to trace the copyright holders of all material reproduced in this publication and apologize to copyright holders if permission to publish in this form has not been obtained. If any copyright material has not been acknowledged please write and let us know so we may rectify in any future reprint.

Except as permitted under U.S. Copyright Law, no part of this book may be reprinted, reproduced, transmitted, or utilized in any form by any electronic, mechanical, or other means, now known or hereafter invented, including photocopying, microfilming, and recording, or in any information storage or retrieval system, without written permission from the publishers.

For permission to photocopy or use material electronically from this work, please access www.copyright.com (http://www.copyright.com/) or contact the Copyright Clearance Center, Inc. (CCC), 222 Rosewood Drive, Danvers, MA 01923, 978-750-8400. CCC is a not-for-profit organization that provides licenses and registration for a variety of users. For organizations that have been granted a photocopy license by the CCC, a separate system of payment has been arranged.

Trademark Notice: Product or corporate names may be trademarks or registered trademarks, and are used only for identification and explanation without intent to infringe.

Visit the Taylor & Francis Web site at
http://www.taylorandfrancis.com

and the CRC Press Web site at
http://www.crcpress.com

Contents

Preface

Global warming, rising environmental awareness, waste management issues, dwindling fossil resources, and rising oil prices are some of the reasons why green materials obtained from renewable resources are increasingly being promoted for sustainable development. Various kinds of renewable green materials, such as starchy and cellulosic polymers including natural fibers, vegetable oils, wood bark, cotton, wool, and silk, have been used for thousands of years for food, furniture, and clothing. However, it is only in the past two decades they have experienced a renaissance as one of the most feasible alternatives to synthetic polymers for a variety of industrial applications, such as building, construction, automotive, packaging, films, and paper coating, as well as in biomedical applications. The prime disadvantages of synthetic polymers, such as release of toxic gases and vapors as a result of incineration and difficulty in their disposal, have led to intense investigations in the field of new green polymeric materials with a particular interest in the use of biopolymers obtained from renewable resources for green composite applications. This book is the outcome of contributions by world-renowned experts in the field of green polymer materials from different disciplines, and various backgrounds and expertise. The material enclosed in the book gives a true reflection of the vast area of research in green composites, which is also applicable to a number of industries.

This book contains precisely referenced chapters, emphasizing green composite materials from different natural resources with eco-friendly advantages that can be utilized as alternatives to synthetic polymers through detailed reviews of various lignocellulosic reinforcing materials and their property control using different approaches. Each chapter in this book covers a significant amount of basic concepts and their development until its current status of development. The book aims at explaining basic characteristics of green composite materials, their synthesis, and applications for these renewable materials obtained from different natural resources that present future directions in a number of industrial applications including the automotive industry. The book attempts to present emerging low-cost and eco-friendly green composite materials. I hope this book will contribute significantly to the basic knowledge of students and researchers all around the globe working in the field of green materials. I thank all the contributors for their innovative contributions and Laurie Schlags (project coordinator) along with Allison Shatkin (senior editor) for their invaluable help in the editing process.

Vijay Kumar Thakur
Iowa State University

Editor

Vijay Kumar Thakur, PhD, graduated with a BSc in chemistry (nonmedical), physics (nonmedical), and mathematics (nonmedical); BEd; and MSc in organic chemistry from Himachal Pradesh University, Shimla, India, in 2006. He then moved to the National Institute of Technology, Hamirpur, India, where he obtained his doctoral degree in polymer chemistry from the Chemistry Department in 2009. After a brief stay in the Department of Chemical and Materials Engineering at Lunghwa University of Science and Technology, Taiwan, he joined Temasek Laboratories at Nanyang Technological University, Singapore, as a research scientist in October 2009 and worked there until 2012. He has a general research interest in the synthesis of polymers, nanomaterials, nanocomposites, biocomposites, graft copolymers, high-performance capacitors, and electrochromic materials. He has coauthored five books, 20 book chapters, one U.S. patent, and has published more than 60 research papers in reputed international peer-reviewed journals, including *Advanced Materials*, *Journal of Materials Chemistry*, *Polymer Chemistry*, and *RSC Advances*, along with 40 publications in proceedings of international/national conferences. He has been included in the *Marquis Who's Who in the World* in the field of science and engineering for the year 2011. He is a reviewer for more than 37 international journals and currently serves as a member on the steering committee of the WAP Conference Series: Engineering and Technology Frontier. He also serves on the editorial board of 22 international journals including *Advanced Chemistry Letters*, *Lignocelluloses*, *Drug Inventions Today* (Elsevier), *International Journal of Energy Engineering*, and *Journal of Textile Science & Engineering* (USA) being published in the fields of natural/synthetic polymers, composites, energy storage materials, and nanomaterials.

Contributors

Eldho Abraham
Department of Chemistry
Bishop Moore College
and
Department of Chemistry
C.M.S. College
Kerala, India

Jou-Hyeon Ahn
Department of Chemical and Biological
    Engineering and Research Institute
    for Green Energy Convergence
    Technology
Gyeongsang National University
Jinju, Republic of Korea

Abdullah Alhuthali
Department of Imaging and Applied
    Physics
Curtin University
Perth, Australia

Pramendra K. Bajpai
Department of Mechanical and
    Industrial Engineering
Indian Institute of Technology
Roorkee, India

Jean-Charles Benezet
Ecole des Mines d'Alès
Materials Research Centre
Alès, France

Anne Bergeret
Ecole des Mines d'Alès
Materials Research Centre
Alès, France

Marcia R.M. Chaves
Centre of Applied Sciences
University of Sagrado Coração
São Paulo, Brazil

Bibin M. Cherian
Department of Rural Engineering
São Paulo State University
São Paulo, Brazil

Deepa B.
Department of Chemistry
Bishop Moore College
and
Department of Chemistry
C.M.S. College
Kerala, India

Hom N. Dhakal
Advance Polymer and Composites
    Research Group, School of
    Engineering
University of Portsmouth
Portsmouth, United Kingdom

Anna Dilfi
Department of Electronics and
    Communication Engineering
SNS College of Engineering
Tamil Nadu, India

Nilmini P.J. Dissanayake
College of Engineering, Mathematics
    and Physical Sciences
University of Exeter
Exeter, United Kingdom

Mizi Fan
Civil Engineering Material Research
   Laboratory
Brunel University
Uxbridge, United Kingdom

Raju K. Gupta
Department of Chemical Engineering
Indian Institute of Technology
Kanpur, India

Michael R. Kessler
Department of Materials Science and
   Engineering
Iowa State University
Ames, Iowa

Anastasia Koutsomitopoulou
Composite Materials Group
University of Patras
Patras, Greece

Alan K.T. Lau
Department of Mechanical Engineering
The Hong Kong Polytechnic University
Hung Hom, Hong Kong

Alcides L. Leão
Department of Rural Engineering
São Paulo State University
São Paulo, Brazil

It Meng Low
Department of Imaging and Applied
   Physics
Curtin University
Perth, Australia

George C. Papanicolaou
Composite Materials Group
University of Patras
Patras, Greece

Bessy M. Philip
Department of Chemistry
Bishop Moore College
Kerala, India

Laly A. Pothan
Department of Chemistry
Bishop Moore College
Kerala, India

Raghavan Prasanth
School of Materials Science and
   Engineering, and Energy Research
   Institute
Nanyang Technological University
Singapore

and

Department of Chemical and Biological
   Engineering and Research Institute
   for Green Energy Convergence
   Technology
Gyeongsang National University
Jinju, Republic of Korea

Mohini Sain
Centre for Biocomposites and
   Biomaterials Processing
University of Toronto
Toronto, Ontario, Canada

Ravi Shankar
South Dakota School of Mines and
   Technology
Rapid City, South Dakota

Robert A. Shanks
School of Applied Sciences
RMIT University
Melbourne, Victoria, Australia

Inderdeep Singh
Department of Mechanical and
   Industrial Engineering
Indian Institute of Technology
Roorkee, India

Sivoney F. de Souza
Centre for Science and Humanities
Universidade Federal do ABC
São Paulo, Brazil

John Summerscales
Advanced Composites Manufacturing
  Centre, School of Marine Science
  and Engineering
University of Plymouth
Plymouth, England

Marcelo Telascrea
Department of Rural Engineering
São Paulo State University
São Paulo, Brazil

Manju K. Thakur
Division of Chemistry
Government Degree College Sarkaghat
Himachal Pradesh University
Shimla, India

Vijay K. Thakur
Temasek Laboratories
Nanyang Technological University
Singapore

and

Department of Materials Science and
  Engineering
Iowa State University
Ames, Iowa

Sabu Thomas
School of Chemical Sciences
Mahatma Gandhi University
Kerala, India

Mahendra Thunga
Department of Materials Science and
  Engineering
Iowa State University
Ames, Iowa

Thi-Phuong-Thao Tran
Ecole des Mines d'Alès
Materials Research Centre
Alès, France

Hao Wang
Centre of Excellence in
  Engineered Fibre
  Composites
University of Southern
  Queensland
Toowoomba, Australia

Bartosz T. Weclawski
Civil Engineering Material Research
  Laboratory
Brunel University
Uxbridge, United Kingdom

Zhong Y. Zhang
Advance Polymer and Composites
  Research Group, School of
  Engineering
University of Portsmouth
Portsmouth, United Kingdom

1 Green Composites
An Introduction

Vijay K. Thakur, Manju K. Thakur, Raju K. Gupta, Raghavan Prasanth, and Michael R. Kessler

CONTENTS

1.1 INTRODUCTION

Global environmental concerns, such as rising sea levels, rising average global temperatures, decreasing polar ice caps, and rapidly depleting petroleum resources, have intensified pressure on humans and industries. How to respect the environment and improve living conditions for the benefit of all living organisms are key global issues. These concerns and an increased awareness of renewable "green" materials have initiated efforts in many industries to mitigate their impact on the environment. With an emphasis on reduction of greenhouse gas emissions and carbon footprint, there is a demand for sustainably produced green materials with improved performance (Luo and Netravali 1999).

Sustainable development has become a major issue in recent years, and the foreseeable depletion of oil-based resources will require the use of biopolymer materials from renewable resources (Bledzki and Gassan 1999; Singha and Thakur 2012; Thakur et al. 2011). The generally accepted definition of sustainable development is "development that meets the needs of the present without compromising the ability of future generations to meet their own need" (Brundtland Commission 1987). In a broader approach, sustainable development is defined to be comprised of three components: society, environment, and economy. Biopolymeric materials obtained from different natural resources offer the potential to aid the transition toward sustainable and green development (Zain et al. 2011). One of the most significant advantages of some biopolymeric, green materials is that they easily decompose into environmentally benign components, such as carbon dioxide, water, and humus-like matter (Klemm et al. 2005; Scott 2000; Wambua et al. 2003). Biodegradation of biobased, biodegradable polymers can be achieved by exposing them to environmental

1

influences (such as UV, oxygen, water, and heat) or microorganisms that will metab-
olize the polymer and produce an inert, humus-like material that is not harmful to
the environment and can be easily mixed with natural soil.

The effective use of eco-friendly, green materials in a variety of applications, with
a particular focus on energy-efficient, cost-effective materials, is one of the daunt-
ing challenges of the twenty-first century. This brings natural polymeric materials,
such as cellulosic fibers, vegetable oils, wood bark, cotton, wool, and silk, into focus
as feasible alternatives to traditional synthetic polymeric materials for a variety of
industrial applications, such as building construction, automobiles, packaging, films,
and paper coatings (Klyosov 2008). Archeological artifacts suggest that human
beings used natural fibrous materials in fabrics several thousand years ago (Nikolaos
2011). Natural fibers have been used in ropes, lines, and other one-dimensional products. Some other applications of natural fibers in earlier times included suspension
bridges for on-foot passage of rivers and rigging for naval ships.

Biopolymers have already made inroads in biomedical applications, such as
surgical sutures, implants, and controlled drug-delivery devices (Averous 2004).
However, most commercial polymers used in everyday applications are still prepared
from nonbiodegradable/nonrenewable constituents (Hon 1996). These polymers are
generally derived from petroleum, by definition a nonsustainable resource. We are
currently consuming petroleum at an "unsustainable" rate: nearly 100,000 times
faster than nature can create it.

These concerns have led to intense investigations in the field of environmentally friendly, green polymeric materials with a particular interest in the use of
biopolymers for green composite applications (Bledzki et al. 2010). Humans have
been using biopolymeric, natural, renewable materials, such as wood, amber, silk,
natural rubber, celluloid, shellack, for many centuries (Cipriani et al. 2010; Davim
2011). Naturally occurring biopolymeric materials were used beginning in early
civilizations and have been increasingly utilized since the industrial revolution
in the late nineteenth century to meet specific needs. Today, the demand for the
effective utilization of biopolymeric materials is increasing significantly. In particular, the idea of using biopolymer-based materials as one of the components in
advanced green composite materials is gaining more and more interest both in
academia and in industry (Hamad 2002; Thakur et al. 2011). This chapter covers
a brief introduction of green composites/natural fibers and does not cover traditional synthetic composites/animal fibers.

1.2   OVERVIEW OF COMPOSITES

Composites are made of individual materials referred to as constituent materials. The
properties of composites are typically determined by the combination of the respective properties of the constituent materials (Klyosov 2008). The term "composite"
comes from the Latin word *compositus*, stemming from the root word *componere*,
which means "to bring together." Different researchers have defined composite materials in different ways (Singha and Thakur 2012). In general, composites are defined
as engineering materials made from two or more constituents with significantly different physical or chemical properties. Composites are typically stronger than the

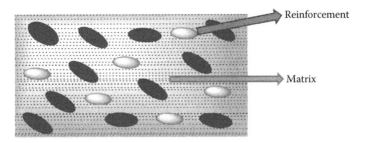

FIGURE 1.1   General schematic of a composite.

individual materials and may exhibit entirely different properties than the individual constituents (Thakur et al. 2011). According to American Society for Materials Handbook, composites can be defined as "a macroscopic combination of two or more distinct materials having a recognizable interface between them" (Hazizan and Cantwell 2003). A composite is made of at least two materials, where one material essentially acts as the binding material (the matrix), while the other acts as reinforcement (Chapter 10 provides a detailed description of composites, classification of green composites, and factors affecting composite properties). Figure 1.1 shows the general schematic of a composite.

The matrix material surrounds the reinforcement; it is also referred to as the "continuous phase." The type and level of reinforcement determines the mechanical and physical properties of the composite; it is referred to as the "discontinuous phase." The discontinuous phase is generally harder and exhibits mechanical properties superior to those of the continuous phases. In a given composite material, both the matrix and the reinforcement are distinguishable. To a considerable extent, each of the materials maintains its distinctive characteristics, which enhance the properties of the resulting composite material. In particular, mechanical and physicochemical properties of the composite are superior to those of the matrix material (Thakur and Singha 2012). Composites can be tailored to exhibit a wide range of desired properties. Many composite materials offer advantageous properties in terms of high fatigue strength, low weight, corrosion resistance, and higher specific properties such as tensile, flexural, and impact strengths (Oksman and Sain 2008).

1.2.1   Potential Merits of Composites

The advantages of composite materials have led to their widespread use in a variety of industries. Some of the important features and benefits that some composites provide are as follows:

- Light weight: composites are significantly lighter, especially in comparison to materials such as concrete and metals
- Increased design flexibility
- High strength
- Better damage tolerance

- Increased impact resistance
- Increased chemical resistances
- Increased fracture toughness
- Superior corrosion resistance
- Low coefficient of thermal expansion
- Superior fatigue resistance
- Potentially lower component costs

Composites are generally classified by the type of matrix, such as polymer, metal, or ceramic, or by the type of reinforcement, such as fibers, particulate, flake, or whiskers. Figure 1.2 shows the classification of different types of composites.

Each type of composite material is designed to meet the requirements of specific applications. In this book, the term composites will always refer to *polymer composites*.

Although we find polymers in a wide range of applications, from low-cost/high-volume consumer products to automotive engineering components under the hood to highly complex medical devices, in some applications their stiffness and strength cannot compete with metals. Polymer composites were developed to meet the need for light weight, high stiffness materials that exhibit additional functionalities, such as wear resistance, electrical properties, and thermal stability (Oksman and Sain 2008). Polymer composites consist of two or more distinct phases, including a polymer matrix (continuous phase) and fibrous or particulate reinforcing material (dispersed phase). Based on their reinforcements, polymer composites can be categorized into three main classes: particulate composites, continuous fiber composites, and discontinuous fiber composites.

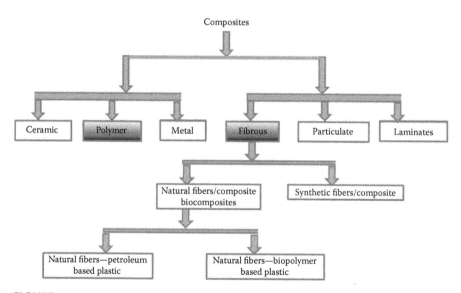

FIGURE 1.2 Classification of different types of composites.

1.3 GREEN COMPOSITES

Increasing environmental pollution, concerns over petroleum supplies, high crude oil prices, and lack of resources have resulted in increased research in materials that are friendly to our health and environment (Luo and Netravali 1999). A number of renewable biopolymer materials, including cellulose, proteins, starch, vegetable oils, and sugar, have been investigated as potential components of green polymer composites (Luo and Netravali 1999).

Green composites are a specific class of composites, where at least one of the components (such as the matrix or the reinforcement) is obtained from natural resources (Netravali and Chabba 2003). The terms green composites, biocomposites, and eco-composites all broadly refer to the same class of materials (Nikolaos 2011; Oksman and Sain 2008). Green composites, especially natural fiber–reinforced composites, have been used by humans since the beginning of human civilization (Richard 1994; Thakur et al. 2011). They were used as a source of energy, and as a material to make shelters, clothes, tools, and more. In ancient Egypt, 3000 years ago, people used straw as the reinforcing component for the mud-based wall materials in houses. These green composites were produced in simple shapes by layering the different structural elements to create the desired design. They made bricks of mud with straw as reinforcement and used these bricks to build walls.

Biopolymer-based green composites can offer green, renewable alternatives to the most widely used petroleum-based polymers with equal or better properties, enabling new applications (Azizi et al. 2005; Pandey et al. 2013; Thakur and Singha 2012). The interest in green composites that contain biopolymers as one of the essential components has increased in recent years due to the renewable and biodegradable nature of biopolymers (Markarian 2005). Biopolymeric composites, such as cellulosic fiber–reinforced polymer composites, are gaining greater acceptance in a number of applications, particularly in structural and packaging applications (Kabir et al. 2012). For a material to be effectively used in packaging applications, the basic raw materials should be renewable and the end products should be compostable to reduce the use of fossil fuels and limit the cost and environmental impact of waste treatment (Rowell 2012). The industrial-scale production processes used to prepare different kinds of packaging materials using biopolymers should be efficient, economically competitive, and environmentally friendly (Dufresne et al. 1999).

Natural fiber–reinforced polymer composites represent an emerging area in polymer science (Ouajai and Shanks 2009a). These composites are both environmentally friendly and sustainable. After decades of high-tech development of artificial fibers, such as aramid, carbon, and glass, natural fibers, such as wood fibers, sisal, pine needles, kenaf, flax, jute, hemp, and others, have attracted renewed interest (Rowell 2012; Thakur and Singha 2012). These natural cellulosic fibers have shown great potential as substitutes for synthetic fibers, in particular glass fibers in composites that are extensively used in the automotive and construction industries. The advantages of natural fibers over synthetic fibers are their low cost, eco-friendliness, low density, low abrasion, acceptable specific strength properties, ease of separation, carbon dioxide sequestration, and biodegradability, to name a few (Thakur et al. 2011; Thakur and Singha 2012).

1.3.1   CLASSIFICATION OF GREEN COMPOSITES

Intense research efforts are currently focused on developing "green" composites by combining (natural/bio) fibers with suitable polymer matrices (Schneider and Karmaker 1996; Singha and Thakur 2012). A variety of natural and synthetic polymer matrix resins are available for green composites, including polythene, polypropylene, polystyrene, polyester, epoxy, phenolic resins, starch, and polylactic acid (Garlotta 2001; Ouajai and Shanks 2009b). Depending on the type of reinforcement and polymer matrix, green composites can be divided into three main types:

1. Totally renewable composites, in which both the matrix and reinforcement are from renewable resources
2. Partly renewable composites, in which the matrix is obtained from renewable resources and reinforced with a synthetic material
3. Partly renewable composites, in which a synthetic matrix is reinforced with natural biopolymers

Although the number of green composites made from renewable resources has been increasing, spurred by the growing seriousness of environmental problems (Uma Devi et al. 2010), the processing temperature remains a limiting factor in the choice of a suitable polymer matrix for green composites. Polymer matrices are generally classified into thermosetting, thermoplastic, or biodegradable (Bledzki and Gassan 1999; John and Thomas 2008).

In fiber-reinforced green composites, the fiber serves as reinforcement and provides strength and stiffness to the resulting composite structure (Rowell 2012). Natural biomass (agricultural residues, wood, plant fibers, etc.), which primarily contains cellulose, hemicelluloses, and lignin, represents an abundant source of renewable reinforcement for green composites and is considered one of the most important components of green composites. The numerous advantages of natural fiber–reinforced green composites such as low cost, light weight, eco-friendliness, nonabrasiveness, and biodegradability place them among the high-performance composites having both economic and environmental advantages (Voichita 2011). Although natural fiber reinforcement was used in various applications in the last two decades, extensive research is still required in order to fully understand and explore the potential of natural fibers (Voichita 2006). The effective utilization of natural fibers derived from renewable resources provides environmental benefits with respect to ultimate disposability and utilization of raw material (Singha and Thakur 2012). Therefore, many industrial sectors consider natural fibers as potential substitutes for synthetic reinforcement. Natural fiber–reinforced green composite materials are presently used in various applications, such as door components, furniture, deck surfaces, windows, and automotive components.

The properties of natural fibers vary to a considerable degree, depending on the processing method used to obtain the fibers (Hon 1996). At the present state of technology, wood as well as non-wood fibers such as hemp, kenaf, flax, and sisal have achieved commercial success in green composites automotive sector, employing different polymer matrices (Davim 2011). A number of studies on natural fiber–reinforced composites revealed that their mechanical properties are strongly influenced by parameters such as fraction of the fibers (volume/weight), fiber orientation, fiber aspect ratio, fiber length, fiber–matrix adhesion, and stress transfer at the interface (Hepworth et al. 2000). The term "natural fiber" refers to a broad range of animal and vegetable fibers. These fibers often contribute to the structural performance of the plant and, when used in polymer composites, can provide significant reinforcement (Joshi et al. 2004; Karus and Kaup 2002). Natural fibers have been further classified into two broad categories based on their origins: animal fibers and plant fibers (John and Thomas 2008; Singha and Thakur 2012; Thakur et al. 2011). Figure 1.3 shows a classification of natural fibers.

Animal fibers: Animal fibers generally consist of proteins. Typical examples are hair (from sheep, goats, rabbits, alpaca, and horses) and silk.

Plant fibers: Plant fibers primarily consist of cellulose fibrils embedded in a lignin matrix along with minor amounts of additional extraneous components. Their main components are cellulose, hemicellulose, pectin, and lignin (Baley 2002). The extraneous components include low-molecular-weight organic compounds (extractives) and inorganic matter. Natural fibers are lighter than traditional inorganic reinforcements leading to possible benefits such as fuel savings when their composites are used in transportation applications (Clemons 2009).

Plant fibers can be subdivided into several classes: straw, seed, bast, leaf, and wood fibers (Hamad 2002). Cellulose is the prime constituent of all cellulosic fibers, and most plant fibers (except for cotton) are composed of cellulose, hemicelluloses, lignin, waxes, and some water-soluble compounds (Summerscales et al. 2010). The reinforcing efficiency of natural plant fibers in a given polymer matrix is primarily related to the nature of its cellulosic content and crystallinity (Singha and Thakur 2012; Thakur et al. 2011).

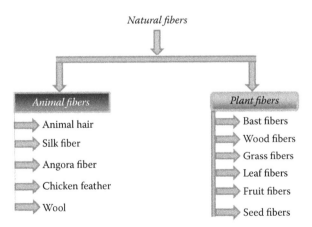

FIGURE 1.3 Classification of natural fibers.

A detailed description of the structure and the advantages/disadvantages of plant fibers is given in Chapter 10.

Fibers are also classified by their length: long fibers are designated as continuous fibers, while short fibers are called discontinuous fibers (sometimes also referred to as chopped fibers). Compared to synthetic fibers, the orientation of short or discontinuous fibers in a green polymer composite cannot be easily controlled (Gowda et al. 1999). In most cases, the fibers are assumed to be randomly oriented in the polymer composite. Natural fiber–reinforced green composites have been prepared by different methods, including extrusion-compression molding, compression molding, structural reaction injection molding, and injection molding with short fiber reinforcement (Singha and Thakur 2012). The most commonly used technique for green composite fabrication is compression molding, and different variations of this technique are suitable for the processing of natural fibers. In general, they differ in the way the reinforcement and the polymer matrices are combined and brought into the mold. Some processes use a premelted polymer, some use a fibrous polymer that is combined with the natural fibers to form hybrid materials prior to compression molding, and others use a polymer powder that is introduced into the fiber materials before compression molding.

Green composites can be economically viable, environmentally friendly materials for a number of applications (Jog and Nabi Saheb 1999; Gacitua et al. 2005). In the building and construction industry, the applications include partition boards, window and door frames, roof tiles, and mobile or prefabricated buildings. They are also found in aerospace, military, and marine applications; in the transportation industry, automobile, and railway coach interiors; in storage devices, such as postboxes, grain storage silos, or biogas containers; and in furniture, showers, bath units, and many other products (Gomes 2004; Shibata et al. 2013).

To conclude, the development of green composites from renewable resources has just begun. Even though these composites offer many advantages, their specific characteristics must be carefully evaluated when choosing a material for a given application. Renewable polymers such as natural fibers are already used in industrial products because of their economic and environmental advantages; however, future work will have to focus on the specific technological properties and advantages of natural fibers.

REFERENCES

Averous, L. 2004. Biodegradable multiphase systems based on plasticized starch: A review. *Journal of Macromolecular Science C Polymer Reviews* 44(3): 231–274.

Azizi, S.M.A.S., Alloin, F., and Dufresne, A. 2005. Review of recent research into cellulosic whiskers, their properties and their applications in nanocomposite field. *Biomacromolecules* 6: 612–626.

Baley, C. 2002. Analysis of the flax fibres tensile behavior and analysis of the tensile stiffness increase. *Composites: Part A* 33: 939–948.

Bledzki, A.K. and Gassan, J. 1999. Composites reinforced with cellulose based fibers. *Progress in Polymer Science* 24: 221–274.

Bledzki, A.K., Mamun, A., and Volk, J. 2010. Barley husk and coconut shell reinforced polypropylene composites: The effect of fibre physical, chemical and surface properties. *Composites Science and Technology* 70: 840–846.

Brundtland Commission. 1987. *Our Common Future*. Report by the Brundtland Commission [formally the World Commission on Environment and Development (WCED)], Oxford University Press: Oxford.

Cipriani, G., Salvini, A., Baglioni, P. and Bucciarelli, E. 2010. Cellulose as a renewable resource for the synthesis of wood consolidants. *Journal of Applied Polymer Science* 118: 2939–2950.

Clemons, C. 2009. Composites from wood and plastics. In: *Wood and Fiber Product Seminar*. VTT and USDA Joint Activity, Finland, September 22–23.

Davim, J. 2011. *Wood Machining*. Hoboken, NJ: John Wiley & Sons.

Dufresne, A., Kellerhals, M.B. and Witholt, B. 1999. Transcrystallization in mcl-HAs/cellulose whiskers composites. *Macromolecules* 32(22): 7396–7401.

Gacitua, W., Ballerini, A. and Zhang, J. 2005. Polymer nanocomposites: Synthetic and natural fillers—A review. *Maderas: Cienc y Tecnologia* 7(3): 159–178.

Garlotta, D. 2001. A literature review of poly (lactic acid). *Journal of Polymers and the Environment* 9(2): 63–84.

Gomes, M.E. 2004. Biodegradable polymers and composites in biomedical applications: From catgut to tissue engineering. Part 1: Available systems and their properties. *International Materials Reviews* 12(4): 65–81.

Gowda, T.M., Naidu, A.C.B., and Chhaya, R. 1999. Some mechanical properties of untreated jute fabric-reinforced polyester composites. *Journal of Composite Part A: Applied Science and Manufacturing* 30(3): 277–284.

Hamad, W. 2002. *Cellulosic Materials: Fibers, Networks, and Composites*. Boston, MA: Kluwer Academic Publishers.

Hazizan, M.A. and Cantwell, W.J. 2003. The low velocity impact response of an aluminium honeycomb sandwich structure. *Composites: Part B* 34: 679–687.

Hepworth, D.G., Hobson, R.N., Bruce, D.M. and Farrent, J.W. 2000. The use of unretted hemp fibre in composite manufacture. *Composites Part A: Applied Science and Manufacturing* 31(11): 1279–1283.

Hon, D.N.-S. 1996. *Chemical Modification of Lignocellulosic Materials*. New York: Marcel Dekker.

Jog, J.P. and Nabi Saheb, D. 1999. Natural fiber polymer composites: A review. *Advanced in Polymer Technology* 18(4): 315–363.

John, M.J. and Thomas, S. 2008. Biofibers and biocomposites. *Carbohydrate Polymers* 71:343–364.

Joshi, S.V., Drzal, L.T., Mohanty, A.K. and Arora, S. 2004. Are natural fiber composites environmentally superior to glass fiber reinforced composites? *Composites Part A: Applied Science and Manufacturing* 35: 371–376.

Kabir, M.M., Wang, H., Lau, K.T., Cardona, F. and Aravinthan, T. 2012. Mechanical properties of chemically-treated hemp fibre reinforced sandwich composites. *Composites Part B: Engineering* 43: 159–169.

Karus, M. and Kaup, M. 2002. Natural fibres in the European automotive industry. *Journal of Industrial Hemp* 7(1): 119–131.

Klemm, D., Heublein, B., Fink, H.-P. and Bohn, A. 2005. Cellulose: Fascinating biopolymer and sustainable raw material. *Angewandte Chemie International Edition* 44(22): 3358–3393.

Klyosov, A.A. 2008. *Wood-Plastic Composites*. Hoboken, NJ: John Wiley & Sons.

Luo, S. and Netravali, A.N. 1999. Mechanical and thermal properties of environmentally friendly green composites made from pineapple leaf fibres and poly(hydroxybutyrate-co-valerate) resin. *Polymer Composite* 20(3): 367–378.

Markarian, J. 2005. Automotive and packaging offer growth opportunities for nanocomposites. *Journal of Plastics Additives and Compounding* 7: 18–25.

Netravali, A.N. and Chabba, S. 2003. Composites get greener. *Materials Today* 6(4): 22–29.

Nikolaos, E.Z. 2011. *Interface Engineering of Natural Fibre Composites for Maximum Performance*. Cambridge, UK: Woodhead Publishing.

Oksman, K. and Sain, M. 2008. *Wood-Polymers Composites*. Boca Raton, FL: CRC Press.

Ouajai, S. and Shanks, R.A. 2009a. Biocomposites of cellulose acetate butyrate with modified hemp cellulose fibres. *Macromolecular Materials and Engineering* 294: 213–221.

Ouajai, S. and Shanks, R.A. 2009b. Preparation, structure and mechanical properties of all-hemp cellulose biocomposites. *Composites Sciences and Technology* 69: 2119–2126.

Pandey, J.K., Nakagaito, A.N. and Takagi, H. 2013. Fabrication and applications of cellulose nanoparticle-based polymer composites. *Polymer Engineering and Science* 53: 1–8.

Richard, D.G. 1994. *Cellulosic Polymers, Blends, and Composites*. Munich, Germany: Hanser Publishers.

Rowell, R.M. 2012. *Handbook of Wood Chemistry and Wood Composites*. Boca Raton, FL: Taylor and Francis.

Schneider, J.P. and Karmaker, A.C. 1996. Mechanical performance of short jute fiber reinforced polypropylene. *Journal of Materials Science* 15: 201.

Scott, G. 2000. Green polymers. *Polymer Degradation Stability* 68(1): 1–7.

Shibata, M., Yamazoe, K., Kuribayashi, M. and Okuyama, Y. 2013. All-wood biocomposites by partial dissolution of wood flour in 1-butyl-3-methylimidazolium chloride. *Journal of Applied Polymer Science* 127: 4802–4808.

Singha, A.S. and Thakur, V.K. 2012. *Green Polymer Materials*. Houston, TX: Studium Press LLC.

Summerscales, J., Nilmini P.J.D., Virk, A.S. and Hall, W. 2010. A review of bast fibres and their composites. Part 2 Composites. *Composites: Part A: Applied Science and Manufacturing* 41: 1336–1344.

Thakur, V.K., Singha, A.S. and Thakur, M.K. 2011. *Green Composites from Natural Cellulosic Fibers*. Germany: GmbH & Co. KG.

Thakur, V.K. and Singha, A.S. 2012. *Nanotechnology in Polymers*. Houston, TX: Studium Press LLC.

Uma Devi, L., Bhagawan S.S. and Thomas, S. 2010. Dynamic mechanical analysis of pineapple leaf/glass hybrid fiber reinforced polyester composites. *Polymer Composites* 31(6): 956–965.

Voichita, B. 2006. *Acoustics of Wood*. Dordrecht: Springer-Verlag Berlin and Heidelberg GmbH & Co. KG.

Voichita, B. 2011. *Delamination in Wood, Wood Products and Wood-Based Composites*. Berlin, Heidelberg, Germany: Springer.

Wambua, P., Ivens, J. and Verpoest, I. 2003. Natural fibres: Can they replace glass in fibre reinforced plastics? *Composites Science and Technology* 63: 1259–1264.

Zain, M.F.M., Islam, M.N., Mahmud, F. and Jamil M. 2011. Production of rice husk ash for use in concrete as a supplementary cementious material. *Construction and Building Materials* 25: 798–805.

2 Valorization of Agricultural By-Products in Poly(Lactic Acid) to Develop Biocomposites

Anne Bergeret, Jean-Charles Benezet,
Thi-Phuong-Thao Tran, George C. Papanicolaou,
and Anastasia Koutsomitopoulou

CONTENTS

2.1 INTRODUCTION

Global warming, the growing awareness of environmental and waste management issues, dwindling fossil resources, and rising oil prices are some of the reasons why green products are increasingly being promoted for sustainable development. These green products, such as starchy and cellulosic polymers, have been used for thousands of years for food, furniture, and clothing. But it is only in the past two decades that they have experienced a renaissance, with substantial commercial production. For example, many old processes have been reinvestigated, such as the chemical dehydration of ethanol to produce green ethylene and therefore green polyethylene, polyvinylchloride, and other plastics.

Poly(lactic acid) or PLA is currently one of the most promising bio-based polymers. During the last decade, PLA has been the subject of abundant literature, with several reviews and book chapters (Auras et al. 2004; Averous 2004; Mehta et al. 2005). Processable by many techniques (blowing films, injection-molded pieces, calendared and thermoformed films, etc.), a wide range of PLA grades is now commercially available. PLA is obtained from lactic acid extracted from starch and converted to a high-molecular-weight polymer through an indirect polymerization route via lactide. This route was first demonstrated by Carothers et al. (1932) but high molecular weights were not obtained until improved purification techniques were developed (Garlotta 2002). The mechanism involved is ring-opening polymerization and may be ionic or coordination insertion depending on the catalytic system used (Mehta et al. 2005; Stridsberg et al. 2002). All properties of PLA depend on the molecular characteristics as well as the presence of ordered structures (crystalline thickness, crystallinity, spherulite size, morphology, and degree of chain orientation). The physical properties of polylactide are related to the enantiomeric purity of the lactic acid stereocopolymers. PLA can be produced totally amorphous or up to 40% crystalline. PLA resins containing more than 93% of L-lactic acid are semicrystalline, while those containing 50%–93% are entirely amorphous. The typical PLA glass transition temperature ranges from 50°C to 80°C, whereas the melting temperature ranges from 130°C to 180°C. The mechanical properties of PLA can vary considerably, ranging from soft elastic materials to stiff high-strength materials, according to various parameters, such as crystallinity, polymer structure, molecular weight, material formulation (plasticizers, blend, composites, etc.), and processing. For instance, commercial PLA with 92% of L-lactide (called PLLA) has a modulus of 2.1 GPa and an elongation at break of 9%. The CO_2 permeability coefficients for PLA polymers are lower than those reported for crystalline polystyrene at 25°C and 0% relative humidity (RH) and higher than those for polyethylene terephtalate. The main abiotic degradation phenomena of PLA involve thermal and hydrolysis degradations.

In addition, natural fibers have established a track record as a reinforcing material in automotive parts and are expanding with high growth rate to packaging, construction, and household utility–based industries because of their light weight, low cost, and environmentally friendly nature. Flax and hemp are among the most widely used natural fibers to date. Natural fibers are commonly used to reinforce PLA, and some products are already on the market for various applications, including automotive, mobile phone, and plant pots (Graupner et al. 2009).

In recent years, a special concern has been devolved toward new waste sources such as lignocellulosic cereal waste by-products, for example husks, which are available in abundant volume throughout the world and which will alleviate the shortage of wood resources. Traditional use of these grain by-products includes bedding for animals and livestock feeding. Many studies have investigated possible uses for these waste products as a fuel and the fuel residue as activated carbon (called ash) in cement and concrete production (Chao-Lung et al. 2011; Zain et al. 2011) as well as in bricks production (Sutas et al. 2012) and zeolithe synthesis (Dalal et al. 1985). Other studies have reported that cereal husk ash obtained from a burning process could be considered as an alternative reinforcement for thermoplastic polymers (e.g., polypropylene (PP) and polyethylene) in comparison with other commercial fillers (e.g., silica) (Ayswarya et al. 2012; Fuad et al. 1995b; Siriwardena et al. 2003; Turmanova et al. 2008). Moreover, cereal husk raw materials, such as rice and wheat husks, could be potential low-cost alternatives replacing wood for making composites with PP (Bledzki et al. 2010, 2012; Premalal et al. 2002; Yang et al. 2004), polyethylene (Favaro et al. 2010), polycaprolactone (Zhao et al. 2008), polyurethane (Rozman et al. 2003), phenol formaldehyde (Ndazi et al. 2007), and PLA (Yussuf et al. 2010).

Among the exploitation of certain agricultural by-products, olive pits obtained during the processing of olives for oil are interesting to considered. Olive oil production especially concerns the Mediterranean regions, among them Greece (Vourdoubas 1999) and Italy (Pattara et al. 2010) that are considered as the main producers with a pit production in Italy from 1999 to 2007 ranging between 277 and 519 ktons. The use of olive pits as a biofuel offers an alternative in the agriculture industry to the use of fossil fuels, contributing to a reduction in CO_2 emissions. Olive pits have high combustion power, equivalent to hard wood, allowing them to be used as wood briquettes to generate heat or power (Vlyssides et al. 2008). Recent developments investigated the use of the pyrolysis char from an olive pit, which is a carbon-rich material (about 67 wt%) with significant concentration in metals (mainly iron), for the production of green materials when incorporated within an epoxy resin (Papanicolaou et al. 2010). A bending modulus 60% higher than that of the pure resin was obtained when 35 wt% of olive kernel pyrolytic char was incorporated. Even with filler loadings on the order of 5 wt%, a 27% increase in the bending modulus was obtained. The same authors (Papanicolaou et al. 2012) focused their works on the development of epoxy composites reinforced with grinded olive pits (powder 10–30 µm diameter). An increase in the bending modulus of 48% was achieved when 45 wt% of olive pit powder are incorporated. Earlier Tserki et al. (2006) reported equivalent mechanical properties for a polybutylene succinate (PBS) composite filled with olive husk flour.

The aim of this chapter is to investigate whether the use of rice and wheat husks on the one hand and olive pit powders on the other hand can lead to challenging performances (mainly mechanical and thermal properties, and aging resistance) when they are introduced in a biosourced and biodegradable polymer, such as PLA.

2.2   VALORIZATION OF DIFFERENT AGRICULTURAL
BY-PRODUCTS IN POLY(LACTIC ACID)

2.2.1   POLY(LACTIC ACID) BIOCOMPOSITES REINFORCED
BY RICE AND WHEAT HUSKS

2.2.1.1   Materials and Processing

2.2.1.1.1   Materials

PLA 7000D© was supplied by Nature-Works LLC (M_n = 179,200 Da, polydispersity index, I = 1.75 [steric exclusion chromatography (SEC), tetrahydrofurane (THF), T = 25°C], T_g = 58°C, and T_m = 152°C).

Rice belongs to the grass family Poaceae and is cultivated in tropical, subtropical, and warm temperate regions for its fruit, or caryopsis, rich in starch. It refers to the genus *Oryza*, of which two species are cultivated, *Oryza sativa* and *Oryza glaberrima*. Depending on the cultivar, the result is white rice (China, India, and France), brown rice (China), red rice (Madagascar), yellow rice (Iran), purple rice (Laos), or sticky rice (China, Indonesia, and Laos), which corresponds to a large level of production, approximately 600 million tons/year. Rice is normally grown as a half annual plant, although in tropical areas its life cycle lasts 3–4 months. Unmilled rice, known as paddy, is usually harvested manually or mechanically when the grains have a moisture content of around 25%. Harvesting is followed by threshing either immediately or within a day or two. Subsequently, paddy needs to be dried to bring down the moisture content to no more than 20% for milling. This operation is performed, after threshing, using a machine with two horizontal discs coated with an abrasive material. The upper disc is stationary while the lower, set at an appropriate distance, is rotating. This action has the effect of separating the glumes and lemmas. The husking rubber roller rotating at variable speeds represents a shift reducing the risk of breaking the rice grain. The proportion of the resulting ball husking paddy (rough rice) fluctuates between 17% and 25% depending on the variety (Ruseckaite et al. 2007). The rice grains used in this study are long-grain rice husks (LRH, high amylase content) that tend to remain intact after cooking, as medium-grain rice (high amylopectin content) become stickier. They were supplied by SOUFFLET S.A. (France).

Wheat is one of the most common and important human food grains and ranks second in total production as a cereal crop with 651 million tons in the world in 2012 (Vocke and Liefert 2012). It also belongs to the grass family *Poaceae*. Wheat refers to genus *Titricum*, of which a greater number of species are cultivated. Einkorn wheat is one of them and refers to the genus *Triticum monococcum*. It is one of the earliest cultivated forms of wheat and its husks enclose the grains tightly. The husking process of Einkorn wheat is shown in Figure 2.1. After the first threshing, Einkorn wheat husks 1 (named WH1) are separated from the grain by winnowing. WH1 includes broken husk (lemmas, paleas, and glumes), entire husk (WH2), and nonthreshing Einkorn wheat. After the second winnowing, entire husks (named WH2) are separated from other compositions. Only results performed with WH2 are exposed in this study. The husk proportion varies between 40% and 50% depending on the husking process used. Einkorn wheat husks are supplied by TOFAGNE S.A (France).

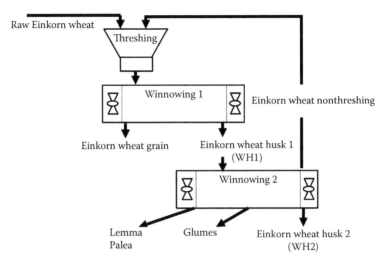

FIGURE 2.1   Husking process of Einkorn wheat.

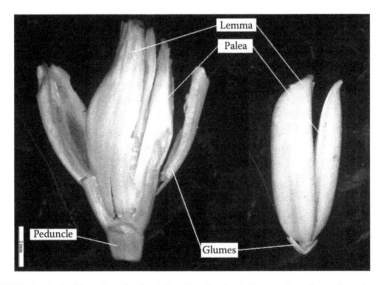

FIGURE 2.2   Left side: wheat husk; right side: rice husk. Bar scale = 2 mm length.

Rice and wheat are characteristically arranged in spikelet, each spikelet having one or more florets. A spikelet consists of two (or sometimes fewer) bracts at the base, called glumes, followed by one or more florets. A floret consists of the flower surrounded by two bracts called the lemma (the external one) and the palea (the internal). In the several lemmas present in the Einkorn wheat husks, paleas entirely cover the grain while two large hard glumes half cover external. Rice husk structures are totally different; their glumes are very tiny, it is difficult to observe them by eye, and they often break in the husking process. Rice husks have only one lemma and one palea. The lemma and palea of rice are harder than those of Einkorn wheat; they cover the grain entirely and protect it. This complex structure is shown in Figure 2.2.

Specific properties of both husks, such as chemical composition, morphology, and surface and thermal properties, are presented in Section 2.2.1.2.

2.2.1.1.2   Biocomposite Processing

Before processing, all materials (PLA, LRH, and WH2) are dried under vacuum at 60°C for 24 hours. Compounding is achieved by twin screw extruder (Clextral BC21; 900 mm length). Screw temperatures are set up at 170°C for the first zone and 180°C for the other zones. Screw speed rate is 250 rpm and feed rate is 4 kg/h. After extrusion, injection molding is then carried out on a Krauss Maffei KM50-T180CX. Screw temperature is set at 200°C and the mold temperature is maintained at 40°C. The injection cycle time is fixed at 60 seconds. LRH.10, LRH.20, and LRH.25 are PLA-reinforced biocomposites with 10%, 20%, and 25% in weight of LRH, respectively. WH2.10, WH2.20, and WH2.25 are PLA-reinforced biocomposites with 10%, 20%, and 25% in weight of Einkorn wheat husk (WH2), respectively.

The morphology evolution of husks after processing is detailed in Section 2.2.1.3. Biocomposite properties, such as thermal and mechanical properties, are presented in Section 2.2.1.4.

2.2.1.2   Specific Properties of Raw Rice and Wheat Husks

2.2.1.2.1   Husk Dimensions

The length of the entire LRH is 9.86 ± 0.41 mm and that of the Einkorn wheat husk is 10.53 ± 1.48 mm.

2.2.1.2.2   Husk Chemical Composition

Waxes are extracted through a commonly used Soxhlet extraction with an etha-nol/toluene 50/50 v/v solution for 18 hours and then with pure ethanol for 6 hours. After that, lignin and hollocellulose are determined after an extraction in the presence of sodium chlorite and acetic acid. Inorganic content is related to ash content determined after pyrolysis at 600°C for 18 hours. Ash composition is determined through X-ray diffraction (EDX) analysis (Oxford Co., Inca 350) in the chamber of an Environmental Scanning Electron Microscope (ESEM) (FEI Co., Quanta 220 FEG). Husk chemical composition is presented in Table 2.1. The wax content of both husks is very low, about 2%. Higher lignin content is obtained for rice husk (37.1%)

TABLE 2.1

Husk Chemical Composition

Component (wt%)	LRH	WH2
Wax	2.2	2.0
Lignin	37.1	28.3
Hollocellulose	46.3	63.0
Ash	14.4	6.7

LRH, long-grain rice husk; WH2, entire wheat husk.

when compared with wheat husk (28.3%), which induces higher hydrophobicity and stiffness. As a balance, hollocellulose contents are about 63% and 43% for WH2 and LRH, respectively. The ash content of rice husk (14.4%) is higher than that of wheat husk (6.7%). Equivalent data were reported in the literature (Yang et al. 2004) for rice husk (3% wax, 20% lignin, 60% holocellulose, and 17% ash, among a major part of silica). The ash composition shows equivalent composition for both husks with a high atomic percentage of Si (about 36%) and O (about 52%), which may suggest the presence of silica.

2.2.1.2.3 Husk Microstructure

The outer surface and cross section of WH2 and LRH are observed by ESEM. Figure 2.3 shows that the outer surface of rice husk is relatively rougher than that of wheat husk. The outer surface of rice husk is highly ridged, and the ridges include regularly interlaid peaks and valleys (Figure 2.3a). In Figure 2.3c, the cross section of rice husk shows clearly that the outer surface is rougher than the inner surface.

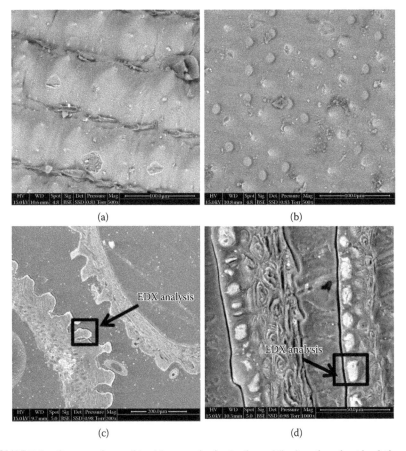

(a)   (b)

(c)   (d)

FIGURE 2.3 Outer surfaces of (a, c) long-grain rice husks and (b, d) entire wheat husk through ESEM observations: (a, b) magnitude ×500; (c) magnitude ×200; (d) magnitude ×1000.

The outer border region of rice husk presents a high peak of Si (EDX analysis) and it can be concluded that there is a significant amount of silica. Silica exists on the outer surface of rice husks in the form of silicon–cellulose membrane that forms a natural protective layer against termites and other microorganisms that attack the paddy. But this component has been alleged to be responsible for insufficient adhesion between accessible functional groups on rice husk surfaces and various matrix binders. On the contrary, wheat husk outer surface is flat and presents tiny scattered knots (Figure 2.3b). From Figure 2.3d, it can be observed that no silicon–cellulose membrane covers the outer surface of wheat husk. However, the EDX spectrum of the white knots near the outer surface reveals the presence of silica. These observations are in perfect agreement with other authors (Ndazi et al. 2007; Park et al. 2003; Yoshida et al. 1962). The heavily silicified outer epidermal cells provide strength, rigidity, and stiffness to husks so they can be used as reinforcement objects for composite products.

2.2.1.2.4 Surface Properties of Husks

Surface properties are consistent with the interfacial phenomena occurring between husks and the host polymer, that is, PLA in this study, during composite processing. It is well known that the dispersive component is related to the ability to create van der Waals interfacial interactions at the solid surface and the polar component to the ability to settle acid–base interfacial interactions at the solid surface. Therefore, the closer the polar and dispersive components of husk and PLA are, the more efficient the husk/matrix adhesion is. Polar and dispersive components are hereby determined through the contact angle method (Digidrop, GBX Co.) with three well-known liquids (di-iodomethane, formamide, and distilled water) using the Owens and Wendt model for measurement. For PLA, the polar and dispersive components are found to be 5.7 and 45.1 mJ/m^2, respectively. The corresponding values are 4.9 and 41.9 mJ/m^2 for rice husk and 3.8 and 34.4 mJ/m^2 for wheat husk. As the polar and dispersive components of rice husk are closer to those of PLA than those of wheat husk, it can be said that the rice husk/PLA interfacial adhesion is better as compared with wheat husk. Moreover, as the difference between the surface free energy of the husks and of PLA is greater for the wheat husk than for the rice husk, it can be concluded that the wetting by PLA is better for wheat than for rice husk.

2.2.1.2.5 Husk Hydrophily

Water uptake is evaluated after conditioning husks at 65% RH and 20°C, until no change in weight is observed. Water contents are 19.28% ± 0.07% and 16.88% ± 0.35% for wheat and rice husks, respectively. This may be due to the differing surface morphologies and chemical composition of the husks. The higher hydrophobicity obtained for rice husk could therefore be related to the higher lignin content when compared with that in wheat husk. Equivalent data were obtained by Bledzki et al. (2012) with water contents ranging from about 12% and 20% after conditioning rice husk at 23°C and 65% and 95% RH, respectively. The moisture uptake reached an equilibrium state after 45 and 55 days of conditioning periods at 65% and 95% RH, respectively.

2.2.1.2.6   Thermal Properties

Thermal stability of rice and wheat husks is investigated through thermogravimetry analysis (TGA) (Perkin-Elmer, Pyris 1, 10°C/min, nitrogen) experiments (Figure 2.4) to determine the feasibility of these husks to be processed at high temperature. It can be observed that husk degrades through different stages. Below 250°C, about 6%–7% of weight loss occurs for both husks and is related to the dehydration of the husks. Equivalent losses are reported in the literature with losses of 6% and 2%–6% for wheat husk and wheat straw, respectively. The husk decomposition starts as low as 250°C, suggesting that these agroresidues are suitable for processing with

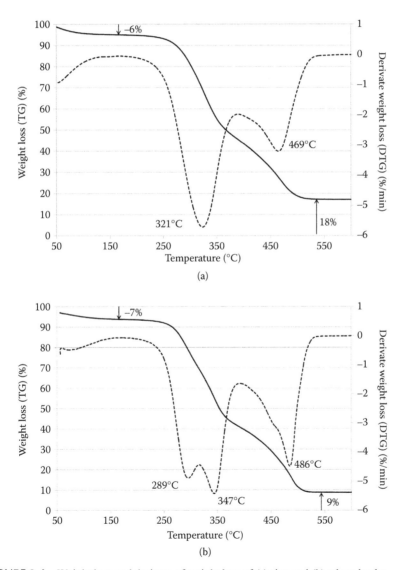

FIGURE 2.4   Weight loss and derivate of weight loss of (a) rice and (b) wheat husks.

polymers having a melting temperature below 250°C, as is the case for PLA. As far as wheat husks are concerned, the derivate of the TGA curve (called DTG) exhibits two decomposition steps with decomposition temperatures of 289 and 347°C. The peak at 289°C is related to the thermal decomposition of hemicelluloses and the glycosidic linkage of cellulose, and the peak at 347°C is due to α-cellulose decomposition. This two-step decomposition has been also reported by Bledzki et al. (2012) for wheat husk. In the case of rice husk, the DTG curve exhibits a single decomposition step at 321°C corresponding to hemicelluloses and cellulose simultaneous decompositions. The overlapping between these two decomposition peaks may depend on the heating rate and the rice husk variety (Mansaray and Ghaly 1998; Stefani et al. 2005). Finally, the last peak in the range 470–490°C is attributed to lignin conversion that is mainly responsible for the char formation. A char yield of 18% for rice husk at 600°C is obtained, which is slightly higher than that for wheat husk (9%). This result is consistent with the ash content evaluation reported in Table 2.1. Differences in the values could be attributed to different conditions for both tests. Nevertheless, it can be seen that rice husk is more thermally stable than wheat husks.

2.2.1.3 Specific Properties of Rice and Wheat Husks after Biocomposite Processing

Biocomposite compounds are achieved by twin screw extrusion processing. During this extrusion, initial husk dimensions are reduced due to shear stresses within the extruder. As the final properties of biocomposite are related to the morphology of husks after compounding, it seems important to analyze it. The distributions of husk areas (laser diffraction particle size analyzer, LS™ 200, Beckman Coulter) are shown in Figure 2.5 and the corresponding pictures are in Figure 2.6. WH2 pieces have smaller sizes after extrusion than LRH pieces as (1) WH2 contains 48.6% of husk

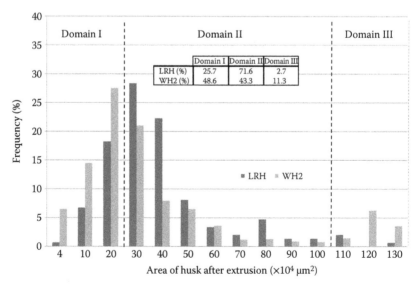

FIGURE 2.5 Area distribution of long-grain rice husks (LRH) and entire wheat husk (WH2) after extrusion (in percentage for each domain in inset table).

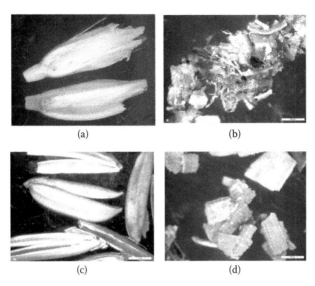

FIGURE 2.6 Rice husk (a) and wheat husk (c) before composite processing (bar scale = 2 mm length); rice husk (b) and wheat husk (d) after composite processing (bar scale = 500 µm length).

pieces have areas below 30×10^4 µm^2 (Domain I) whereas LRH contains only 25.7% and (2) 71.6% of LRH husk pieces have areas between 30×10^4 and 100×10^4 µm^2 (Domain II) whereas WH2 contains only 43.3%. This could be due to the higher stiffness of wheat peduncles when compared with rice. The higher initial length of WH2 could explain the higher number of large pieces more than 100×10^4 µm^2 corresponding to (Domain III) compared with LRH.

2.2.1.4 Biocomposite Properties

2.2.1.4.1 Thermal Properties

Study of the thermal behavior of biocomposites is performed using a Perkin-Elmer differential scanning calorimeter (DSC) (Pyris Diamond) by heating specimens from 25 to 200°C at 10°C/min. It can be observed (Figure 2.7) that the addition of the husk induces a significant shift of PLA cold-crystallization peak to lower temperatures due to an increased melt shear rate during compounding leading to a decrease in the average molecular weight (Figure 2.8), as also shown by other authors (Le Marec 2011; Plackett et al. 2003). At the same time, a shift of the melting temperature to higher temperatures is obtained, which could be related to the slight increase in crystallinity ratio (from about 1% to 5%). No variation in glass transition temperature is observed in presence of husks.

Thermal stability of biocomposites measured through TGA experiments is also investigated as a function of the husk content (Figure 2.9). Results show equivalent behavior for both husks, that is a slight decrease in the thermal stability for increased husk contents due to the decrease in the average molecular weight. The char content increase with the husk content is related to the ash content determined in Section 2.2.1.2.

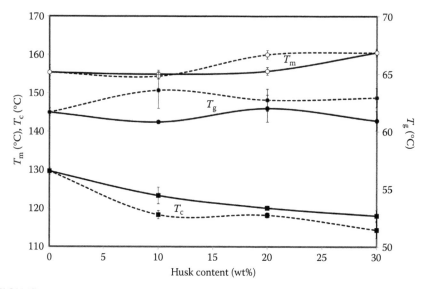

FIGURE 2.7 Evolution of cold crystallization (■), melting (○), and glass transition (●) temperatures as a function of the (full line) rice and (dashed line) wheat husk content.

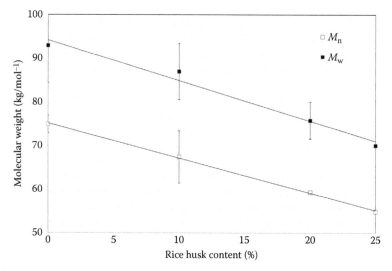

FIGURE 2.8 Evolution of molecular weight in weight (M_w) and in number (M_n) as a function of long-grain rice husk weight content (THF, 45°C, 0.65 ml/min, 100 μL).

2.2.1.4.2 Mechanical Properties

The influence of the husk content on the bending (modulus, stress) (Zwick Z010, ISO178:2001) and the unnotched impact (Zwick, ISO179:2000) properties are shown in Figure 2.10. An increase in the bending modulus with an increase in husk loading is observed, irrespective of the nature of the husk (4566 ± 50 MPa for WH2.25 and 4476 ± 660 for LRH.25 compared to 3449 ± 25 MPa for PLA). A decrease in the

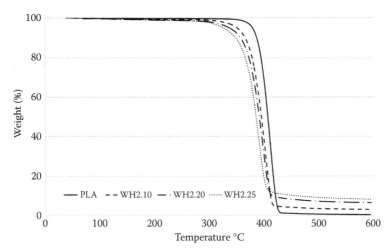

FIGURE 2.9   Weight loss of wheat husk–reinforced biocomposites (10, 20, and 25 wt%) compared with that of neat poly(lactic acid).

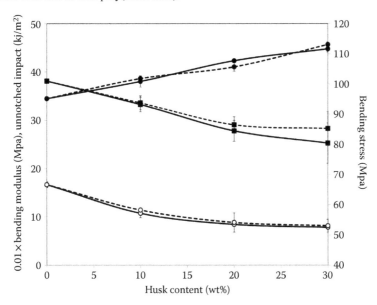

FIGURE 2.10   Bending modulus (●), bending stress (■), and unnotched impact (○) properties of biocomposites as a function of the (full line) rice and (dashed line) wheat husk content.

bending stress (−16% for WH2.25 and −20% for LRH.25) and the impact resistance (−51% for WH2.25 and −53% for LRH.25) with an increase in husk content is also obtained. Same evolutions were reported by other authors. Yang et al. (2004) observed a decrease of about 33% and 57% in tensile and unnotched impact strengths, respectively, for a grinded rice husk filled polypropylene (PP) biocomposite (40% in weight of husk). The investigations by Premalal et al. (2002) revealed a significant increase in the Young's and flexural moduli (about 52% and 32%, respectively), whereas the elongation at break and impact strength decreased (−72% and −77%, respectively) for a rice husk powder filled PP biocomposite (60% in weight of husk). They related these results to more or less strain transfer at the interface between husk and PP and to matrix degradation due to an increased melt viscosity during processing.

2.2.1.4.3 Aging Resistance

A lot of studies concern the aging resistance of PLA-based biocomposites. Nevertheless, none of them deal with rice and wheat husk filled composites, irrespective of the nature of the matrix. Therefore, in accordance with potential final applications of the studied biocomposites, three aging conditions are investigated: (1) water immersion at 45°C for 7 days; (2) UV exposition at 65°C for 1000 hours (SEPAP 12/24 photo-aging chamber); and (3) cyclic hygrothermal aging (one cycle corresponding to age of the biocomposites for 16 hours at 85°C and 45% RH, then for 8 hours at 40°C and 95% RH, this cycle being repeated three times) (WEISS climatic chamber). Molecular weight in weight and in number (Figure 2.11), flexural modulus and stress (Figure 2.12), and thermal properties (Table 2.2) are determined as a function of the rice husk content and of the aging conditions compared to unaged corresponding biocomposites.

Results for unreinforced PLA are first analyzed as the three aging conditions induce different effects on pure PLA characteristics. A drastic decrease in molecular weight is observed after a cyclic hygrothermal aging of PLA as no variation is obtained after other aging tests. The crystallinity rate significantly goes up after cyclic hygrothermal aging (38.64% ± 0.50%) and UV aging (28.78% ± 2.95%) compared to unaged PLA (3.23% ± 0.30%) and water immersed PLA (5.45% ± 1.24%). This increase is due to the fact that the aging temperature is close to the glass transition temperature. The highest crystallinity rate obtained for cyclic hygrothermal aged PLA is explained by involving a chimicrystallization mechanism due to the presence of shorter macromolecular chains as indicated by the decrease in molecular weight. A double melting peak is also observed and could be linked to different phenomena according to the literature, that is (1) to a phase transition from a metastable crystalline form α′ to a stable crystalline form α (Pan et al. 2007), (2) to a melt-recrystallization process (Yasuniwa et al. 2004), or (3) to a solid–solid phase transition (Kawai et al. 2007). In the case of mechanical properties, a slight increase in the flexural modulus is observed irrespective of the aging conditions, showing that the increase in crystallinity may compensate for the PLA degradation. On the contrary, except for the water-immersed PLA (for which a lower crystallinity ratio is obtained), a decrease in strength is observed, indicating the brittleness of PLA after aging.

Let us now analyze the results concerning aged rice husk–reinforced biocomposites. First, it can be said that the crystallinity ratio of aged biocomposites is close to those of aged PLA except in the case of the cyclic hygrothermal aging. For this last aging condition, a slight increase in the crystallinity ratio, a decrease in the melting temperature, and the disappearance of the cold crystallization peak are observed for aged LRH.20 biocomposites compared to pure aged PLA (38.64% ± 0.50% and 163.89 ± 1.93°C for aged PLA in comparison with 45.87% ± 2.44% and 152.05 ± 3.34°C for aged LRH.20 biocomposite). Moreover, a double melting peak located around 150°C was also revealed for biocomposites immersed in water and aged under cyclic hygrothermal conditions. Moreover, it can be noticed that the aging resistance of PLA/rice husks biocomposites is different from those of unreinforced PLA with regard to the molecular weight variations. Indeed, M_w and M_n values are always lower for aged biocomposites than for PLA, irrespective of the aging conditions, the

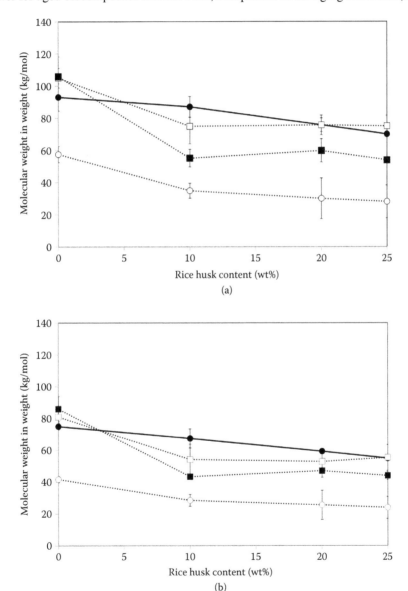

(a)

(b)

FIGURE 2.11 Evolution of molecular weight (a) in weight and (b) in number of poly(lactic acid)/long-grain rice husk biocomposites as a function of the rice husk content and the aging conditions: (•) unaged; (■) water immersion at 45°C for 7 days; (○) humidity–temperature cycled aging; (□) UV exposure for 1000 hours.

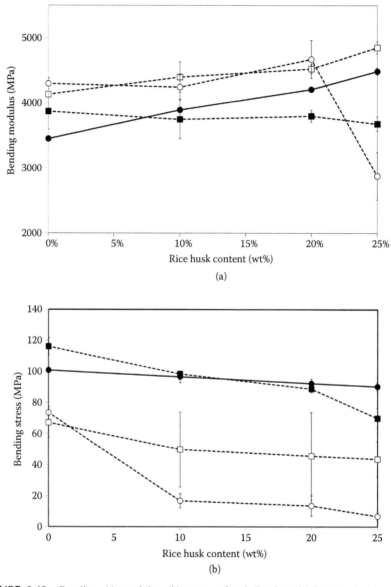

FIGURE 2.12 Bending (a) modulus (b) stress of poly(lactic acid)/long-grain rice husk biocomposites as a function of the rice husk content and the aging conditions: (•) unaged; (■) water immersion at 45°C for 7 days; (○) humidity–temperature cycled aging; (□) UV exposure for 1000 hours.

cyclic hygrothermal aging conditions being more severe than the water immersion conditions, which are also more severe than the UV aging conditions. Finally, the bending strength of biocomposites is significantly decreased after aging. The cyclic hygrothermal aging conditions are once again the most severe aging conditions. Panthapulakkal and Sain (2007) also observed a decrease in bending modulus and

TABLE 2.2

Evolution of Thermal Characteristics of PLA/LRH Biocomposites as a Function of the Rice Husk Content and the Aging Conditions

Aging Conditions		T_g (°C)	T_c (°C)	T_{m1} (°C)	T_{m2} (°C)	X_c (%)
Unaged	PLA	61.64 ± 1.08	129.64 ± 0.73	155.37 ± 0.50	–	3.23 ± 0.30
	LRH.10	61.55 ± 0.23	127.08 ± 2.55	154.84 ± 0.42	–	4.32 ± 1.58
	LRH.20	61.11 ± 1.17	126.39 ± 2.14	156.365 ± 0.11	–	2.77 ± 0.30
	LRH.25	62.84 ± 0.74	124.965 ± 0.05	158.31 ± 2.26	–	3.71 ± 3.17
UV aging	PLA	65.44 ± 1.56	135.32 ± 0.21	159.17 ± 0.94	–	28.78 ± 2.95
	LRH.10	62.78 ± 0.75	135.65 ± 0.30	155.97 ± 0.73	–	29.15 ± 0.39
	LRH.20	63.21 ± 1.30	136.05 ± 0.13	154.31 ± 0.02	–	25.86 ± 0.63
	LRH.25	63.19 ± 0.97	135.96 ± 0.04	155.63 ± 0.11	–	26.02 ± 0.11
Water immersion	PLA	69.12 ± 0.78	127.31 ± 0.13	158.93 ± 1.10	–	5.45 ± 1.24
	LRH.10	67.18 ± 0.34	113.77 ± 4.09	156.17 ± 1.15	–	0.62 ± 0.39
	LRH.20	64.67 ± 0.22	108.57 ± 0.30	155.05 ± 0.03	148.33 ± 0.35	2.95 ± 2.61
	LRH.25	64.39 ± 1.65	107.44 ± 0.43	156.28 ± 0.49	148.57 ± 0.16	3.94 ± 0.15
Cyclic hygrothermal aging	PLA	66.69 ± 1.46	–	163.89 ± 1.93	153.71 ± 1.25	38.64 ± 0.50
	LRH.20	–	–	152.05 ± 3.34	145.74 ± 2.86	45.87 ± 2.44

T_g, glass transition temperature; T_c, crystallization temperature; T_{m1} and T_{m2}, melting temperatures; X_c, crystallinity ratio; LRH, long-grain rice husks; PLA, poly(lactic acid).

strength of 38% and 20%, respectively, for PP/65 wt% of wheat straw biocomposites immersed in water at room temperature for 40 days.

2.2.1.5 Conclusion

Rice and wheat husks can be used as reinforcements in PLA so that a full bio-based composite is available. The bending moduli of biocomposites are improved with increased husk contents. The lowered flexural strength and impact resistance may be due to poor interfacial bonding between husks and PLA.

2.2.2 Poly(Lactic Acid) Biocomposites Reinforced by Milled Olive Pits

2.2.2.1 Materials and Processing

2.2.2.1.1 Materials

PLA 7000D© supplied by Nature-Works LLC is also used in this study. Olive pits are obtained from Greek olive production without any physical or chemical modifications. They are just cleaned and washed to remove impurities produced during the olive oil process. The different steps for obtaining different olive pit powders and their characterization are described in Section 2.2.2.2.

2.2.2.1.2 Biocomposite Processing

Compounding is achieved by twin screw extruder, and injection molding is then carried out on a Krauss Maffei KM50-T180CX. The processing conditions are as previously described in Section 2.2.1.1 Olive pits powder contents vary in the 0–20 wt% range in PLA. Biocomposite properties, such as thermal and mechanical properties, are presented in Section 2.2.2.3.

2.2.2.2 Olive Pit Powder Characterization

2.2.2.2.1 Olive Pit Grinding Process

Olive pits are first placed in an oven at 50°C for 24 hours to remove moisture. Then, three different olive pit powders (called A, B, and C from the finest to the coarsest) are produced using different milling conditions. For milled olive pits A, olive pits are (1) first grinded using a conventional coffee electric grinder for 10 min and then (2) milled successively three times using an agate bowl with agate balls (2 mm diameter) at 600 rpm torque motor speed for 10 min and a Planetary Mono Mill Pulverisette 6 (Fritsch) at a speed of 350–400 rpm for 1 min. The resulting particle size (laser diffraction particle size analyzer) is around 34 μm. Olive pit powders B and C are created using an Ultra Centrifugal Mill ZM 200 (Retsch). In this case, the influence of the speed of the rotor (ranging between 6,000 and 12,000 rpm) and of the ring sieve diameter (0.2 or 0.5 mm) for a milling duration of 15 min on the size of the particle is studied. As shown in Figure 2.13, no significant influence by either parameters is revealed. A modal size of about 60 μm is obtained corresponding to milled olive pits B. Olive pit powder C corresponds to that was sieved using a Vibratory Sieve Shaker (Retsch). Olive pit powder C has a mean size of about 170 μm. Figure 2.14 shows ESEM pictures of each milled olive pit, showing that the powders are particles with irregular shapes.

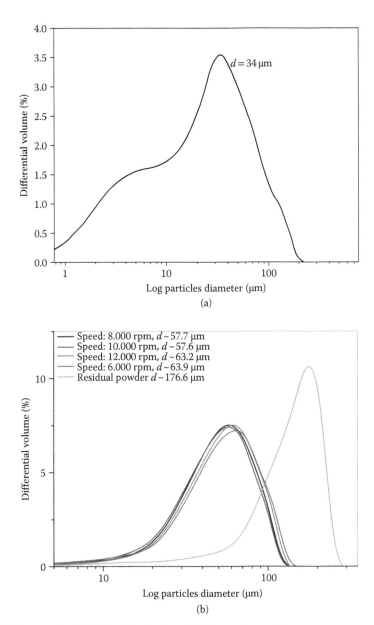

FIGURE 2.13   (a) and (b) Size distribution of different pits powder.

2.2.2.2.2   *Chemical Composition of Olive Pit Powders*

Olive kernel is a lignocellulosic biomass that is composed of cellulose (37.5 wt%), hemi-celluloses (26 wt%), and lignin (21.5 wt%). Other components include moisture (8 wt%), minerals (1 wt%), and proteins (pectines, tannins, etc.) (Rodriguez et al. 2008; Tserki et al. 2005). The ultimate analysis is carried out using a CHN-LECO 800 Analyzer. Results show mainly carbon and oxygen (49 and 31 wt%, respectively) in agreement with

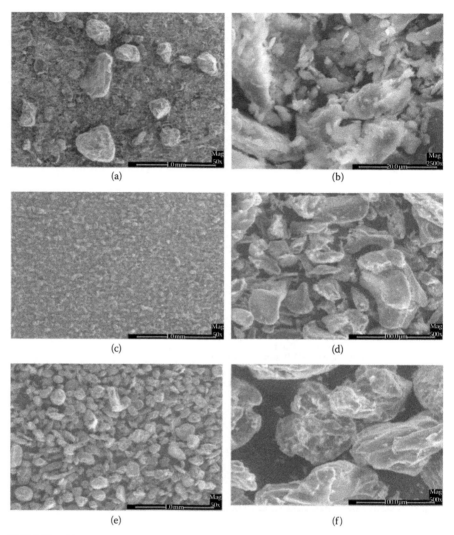

FIGURE 2.14 Microscopic observations of (a, b) powder A, (c, d) powder B, and (e, f) powder C—(a, c, e) magnitude ×50, (d, f) magnitude ×500, (b) magnitude ×2500.

Gonzalez et al.'s (2004) experiments. Metal analysis conducted on the char shows mainly the presence of iron (1236 mg/kg of olive pit) and aluminum (463 mg/kg). Other metals are Zn, Cu, Ni, Cr, and Mn with amounts in the range 12–29 mg/kg of olive kernel.

2.2.2.2.3 Thermal Properties of Olive Pit Powders

The thermal stability of milled olive pits is investigated through TGA experiments (Figure 2.15). Results show that the degradation occurs through three main steps. Up to 100°C, the decrease (5%) is related to moisture extraction. Between 100 and 500°C, two peaks are depicted. The first peak is due to light molecule decomposition (hemicellulose and cellulose) and the second is attributed to heavy molecule decomposition (lignin). After 500°C, char oxidation occurred with a char amount of around 25%. The same decomposition mechanisms were reported in the literature (Tserki et al. 2005) for olive husk flour.

2.2.2.3 Biocomposite Properties

2.2.2.3.1 Thermal Properties

DSC experiments (Perkin-Elmer, Pyris Diamond, 10°C/min, nitrogen) are investigated first to study the thermal behavior of PLA/olive pit powder biocomposites (Table 2.3). An increase in the crystallinity ratio (+4%–8%) with the presence of

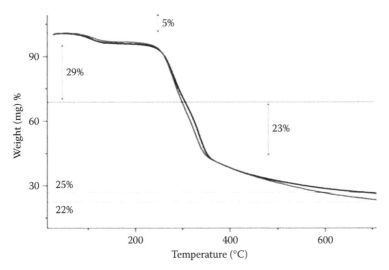

FIGURE 2.15 TGA curves of A and B olive pit powders.

TABLE 2.3

Evolution of Thermal Characteristics of PLA/Olive Pit Powder Biocomposites

Olive Pits (wt%)		T_g (°C)	T_c (°C)	T_m (°C)	X_c (%)
0		65.8	125.1	155.1	8.5
5	A	66.9	125.4	154.2	8.2
	B	65.8	127.3	155.2	6.9
	C	63.1	121.1	153.6	15.9
10	A	66.8	124.4	155.6	12.5
	B	62.6	123.3	153.2	12.5
	C	62.2	121.0	153.9	14.0

T_g, glass transition temperature; T_c, crystallization temperature; T_m, melting temperature; X_c, crystallinity ratio; PLA, poly(lactic acid).

milled olive pits is observed, suggesting that the powders act as nucleating agents. The increase in crystallinity occurs for 5 wt% for larger particles (olive pit powder C) against 10 wt% for the finest particles (olive pit powders A and B). This result is in good agreement with those of Perinovic et al. (2010), who found that olive pit powders act, at the same time, as nucleating agents at lower concentrations and as obstacles for PLA chains to make crystals at higher concentrations. A slight decrease in all temperatures (T_g, T_m, and T_c) is observed in the presence of olive pit powders that is less effective for the finest powder (olive pit powder A).

The thermal stability of PLA/milled olive pit biocomposites is also investigated through TGA experiments (Table 2.4). Results show that the thermal stability of PLA is decreased as the olive pit powder content increases, irrespective of the powder granulometry, as observed by other authors on PLA/wood fiber biocomposites (Huda et al. 2006).

2.2.2.3.2 Mechanical Properties

Figure 2.16 shows the tensile and bending properties of PLA/milled olive pit biocomposites as a function of the content of the milled olive pits and of their size. Moduli increase with increase in the olive pit powder content, especially for milled olive pits B and C. Nevertheless, when smaller particles are used (olive pit powder A), a reduction in both tensile and flexural moduli is observed for weight contents equal to 10%, which may be relative to some agglomeration phenomena. Tserki et al. (2006) also observed an increase in Young's modulus upon olive husk flour (125 μm) addition in a polubutylene succinate (PBS) biodegradable biopolyester from about 375 MPa for unfilled PBS to 750 MPa for 30 wt% filled PBS (100% increase). Moreover, whatever

TABLE 2.4

Degradation Temperature and Mass Loss of PLA/Olive Pit Powder Biocomposites

Olive Pits (wt%)		Degradation Temperature (°C)	Mass Loss (wt%)
0		306	98.6
5	A	286	94.7
	B	283	96.7
	C	290	97.3
10	A	260	91.9
	B	255	93.5
	C	270	94.8
15	B	248	92.1
	C	249	92.4
20	B	256	94.0

PLA, poly(lactic acid).

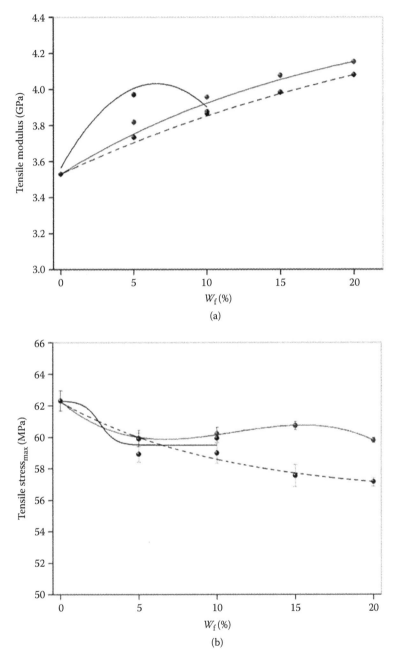

FIGURE 2.16   (a)(b) Tensile modulus and strength: (black line) olive pits A, (gray line) olive pits B, and (dashed line) olive pits C.

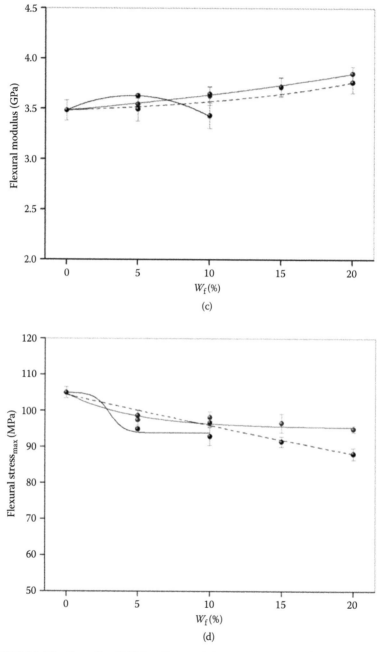

FIGURE 2.16 (*Continued*) (c)(d) bending modulus and strength of poly(lactic acid)/olive pit powder biocomposites as a function of the milled olive pits content: (black line) olive pits A, (gray line) olive pits B, and (dashed line) olive pits C.

the milled olive pit size, a decrease in strength with increase in filler content is observed indicating the presence of a poor interfacial adhesion between the hydrophilic ligno-cellulosic powder and the hydrophobic PLA, which does not allow efficient stress transfer between the two phases of the biocomposite. This lack of adhesion is con-firmed through ESEM observations (Figure 2.17) especially in the case of olive pits B and C. Indeed, in several locations of the inner structure and especially in the inter-facial area extended between particle fillers and the matrix, some gaps or entrapped air can be observed to a great extent, which indicates bad inclusion–matrix adhesion. For olive pits A, agglomerates are observed, which may also induce a decrease in strength. Such a decrease was also described in the literature (Tserki et al. 2006) for PBS/30 wt% olive husk flour biocomposite (46% decrease). Papanicolaou et al. (2012) reported a 48% increase in bending modulus and a 45% reduction in tensile strength in the case of an epoxy/olive pit powder (45 wt%) composite.

2.2.2.4 Conclusion

Milled olive pits can be used as reinforcements in PLA so that a full bio-based composite is available. Bending moduli of biocomposites are improved as powder content increases. The lowered flexural strength may be due to a poor interfacial

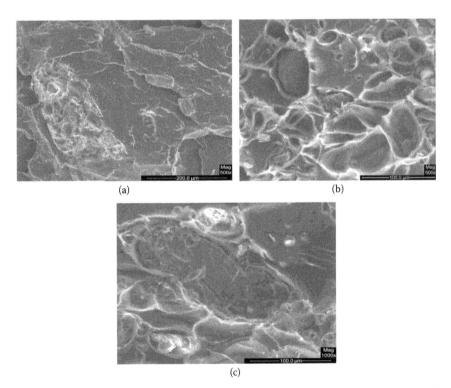

(a)   (b)

(c)

FIGURE 2.17   ESEM cross section analysis of fracture surface under flexion of poly(lactic acid)/olive pit powder biocomposites: (a) magnitude ×500—10 wt% olive pit powder A, (b) magnitude ×500—15 wt% olive pit powder B, and (c) magnitude ×1000—15 wt% olive pit powder C.

bonding between olive pit powder and PLA. From a general point of view, this is the main disadvantage encountered during the incorporation of natural lignocellulosic materials into polymers. Therefore, suggestions to achieve better interfacial bonding are reported in the literature. Tserki et al. (2006) suggested promoting olive husk flour/PBS interfacial adhesion by means of surface treatment, acetylation, and propionylation, as well as the addition of maleic-grafted PBS as a compatibilizer. Other alternatives were extensively depicted (Frisoni et al. 2001; Mahlberg et al. 1998) and are described in Section 2.2.3.

2.2.3 PROPERTIES IMPROVEMENT THROUGH SPECIFIC SURFACE TREATMENTS OF RICE AND WHEAT HUSKS

2.2.3.1 Introduction

It was previously shown that the performance of PLA-reinforced biocomposites is altered because of the high hydrophily of the agricultural by-products used such as rice and wheat husks due to the high number of hydroxyl groups on their surface. Furthermore, the presence of noncellulosic components in such by-products (e.g., lignin and wax) prevents them from obtaining good interfacial adhesion with the host polymer. Therefore, chemical treatments of their surface is a commonly used alternative. Many chemicals, such as alkali treatment, acetylation, organofunctional metals, and maleation, have been screened until now at a laboratory scale.

It is well known that alkali treatments remove noncellulosic components away from the fiber surface and therefore activate hydroxyl groups and water sensitivity. However, the surface properties of treated fibers depend on various factors such as the alkali concentration, the immersion time, and temperature (Alawar et al. 2009; Oh et al. 2005a,b). In major cases, these treatments lead to an increase in the mechanical properties (Huda et al. 2008; Valadez-Gonzalez et al. 1999). Nevertheless, to reduce the hydrophilic character of fiber, bifunctional molecules are mainly supplied on the fiber surface, such as organofunctional metals, leaving one of the functions available for further grafting (Abdelmouleh et al. 2004; Seki 2009). These coupling agents may be simplified to the general formula of $R-M-X_3$, where R is the organofunctional group, M is the tetravalent base metal (Si, Ti, and Zr), and X is a hydrolysable group (e.g., ethoxy for the silane-based coupling agent). When coupling to silica, the Si–O–Si bond from the silane-based coupling agent being stronger than the Ti–O–Ti bond is likely to impart better mechanical properties than titanate-based system. But as Ti tends to scavenge free radicals, the titanate-based coupling agent is likely used because of an acceleration of the curing process. Organozirconate systems are less readily available due to higher production costs (Fuad et al. 1995a).

2.2.3.2 Surface Treatments Used on Husks

Husks are treated by sodium hydroxide (NaOH) at three concentrations (0.5, 1.25, and 2.5 M) at room temperature for between 6 and 48 hours. The influence of NaOH concentration and treatment duration on the global weight loss of husks, and their chemical composition, moisture sensitivity, surface energy, and intrinsic color is investigated and results in the choice of a concentration of 1.25 M and duration of 24 hours

(Tran et al. 2012). Afterward silane treatments are performed on alkali-treated and raw husks. Various parameters, such as the chemical nature of the organosilane (γ-aminopropyltriethoxysilane named APS or γ-glycidoxypropyltrimethoxysilane named GPS), their concentrations (in a 1–10 wt% range, compared to fillers), the solvent used (acetone, methanol, ethanol, or water), the immersion time (1–24 h), and temperature (60–120°C) are studied (Le Moigne et al. 2012). For this study, 1 wt% of each silane is dissolved in a water/ethanol 30/70 w/w solution at pH 4 under continuous stirring for 30 minutes. Husks are then soaked in this silanol solution for 3 hours, washed, and finally dried at 60°C for 24 hours. AAPS (corresponding to NaOH + APS treatments), AGPS (corresponding to NaOH + GPS treatments), APS, and GPS correspond to the different treatments realized. Significant variations in the husk properties, such as chemical composition, surface properties, and water sensitivity, are given as evidence (Tran et al. 2012). Finally, treated husks are incorporated in PLA (20 wt%) and the properties of obtained biocomposites were determined with the purpose of observing enhanced husk/PLA adhesion due to surface treatments.

2.2.3.3 Biocomposite Properties

2.2.3.3.1 Mechanical Properties

Figure 2.18 aims to present the influence of different surface treatments performed on rice and wheat husks on mechanical properties of PLA-based biocomposites. An increase in flexural modulus of about 6%–8% is obtained for wheat husk–reinforced biocomposites (WH2.20), irrespective of the surface treatment applied, compared to untreated husks. No significant difference between the five treatments is observed even if the presence of a silane agent seems to have a more positive effect. An increase of about 5% is also obtained for rice husk–reinforced biocomposites (LRH.20) but only when rice husks are treated by NaOH alone. In the other case, such as in the presence of a silane agent, a decrease in modulus ranging between 3% and 6% is revealed. In the case of the bending strength, an increase is observed for both treated husks compared to untreated husks. This result is independent of the nature of the surface treatment. Improvements in moduli ranging between 8% and 20% for WH2.20 biocomposites and 8% and 10% for LRH.20 biocomposites are estimated. Higher bending strengths are obtained when both alkaline and silane treatments are applied on husks with slightly higher strengths in the case of APS agent (AAPS), suggesting a better husk/PLA adhesion. Only alkali treatment (A) leads to a drastic decrease in the bending strength (more than −50%) for both husks because of increasing husk hydrophily. Fuad et al. (1995a) compared the effect of different silane coupling agents applied on rice husk ash used as fillers in PP and also showed an improvement in the tensile strength and impact properties. Nevertheless, none of them increased the stiffness of the composites.

2.2.3.3.2 Aging Resistance

Figure 2.19 aims to present the influence of different surface treatments performed on rice and wheat husks on the aging resistance of PLA-based biocomposites at 65°C for 7 days in a ventilated oven. An increase in flexural modulus is obtained for aged biocomposites when husks are treated compared to untreated corresponding husks. The effect is more important for wheat husks (increase of about

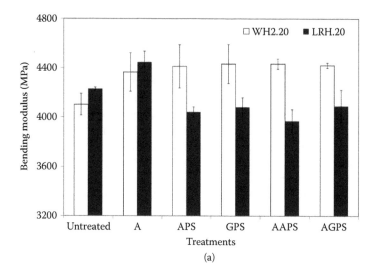

(a)

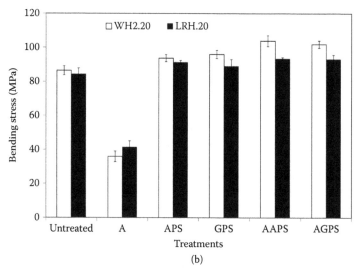

(b)

FIGURE 2.18 Influence of different surface treatments (A: NaOH; APS: γ-aminopropyltri ethoxysilane; GPS: γ-glycidoxypropyltrimethoxysilane; AAPS: NaOH + APS; AGPS: NaOH + GPS) applied on rice and wheat husks on mechanical properties: (a) bending modulus and (b) bending stress of poly(lactic acid) biocomposites reinforced by 20 wt% of husks.

18%–24%) when compared with rice husks (maximum increase of 7%). Moreover, a better aging resistance is obtained for husks treated only by the silane agent without alkaline pretreatment. In the case of the bending strength, an increase of about 33%–39% for husks treated by the silane agent alone (APS and GPS) and about 39%–58% for wheat husks treated by the silane agent with a preliminary alkaline treatment (AAPS and AGPS) is determined. For rice husks, no influence of AAPS and AGPS treatments is revealed as a decrease is given in evidence in

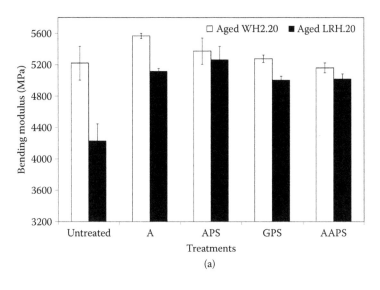

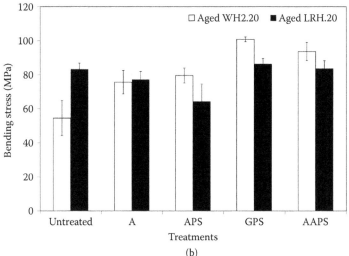

FIGURE 2.19 Influence of different surface treatments (A: NaOH; APS: γ-aminopropyltri ethoxysilane; GPS: γ-glycidoxypropyltrimethoxysilane; AAPS: NaOH + APS; AGPS: NaOH + GPS) applied on rice and wheat husks on aging resistance (thermal aging at 65°C for 7 days): (a) bending modulus and (b) bending stress of poly(lactic acid) biocomposites reinforced by 20 wt% of husks.

the case of APS and GPS treatments. As it is evident that the presence of a surface treatment improves the thermal aging resistance of husk-reinforced biocomposites, it is important to notify that the effect is highly dependent on the nature of the husk, with a greater effect of the surface treatment on wheat husks than on rice husks. This result is related to the presence of a silicon–cellulose membrane that covers the outer surface of rice husk (see Section 2.2.1.2.3) and that is responsible for the insufficient adhesion between accessible functional groups on rice husk surfaces and PLA. Finally, it can be concluded that a combined treatment of NaOH + silane or a single silane treatment is the best treatment to improve according to the mechanical characteristics.

2.3 CONCLUSIONS

Different by-products of local agriculture, such as rice and wheat husks produced in the south of France and milled olive pits coming from Greek olive oil production, are characterized in detail to determine their potential to be used as reinforcement systems in a well-known bio-sourced and biodegradable polymer, such as PLA. As a significant reinforcement effect is obtained (increase in modulus), a decrease in strength and impact resistance are produced in many cases, due to a lack of interfacial adhesion between the lignocellulosic agroresidues and the host polymer. Therefore, extensive investigation is carried out to propose efficient surface treatments, especially in the case of rice and wheat husks. An interesting improvement in mechanical properties as well as in aging resistance in different aging conditions (UV exposure, humidity-moisture cycling, water immersion) is obtained when combining an alkali treatment with silane grafting.

REFERENCES

Abdelmouleh, M., Boufi, S., Belgacem, M.N., Duarte, A.P., Ben Salah, A. and Gandini, A. 2004. Modification of cellulosic fibres with functionalised silanes: Development of surface properties. *International Journal of Adhesion and Adhesives* 24(1): 43–54.
Alawar, A., Hamed, A.M. and Al-Kaabi, K. 2009. Characterization of treated date palm tree fiber as composite reinforcement. *Composites Part B* 40(7): 601–606.
Auras, R., Harte, B. and Selke, S. 2004. An overview of polylactides as packaging materials. *Macromolecular Bioscience* 4(9): 835–864.
Averous, L. 2004. Biodegradable multiphase systems based on plasticized starch: A review. *Journal of Macromolecular Science. Polymer Reviews* C44(3): 231–274.
Ayswarya, E.P., Vidya Francis, K.F., Renju, V.S. and Thatchil, E.T. 2012. Rice husk ash—A valuable reinforcement for high density polypropylene. *Materials and Design* 41: 1–2.
Bledzki, A.K., Mamun, A. and Volk, J. 2010. Barley husk and coconut shell reinforced polypropylene composites: The effect of fibre physical, chemical and surface properties. *Composites Science and Technology* 70: 840–846.
Bledzki, A.K., Mamum, A., Bonnia, N.N. and Ahmad, S. 2012. Basic properties of grain by-products and their viability in polypropylene composites. *Industrial Crops and Products* 37: 427–434.
Carothers, H., Dorough, G.L. and Van Natta, F.J. 1932. Studies of polymerization and ring formation. X. The reversible polymerization of six membered cyclic esters. *Journal of American Chemical Society* 54(2): 761–772.
Chao-Lung, H., Le Anh-Tuan, B. and Chun-Tsun, C. 2011. Effect of rice husk ash on the strength and durability characteristics of concrete. *Construction and Building Materials* 25: 3768–3772.
Dalal, A.K., Rao, M.S. and Gokhale, K.V.G.K. 1985. Synthesis of NaX zeolite using silica from rice husk ash. *Industrial Engineering Chemistry and Product Research and Development* 24(3): 465–468.

Favaro, S.L., Lopes, M.S., Vieira de Carvalho Neto, A.G., Rogerio de Santana, R. and Radovanovic, E. 2010. Chemical, morphological and mechanical analysis of rice husk/ post-consumer polyethylene composites. *Composites Part A* 41: 154–160.

Frisoni, G., Baiardo, M., Scandola, M., Lednicka, D., Cnockaert, M.C., Mergaert, J. and Swings, J. 2001. Natural cellulose fibers: Heterogeneous acetylation kinetics and bio-degradation behavior. *Biomacromolecules* 2: 476–482.

Fuad, M.Y.A., Ismail, Z., Ishak, Z.A.M. and Omar, A.K.M. 1995a. Application of rice husk ash in polypropylene: Effect of titanate, zirconate and silane coupling agents. *European Polymer Journal* 31(9): 885–893.

Fuad, M.Y.A., Ismail, Z., Mansor, M.S., Ishak, M.Z.A. and Omar, M.A.K. 1995b. Mechanical properties of rice husk ash/polypropylene composites. *Polymer Journal* 27: 1002–1015.

Garlotta, D. 2002. A literature review of poly(lactic acid). *Journal of Polymer and the Environment* 9(2): 63–84.

Gonzalez, J.F., Gonzia, C.M., Ramiro, A., Gonzalez, J., Sabio, E., Ganan, J. and Rodríguez, M.A. 2004. Combustion optimization of biomass residue pellets for domestic heating with a mural boiler. *Biomass Energy* 27: 145–154.

Graupner, N., Herrmann, A.S. and Müssig, J. 2009. Natural and man-made cellulose fibre-reinforced poly(lactic acid) (PLA) composites: An overview about mechanical characteristics and application areas. *Composites Part A* 40(6–7): 810–821.

Huda, M.S., Drzal, L.T, Misra, M. and Mohanty, A.K. 2006. Wood-fiber-reinforced poly (lactic acid) composites: Evaluation of the physic-mechanical and morphological properties. *Journal of Applied Polymer Science* 102(5): 4856–4869.

Huda, M.S., Drzal, L.T., Mohanty, A.K. and Misra, M. 2008. Effect of fiber surface-treatments on the properties of laminated biocomposites from poly(lactic acid) (PLA) and kenaf fibers. *Composites Science and Technology* 68(2): 424–432.

Kawai, T., Rahman, N., Matsuba, G., Nishida, K., Kanaya, T., Nakano, M., Okamoto, H. et al. 2007. Crystallization and melting behavior of poly(lactic acid). *Macromolecules* 40: 9463–9469.

Le Marec, P.E. 2011. Modelling of mixing in the melt of PLA/cellulose fibers biocomposites. PhD Dissertation, Ecole Nationale Supérieure d'Agronomie de Montpellier. Montpellier, France.

Le Moigne, N., Longerey, M., Taulemesse, J.M. Benezet, J.C. and Bergeret, A. 2012. Improving the interface in natural fibres reinforced PLA biocomposites by optimized organosilane treatments. *Composites Part A*. Under submission.

Mahlberg, R., Niemi, H.E.M., Denes, F. and Rowell, R.M. 1998. Effect of oxygen and hexamethyldisiloxane plasma on morphology, wettability and adhesion properties of polypropylene and lignocellulosics. *International Journal of Adhesion and Adhesives* 18: 283–297.

Mansaray, K.G. and Ghaly, A. 1998. Thermal degradation of rice husks in nitrogen atmosphere. *Bioresource Technology* 65: 13–20.

Mehta, R., Kumar, V., Bhunia, H. and Upahyay, S.N. 2005. Synthesis of poly(lactic acid): A review. *Journal of Macromolecular Science. Polymer Reviews* C45(4): 325–349.

Ndazi, B.S., Karlsson, S., Tesha, J.V. and Nyahumwa, C.W. 2007. Chemical and physical modifications of rice husks for use as composite panels. *Composites Part A* 38: 925–935.

Oh, S.Y., Yoo, D.I., Shin, Y., Kim, H.C., Kim, H.Y., Chung, Y.S., Park, W.H. and Youk, J.H. 2005a. Crystalline structure analysis of cellulose treated with sodium hydroxide and carbon dioxide by means of X-ray diffraction and FTIR spectroscopy. *Carbohydrate Research* 340(15): 2376–2391.

Oh, S.Y., Yoo, D.I., Shin, Y. and Seo, G. 2005b. FTIR analysis of cellulose treated with sodium hydroxide and carbon dioxide. *Carbohydrate Research* 340(3): 417–428.

Pan, P., Kai, W., Zhu, B., Dong, T. and Inoue, Y. 2007. Polymorphous crystallization and multiple melting behavior of poly(L-lactide): Molecular weight dependence. *Macromolecules* 40: 6898–6905.

Panthapulakkal, S. and Sain, M. 2007. Agro-residue reinforced high-density polyethylene composites: Fiber characterization and analysis of composite properties. *Composites: Part A* 38: 1445–1454.

Papanicolaou, G.C., Koutsomitopoulou, A.F. and Sfakianakis, A. 2012. Effect of thermal fatigue on the mechanical properties of epoxy matrix composites reinforced with olive pits powder. *Journal of Applied Polymer Science* 124: 67–76.

Papanicolaou, G.C., Xepapadaki, A.G., Angelakopoulos, G.C., Zabaniotou, A. and Ioannidou, O. 2010. Use of solid residue from olive kernel pyrolysis for polymer matrix composite manufacturing: Physical and mechanical characterization. *Journal of Applied Polymer Science* 119: 2167–2173.

Park, B.D., Wi, S.G., Lee, K.H., Singh, A.P., Yoon, T.H. and Kim, Y.S. 2003. Characterization of anatomical features and silica distribution in rice husk using microscopic and microanalytical techniques. *Biomass and Bioenergy* 25: 319–327.

Pattara, C., Cappelletti, G.M. and Cichelli, A. 2010. Recovery and use of olive stones: Commodity, environmental and economic assessment. *Renewable and Sustainable Energy Reviews* 14: 1484–1489.

Perinovic, S., Andricic, B. and Erceg, M. 2010 Thermal properties of poly(L-lactide)/olive stone flour composites. *Thermochimica Acta* 510: 97–102.

Plackett, D., Løgstrup Andersen, T., Batsberg Pedersen, W. and Nielsen, L. 2003. Biodegradable composites based on l-polylactide and jute fibres. *Composites Science and Technology* 63(9): 1287–1296.

Premalal, H.G.B., Ismail, H. and Baharin, A. 2002. Comparison of the mechanical properties of rice husk powder filled polypropylene composites with talc filled polypropylene composites. *Polymer Testing* 21(7): 833–839.

Rodriguez, G., Lama, A., Rodriguez, R., Jimenez, A., Guillen, R. and Fernandez-Bolanos, J. 2008. Olive stone an attractive source of bioactive and valuable compounds. *Bioresource Technology* 99: 5261–5269.

Rozman, H.D., Yeo, Y.S. and Abubakar, A. 2003. The mechanical and physical properties of polyurethane composites based on rice husk and polyethylene glycol. *Polymer Testing* 22: 617–623.

Ruseckaite, R.A., Ciannamea, E., Leiva, P. and Stefani, P.M. 2007. Particle boards based on rice husk. In: *Polymer and Biopolymer Analysis and Characterization*. Zaikov, G. E. and Jiménez, A., Eds. Nova Science Publishers, 1–12. New York.

Seki, Y. 2009. Innovative multifunctional siloxane treatment of jute fiber surface and its effect on the mechanical properties of jute/thermoset composites. *Materials Science and Engineering: A* 508(1–2): 247–252.

Siriwardena, S., Ismail, H. and Ishiaku, U.S. 2003. A comparison of the mechanical properties and water absorption behavior of white rice husk ash and silica filled polypropylene composites. *Journal of Reinforced Plastics Composites* 22: 1645–1666.

Stefani, P.M., Garcia, D., Lopez, J. and Jimenez, A. 2005. Thermogravimetric analysis of composites obtained from sintering of rice husk-scrap tire mixtures. *Journal of Thermal Analysis and Calorimetry* 81: 315–320.

Stridsberg, K.M., Ryner, M. and Albertsson, A.C. 2002. Controlled ring-opening polymerization: Polymers with designed macromolecular architecture. Degradable aliphatic polyesters. In: *Advances in Polymer Science*. Abe, A., Albertsson, A.-C., Coates, G.W., Genzer, J., Kobayashi, S., Lee, K.-S., Leibler, L., Long, T.E., Manners, I., Möller, M., Okay, O., Tang, B.Z., Terentjev, E.M., Vicent, M.J., Voit, B., Wiesner, U. and Zhang, X., Eds. Springer Verlag, 157: 41–65. Berlin, Germany.

Sutas, J., Mana, A. and Pitak, L. 2012. Effect of rice husk and rice husk ash to properties of bricks. *Procedia Engineering* 32: 1061–1067.

Tran, T.P.T., Benezet, J.C. and Bergeret, A. 2012. Effects of alkali and silane treatments on poly(lactic acid) (PLA) reinforced by rice and Einkorn wheat husks mechanical properties. *Industrial Crops and Products*. To be submitted.

Tserki, V., Matzinos, P., Kokkou, S. and Panayiotou, C. 2005. Novel biodegradable composites based on treated lignocellulosic waste flour as filler. Part I. Surface chemical modification and characterization of waste flour. *Composites Part A* 36(7): 965–974.

Tserki, V., Matzinos, P. and Panayiotou, C. 2006. Novel biodegradable composites based on treated lignocellulosic waste flour as filler. Part II. Development of biodegradable composites using treated and compatibilized waste flour. *Composites Part A* 37(9): 1231–1238.

Turmanova, S., Dimitrova, A. and Vlaev, L. 2008. Comparison of water absorption and mechanical behaviors of polypropene composites filled with rice husk ash. *Polymer Plastics Technology and Engineering* 47: 809–818.

Valadez-Gonzalez, A., Cervantes-Uc, J.M., Olayo, R. and Herrera-Franco, P.J. 1999. Effect of fiber surface treatment on the fiber-matrix bond strength of natural fiber reinforced composites. *Composites Part B* 30(3): 309–320.

Vlyssides, A., Barampouti, E. and Mai, S. 2008. Physical characteristics of olive stone wooden residues: Possible bulking material for composting process. *Biodegradation* 19: 209–214.

Vocke, G. and Liefert, O. 2012. *Wheat Outlook*. In United States Department of Agriculture (USDA). A report from the economic research service.

Vourdoubas, J. 1999. Heating greenhouse in Crete using biomass. The case of olive kernel wood. *Energy and Agriculture towards the Third Millennium (AGENERGY'99)*, 101–104. Athens.

Yang, H.S., Kim, H.J., Son, J., Park, H.J., Lee, B.J and Hwang, T.S. 2004. Rice-husk flour filled polypropylene composites: Mechanical and morphological study. *Composite Structures* 63: 305–312.

Yasuniwa, M., Tsubakihara, S., Sugimoto, Y. and Nakafuku, C. 2004. Thermal analysis of the double melting behavior of poly(L-lactic acid). *Journal of Polymer Science: Part B: Polymer Physics* 42: 25–32.

Yoshida, S., Ohnishi, Y. and Kitagishi, K. 1962. The chemical forms, mobility and deposition of silicon on rice plant. *Soil Science and Plant Nutrition* 8(3): 15–21.

Yussuf, A.A., Massoumi, I. and Hassan, A. 2010. Comparison of polylactic acid/kenaf and polylactic acid/rice husk composites: The influence of the natural fibers on the mechanical, thermal and biodegradability properties. *Journal of Polymers and the Environment* 18(3): 422–429.

Zain, M.F.M., Islam, M.N., Mahmud, F. and Jamil, M. 2011. Production of rice husk ash for use in concrete as a supplementary cementious material. *Construction and Building Materials* 25: 798–805.

Zhao, Q., Tao, J., Yam, R.C.M., Mok, R.K.Y. and Song, C. 2008. Biodegradation behavior of polycaprolactone/rice husk ecocomposites in simulated soil medium. *Polymer Degradation and Stability* 93: 1571–1576.

3 Processing Cellulose for Cellulose Fiber and Matrix Composites

Robert A. Shanks

CONTENTS

3.1 INTRODUCTION

Cellulose is abundant as a structural component of plants with properties based on a regular stereochemical structure that leads to strong intra- and intermolecular interactions and packing into crystalline arrays that build into increasing levels of supramolecular structures. Cellulose is a remarkable material from its primary chemical bonding through each structural level to realization of the resultant mechanical properties. Cellulose poses problems for the polymer scientist in that it cannot be melted, it is insoluble in almost all liquids, and it is found accompanied by several other materials including hemicelluloses, lignin, pectin, and waxes. In plants, via its association with other materials, cellulose provides us with an example of the structure and properties of a composite. Craftsman can use cellulosic materials from their natural form by mechanically shaping them. Materials scientists need to separate the cellulose from plant materials and reconstitute it into new materials, independent of the fiber source. The cellulose fibers are then able to be combined with other polymers and additives, and shaped into complex products using normal moulding techniques.

When melting is not possible, separation and processing require dissolving. Early developments were based on soluble cellulose derivatives such as cellulose nitrate, cellulose acetate (CA), CA–butyrate, CA–propionate, methyl cellulose, ethyl cellulose, and carboxymethyl cellulose; these are all soluble in an organic solvent or water. Cellulose acetate butyrate (CAB) composites with hemp fibers were prepared using tributyl citrate as plasticizer. CAB can be dissolved or melt processed, unlike when cellulose is used as a matrix (Ouajai and Shanks 2009a).

Cellulose solution is assisted by complexation to disrupt internal hydrogen bonds. Complexation with tetraamminediaquacopper dihydroxide, or Schweitzer's reagent (Schweizer 1857), can dissolve cellulose, and upon acidification, the cellulose is precipitated or regenerated. Schweitzer's study describes how plant fibers were dissolved in an alkaline copper sulfate solution containing ammonia, and then it is regenerated to form a horn-like, translucent brittle mass. Regeneration from a stream of cellulose solution passed into acid is used to form fibers and films. The process is inefficient; the structure of the complex is imprecise and much waste is created.

A more efficient process is the reaction of the cellulose raw material with carbon disulfide and sodium hydroxide to form a soluble sodium cellulose xanthate derivative. The degree of substitution and molecular conformation, represented by radius of gyration, expansion factor, and effective bond length, of cellulose xanthate solutions were characterized using dilute solution viscometry (Das and Choudhury 1967). Regeneration of the xanthate in acid gives viscose that can be regenerated in the form of fibers or film. Although this is a more efficient process, again there is much waste that cannot be regenerated. The quest for cellulose solvents continues with several recent advances. Dissolution of sodium cellulose xanthate in sodium hydroxide under flow has been investigated using a rheo-optical technique showing that a highly concentrated viscoelastic phase forms that slowly disperses into solution independent of temperature (Le Moigne and Navard 2004).

Cellulose is better described by its chemical name poly(β-1,4-glucopyranose), although this does not fully describe the conformation. The repeat unit is the disaccharide cellobiose that, despite structural complexity, is relatively planar with hydroxyl groups positioned for hydrogen bonding intramolecularly along the axis of the molecule and intermolecularly with other cellulose units that can be adjacently close packed. Figure 3.1 shows a representative energy-minimized cellobiose molecule in two orientations to reveal stereochemistry and another cellobiose molecule enclosed in a solvent-accessible mesh boundary to illustrate the volume occupied by the molecule.

Cellulose is a remarkable biopolymer, where the molecular structure and stereochemistry of the repeat unit has many possible configurations; however, only the one with the most stability due to equatorial conformation of hydroxyl groups and interchain bond is present. The complex bonding results in an almost planar conformation that is highly suited to orientation, intra- and intermolecular hydrogen bonding that leads to fiber formation. Typical of other biomaterials, such as proteins and nucleic acids, the primary molecular bonding establishes the base for a complex, hierarchical, and supramolecular structure.

Cellulose is the most abundant organic substance. It is a renewable resource sourced from all plant materials. Because it is the structural component of plants, it has many applications in structures of the built environment for which it is ideally suited. Cellulose materials differ according to source with regard to fiber density, fiber length, and the proportion of cellulose present. Cellulose content, extractives, and moisture content are variables that influence the nature of wood flour formed

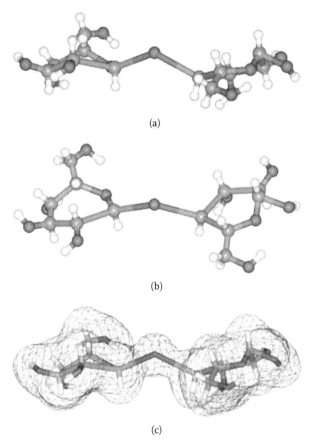

(a)

(b)

(c)

FIGURE 3.1 Cellobiose structures representing a cellulose repeat unit, showing edge and perspective views (a and b) and solvent accessible excluded volume (c).

from or fibers isolated from wood species (Poletto et al. 2012). Another form of very fine cellulose fibers is bacterial cellulose with fibrils where the intensity of a Raman band did not change significantly with rotation angle indicating an in-plane 2D network of fibrils with uniform random orientation, consistent with auxetic behavior (negative Poisson ratio) (Tanpichai et al. 2012).

Cellulose is the structural component of plants. Protein is another green material that can provide strong materials, from sources as diverse as soy fiber products, animal protein, and spider web fiber. Protein-based fibers have been developed, although they have not found the wide interest and applications of cellulose. Starch gives materials with ease of biodegradation either as a disperse phase or as a main component, although not of high performance. Renewable organic materials can be combined with minerals to form composites with a small environmental footprint. Traditional cellulose composites are prepared from or include wood flour, wood chips, waste paper, waste textiles and carpet fibers, and impregnated textiles (Linoleum).

In this chapter, the emphasis is on cellulose as a matrix phase where the dispersed phase can be either cellulose or other materials. Whether reinforcement or matrix, cellulose must be separated from the source plant material. Since cellulose cannot be melt processed, formation of a solution is required for fiber regeneration or matrix phase formation. A solution enables cellulose to be regenerated into a preferred morphology or shaped as needed for a product. These aspects are reviewed and evaluated in this chapter.

3.2   SOURCES OF CELLULOSE

The largest source of cellulose is trees, generally either native forests or plantations. The volume of cellulose available from trees is huge, although trees have a long life cycle inhibiting supply variations in response to changes in demand. Isolation of cellulose from trees is an extensive process. The yield of cellulose from fast growing trees such as eucalypt is about five times that of cotton, with less water, pesticides, and herbicides. Cellulose content of wood is about 40%·w/w. The wood may be reduced to fibers, flour, or pulp and used without separation of cellulose. Alternatively, for highly specified and quality products, cellulose must be separated from other components.

Waste agricultural cellulose fibers are an available resource; however, collection is the barrier to utilization. Crop fibers, particularly bast fibers, are fast growing with ready separation of fibers and high mechanical properties. Crop fibers are waste materials that can be left in soil as fertilizer or value added as a source of cellulose fibers.

Wood is the main source of cellulose for paper, fillers, and regenerated fibers. Bast fibers are increasingly used for composite reinforcement, whereas in the past, they have been prominent in textiles and used for paper. Cellulose fiber waste generated from many agricultural crops is being developed into fiber reinforcements after separation and purification. The source and processes are often associated with a particular crop in a suited region or climate, so research into new materials and applications is often localized. Bamboo is a notable material for high performance and for the breadth of forms in which it can be used. Bamboo fibers regenerated for use in textiles is an application that has recently expanded. The fibers are strong, lustrous, and resist bacterial growth. Other bamboo applications are in composites, with laminated flooring panels being a high value and quality product where hardness and appearance are important.

A material that is economic in one location may not be considered so in another location. Because of the composition and form of the plants, specific separation, extraction, and purification will be required. This means cellulose fiber preparation and processing is source dependent, and composite developments will not be ubiquitous.

Fiber length and quality are important considerations when the fibers are to be used without regeneration. Cellulose fibers from trees are generally superior, although this depends on the part of the tree from which they are derived (Ververis et al. 2004). Hardwood sources are generally superior to softwood. Bast fibers such as flax, hemp, ramie, and kenaf have high modulus and strength that, relative to density, are comparable with glass fibers.

3.3   NATURAL CELLULOSE PURIFICATION

A first step in utilization of cellulose from any source is separation of fibers from the plant source. The physical processes used are retting, decortication, and steam explosion. With retting, the plant materials are left in water for natural enzyme and hydrolytic degradative reactions to occur. Regular highly ordered crystalline cellulose fibers resist degradation compared with more porous amorphous regions of the plant where degradation proceeds faster. This selective degradation separates fibers from extraneous material. Decortication is an alternative mechanical disruption of the plant material where compressive and shear forces of rough surfaces rollers separate strong, high-modulus fibers from other materials. Retting is a natural process that is time dependent, while decortication is a rapid machine-driven process. Steam explosion uses the force of rapidly boiling water to tear the fibers apart from within. The plant materials are immersed in water that is heated under pressure to about 140°C. Pressure-assisted absorption of water into the plant structures occurs, then the pressure is released and the super-heated water rapidly boils, tearing apart the plant material and releasing fibers. The amorphous nonfiber plant material is selectively disrupted freeing the fibers. Steam explosion causes a breakdown of structural lignocellulosic material depending on the time and temperature. The process should be optimized to limit thermal degradation of the cellulose as determined by 5-hydroxymethylfurfural concentration and an increase in char formed during thermogravimetry analysis (TGA) of the product cellulose. TGA showed a decrease in thermal degradation temperatures with severity of the steam explosion process (Jacquet et al. 2011).

After separation of fibers from plant materials, they need to be further purified. Initially, organic solvent extraction removes waxes. For example, a laboratory Soxhlet extraction using acetone will dissolve waxes and leave a more porous structure for further purification. A second extraction can be performed with sodium hydroxide solution, where lignin, pectin, and hemicellulose can be extracted and cellulose swelled but not dissolved.

Enzyme purification using pectinase will selectively hydrolyze and remove pectin. Enzyme treatment, using pectate lyase enzyme, has an optimum pH and temperature, and the suspension of fibers should be adjusted to these conditions for a rapid reaction rate. An increase in the Brunauer–Emmett–Teller surface area was observed (Ouajai and Shanks 2006).

3.4 ENHANCING FIBER STRUCTURE AND PERFORMANCE

Cellulose fibers are swelled in sodium hydroxide solution; typically about 20%·w/v is used. Some hydroxyl groups undergo acid–base equilibria with the sodium alkoxide salt. Cellulose hydroxyl groups are more acidic than typical alcohol hydroxyl groups due to the electron withdrawing influence of neighboring hydroxyl groups. Overall, hydrogen bonding and sodium salt formation cause swelling of cellulose. In the swelled state, the crystal structure slowly changes from metastable native cellulose (type I) to the more stable tensile cellulose (type II). This process is called Mercerization, resulting in more flexible fibers that are preferred for comfort in cellulose fabrics. Cellulose type I has parallel-aligned cellulose molecules forming the crystal structure. Cellulose type II has antiparallel-aligned molecules forming the crystal structure.

After purification to give pure cellulose crystal structures, it can be further perfected by partial hydrolysis. Acid hydrolysis can be performed with sulfuric acid and selectivity obtain partially hydrolyzed cellulose, depending on the hydrolysis time. Amorphous cellulose and defect crystals are hydrolyzed first since diffusion of the acid is faster. The more perfect crystals resist diffusion of the acid and the reaction is limited. After a selected time, the remaining cellulose is washed free of acid. The process may result in half the original cellulose being hydrolyzed; however, the remaining cellulose consists of the more perfect crystals.

Cellulose with larger crystal size has been found to be more resistant to thermal degradation, although activation energy for degradation was constant (Kim et al. 2010). The results indicate that crystal size determines diffusion of degradation products from cellulose rather than degradation mechanism.

In a first stage of selective and progressive diminution of cellulose, microcrystalline cellulose (MCC) is formed and separated by filtration or centrifugation. When dry, MCC is a fine powder. When dispersed in water or polar solvents, MCC gives increased viscosity, shear thinning, and thixotropy. It has application in many products including foods. Further selective hydrolysis leaves cellulose nanofibers that have even more perfect crystal and hence fiber structure than MCC. Cellulose nanofibers are too small to scatter light so dispersion in water is transparent. They are too small to be filtered, so after neutralization with sulfuric acid, the resulting salt solution must be separated through a dialysis membrane. Cellulose nanofibers are finding application in composites due to their superior mechanical properties, high surface/volume ratio, and low density compared with other reinforcing fibers and nanofibers.

Cellulose fibers have been treated with hydrogen bonding compatibilizers to enhance the interface with matrix polymers such as poly(hydroxybutyrate) (PHB) and poly(lactic acid) (PLA). Bisphenol-A and 4,4′-thiodiphenol have been found to be suitable intermediary hydrogen bonding coupling agents. Hyperbranched polyesters are other toughening agents with many hydrogen bonding hydroxyl groups in the outer branching shell. Replacement of water in cellulose by another plasticizer, such as tributyl citrate, will decrease moisture sensitivity and limit embrittlement in dry conditions (Wong et al. 2007).

Hemp fibers were subjected to several treatments including sodium hydroxide acetic anhydride, maleic anhydride, and silane. Sodium hydroxide modified and increased crystallinity of the cellulose giving increased modulus and strength. The other treatments decreased crystallinity and hence strength, although modulus was increased. Increased modulus was related to the removal of noncellulosic components during treatments (Sawpan et al. 2011). Cellulose ethers such as allyl cellulose, allyl carboxymethyl cellulose, and allyl n-hydroxypropyl cellulose were prepared as crosslinkable precursors for cellulose compositions to consolidate wood (Cipriani et al. 2010). Deoxychlorocellulose was prepared by reaction of cellulose with thionyl chloride. Derivatization was performed by reaction of deoxychlorocelluloses with diaminoethane and the aminated products were found to be noncrystalline (da Silva Filho et al. 2010). The amino groups are available for other further reactions.

Graft polymerization within purified and dried hemp fibers was used to create modified fibers with reduced voids that were more water resistant although not brittle when dried. Acrylonitrile with radical initiator was absorbed into the hemp fibers where it was polymerized. Excess polyacrylonitrile not bonded to the fibers was removed by extraction and grafting yields to 10%·w/w were obtained

(Ouajai et al. 2004). Cellulose flax fibers, grafted interstitially with butyl acrylate, 2-ethylhexylacrylate, or methyl methacrylate, were used to form composites with PLA. Monomer was absorbed into voids. Vacuum treatment was followed by initiated monomer addition. The grafted fibers contained acrylate polymer in place of water in voids and limited sensitivity to water or humidity. They showed improved adhesion to the PLA matrix (Shanks et al. 2006).

3.5 CELLULOSE SOLUTIONS

A continuing problem for cellulose processing is its lack of solubility. Figure 3.2 outlines some solvent systems, which initially involved complexation, derivative formation, and then special ionic solvents. The problem presented in dissolving is that the strong internal hydrogen bonding and regularity are greater than the cellulose–solvent interactions.

Cellulose cannot be melted to mix and mould. Cellulose solutions have provided the means to reconstitute cellulose into fibers and other shapes that cannot be obtained from the natural material. However, cellulose is a strongly integrated substance with inter- and intramolecular hydrogen bonds. The challenge in dissolving cellulose has been solved by several processes and new techniques continue to be developed. Complexation with alkaline tetraamminecopper was the first method used. The complex is not stoichiometric, although there is an optimum ratio of reagents. Much waste is formed after the acid regeneration step and better methods were sought.

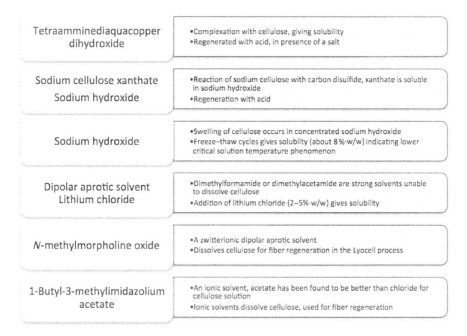

FIGURE 3.2   Solvents for cellulose in order of first use with comments on function.

Derivatization is another approach whereby internal hydrogen bonding is reduced. Cellulose has been converted to cellulose nitrate, acetate, propionate, butyrate, methyl cellulose, ethyl, carboxymethylation, and other derivatives. These derivatives no longer have the properties of cellulose; they are new materials with specific desirable properties. Formation of cellulose materials requires reversible derivatization so cellulose can be regenerated into the desired form. The cuprammonium method was widely used, followed by the sodium cellulose xanthate process. The cellulose raw material such as wood is steeped in sodium hydroxide and then carbon disulfide is added to form the xanthate, which slowly dissolves. Filtration separates other components and cellulose is regenerated by inserting a fine stream of the xanthate solution into an acid bath to produce fibers. The reactions are shown in Figure 3.3. Cellulose fibers formed from the xanthate process are called Rayon or viscose fibers. The fibers have a porous interior with a smooth cellulose skin on the surface (Das and Choudhury 1967).

Sodium hydroxide solution interacts with cellulose in several ways. First, sodium hydroxide solution swells cellulose by disrupting hydrogen bonding. Sodium hydroxide forms sodium alkoxide salts with some of the hydroxyl groups, because they are more acidic than typical alcohol hydroxyl groups. The alkoxide groups facilitate nucleophilic reactions, such as the reactions used for carboxymethylation via chloroacetic acid, and for coupling of some reactive dyes with cellulose via dichlorotriazine groups. As discussed earlier, they initiate reaction with carbon disulfide in the xanthation reaction. With sodium hydroxide concentrations of about 10%–20%·w/v and moderate heating (60°C), the swelling allows rearrangement of crystal structure from type I to type II in the Mercerization process described under fiber structure. Interaction with sodium hydroxide solution is more complex, depending on concentrations and conditions.

Subjecting cellulose and sodium hydroxide solutions to freeze–thaw cycles gradually dissolves some of the cellulose. Using 20%·v/w sodium hydroxide and MCC powder with cooling to −10°C, solutions of 8%·w/v cellulose have been obtained in the author's laboratory. Dissolution by cooling implies that cellulose has a lower critical solution temperature (LCST). This is consistent with cellulose being strongly internally hydrogen bonded. Heating is likely to disrupt hydrogen bonds with water and sodium hydroxide

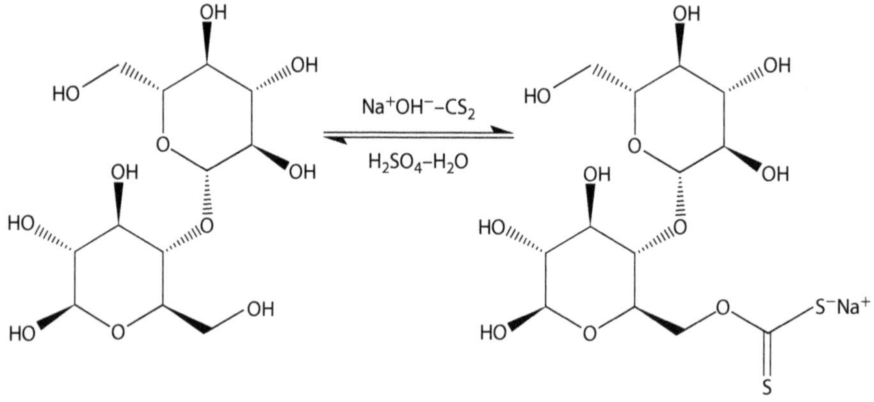

FIGURE 3.3 Sodium cellulose xanthate reaction and regeneration with cellobiose as the example structure, showing preferential formation on C6.

more than the self-hydrogen bonds. Cooling will increase the strength of all hydrogen bonds including those between cellulose and sodium hydroxide solution. Yet, cooling does not change the equilibrium constant for formation of sodium alkoxide salts. Figure 3.4 shows a schematic LCST phase diagram for cellulose in sodium hydroxide.

Figure 3.4 is a schematic based on several assumptions. The first is that dissolving behavior results in a dissolved mass fraction of about 0.1 after cooling to −10°C and that this temperature and concentration is the LCST. Lower concentrations than LCST are expected to be formed at slightly higher temperatures. At higher cellulose concentrations, the solution will be a solution of sodium hydroxide and water in solid swelled cellulose. At ambient temperatures, about 10%–15% w/v water is absorbed into cellulose and referred to as bound water. Further quantities of water are not bound to the cellulose and this excess water is called free water, able to be frozen at temperatures <0°C. As shown in Figure 3.4, cellulose must be heated to at least 100°C and typically higher to remove bound water at cellulose mass fractions approaching 1. Compositions with dissolved cellulose or bound water are below the curve in the one-phase region. Compositions with free water or where cellulose is not dissolved are above the curve in the two-phase region. Although the LCST diagram in Figure 3.4 is a schematic, actual concentrations found in the author's experiments with cellulose have been used to construct the curve. It would be interesting to construct a quantitative LCST phase diagram; however, it is difficult to obtain quantitative data from the freeze-thaw dissolving process thence to measure equilibrium bound water in cellulose.

Hydroxypropyl cellulose (HPC) has been observed to have LCST behavior; at temperatures higher than LCST, HPC becomes self-associated into nanospheres. Formation of nanospheres by heating from LCST has been applied to dispersion of graphene, where the HPC nanospheres were attached to graphene layers thereby stabilizing the layers and enhancing percolation threshold and electrical conductivity (Liao et al. 2012).

Hydrogen bond donors assist in dissolving cellulose by creating stronger hydrogen bonds than the internal cellulose hydrogen bonds. Sodium hydroxide solution combined with a hydrogen bond donor, such as urea or thiourea, forms more concentrated solutions of cellulose. Chitin (2-acetoxyaminocellulose) whiskers were used to reinforce cellulose

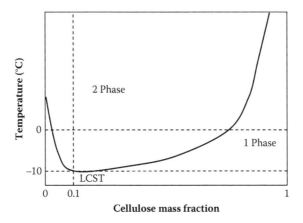

FIGURE 3.4 Schematic representation of cellulose–water–sodium hydroxide LCST phase diagram.

films cast from sodium hydroxide–urea solution that allowed uniform dispersion of the whiskers, enhancing modulus and tensile strength of the films (Huang et al. 2013).

Cellulose can be dissolved in sodium hydroxide (7%·w/w)–urea (12%·w/w) at $-12°C$ within 2 min due to hydrogen bonding between cellulose and the hydrogen bond donor solvent system, and assisted by dispersion of the cellulose (Qin et al. 2012). The low temperature for solubility shows that LCST phenomena exist in the strong hydrogen bond donor systems such as urea. Dispersion was obtained by combinations of stirring time and stirring speed. Solution characteristics were studied with dynamic light scattering and dynamic rheology. Solubility was related to an inclusion complex between cellulose and the hydrogen bond donors.

Regenerated cellulose in aqueous solutions of lithium chloride and urea exhibits strong swelling from 15 to 46 µm diameter, depending on temperature and solution composition. Swelling is limited at temperatures of 120–130°C in a super-heated steam environment (Tatárová et al. 2012). This is typical of LCST behavior, where solvent–solute interactions are less at higher temperature.

Dimethylformamide (DMF) and dimethylacetamide (DMAC) are dipolar aprotic solvents that dissolve many polar self-associating substances although they do not dissolve cellulose directly. DMF and DMAC dissolve cellulose when combined with urea or thiourea that are protic and provide hydrogen donation. Addition of lithium chloride to DMF and DMAC forms Li complexes with cellulose and gives an overall increase in the dielectric constant of the medium and increased solvent power for formation of cellulose solutions. These solvent and additive combinations dissolve cellulose but have the problem of later removal to give a pure cellulose product.

Cellulose solution in sodium hydroxide and thiourea was formed by cooling to 5°C. Multiwalled carbon nanotubes (MWCNT) were dispersed in the cellulose solution and rheology was evaluated to detect a concentration-dependent gel point in the viscoelastic solutions. Mechanical strength of the MWCNT–cellulose composites was relatively high compared with cellulose (Lue and Zhang 2010).

N-methylmorpholine oxide (NMMO) can directly dissolve cellulose due to its highly polar zwitterionic structure. NMMO is used to dissolve cellulose from wood, purify the cellulose by filtration, and regenerate the cellulose as fibers. NMMO is the basis of the Lyocell process that is also used to form Tencel fibers. NMMO is not readily volatile; however, it can be concentrated from water in the regeneration bath and recycled. This results in a single solvent and simple solution process for processing cellulose without derivatives and other additives, and the expensive NMMO is recycled giving an enclosed system.

All-cellulose composites were prepared from purified hemp by adding a solution of cellulose in NMMO to a fiber pad. The NMMO solution contained 12%·w/v cellulose, and after dissolving of the cellulose, the solution was diluted with water to prevent further dissolving of cellulose from the hemp. Impregnation of the fibers was performed at 80°C where viscosity of the NMMO solution was low to ensure efficient fiber wetting. The composite was compression molded at 85°C, then washed in water–ethanol (1:1) until NMMO was removed, dried, and then compression molded again to consolidate the composite. Mechanical properties of the composite depended on size, surface area, crystallinity, and the structural swelling of fibers in the presence of NMMO (Ouajai and Shanks 2009b).

Ionic solvents are currently being used to form cellulose solutions from which the cellulose can be regenerated by dilution and then washing with water. The ionic solvents are similar to NMMO except they have a separate cation and anion. There are many ionic solvents available; however, 1-butyl-3-methylimidazolium acetate (BMIMAc) has been found to be most suitable to dissolving cellulose, the acetate being more effective than the corresponding chloride.

The process of dissolving cellulose in ionic liquids can be indirect such as first swelling the cellulose in a water (2%–5%)–ionic liquid (1-butyl-3-methylimidazolium chloride) mixture, then evaporating water to concentrate the ionic liquid and gradually dissolve the cellulose. Water gives more efficient swelling prior to dissolution and enhanced spinning performance mechanical properties of the regenerated cellulose (Cai et al. 2012). A problem with ionic liquids is their high viscosity and limited miscibility with nonpolar reagents and reaction products when derivatizing cellulose. Co-solvents are used to complement ionic liquids where the solvent mixture can dissolve cellulose, reagents, and reaction products. Suitable co-solvent should have miscibility, low polarity, and relatively high basicity (Gericke et al. 2011).

Wood pulp cellulose has been dissolved in 1-allyl-3-methyl-limidazolium chloride and the rheological properties characterized using steady and dynamic shear. The data showed shear thinning behavior and viscosities that fitted a modified cross equation to obtain zero-shear viscosities (Lu et al. 2012a).

Interactions between cellulose and ionic liquids need understanding to interpret solubility and solution properties. These interactions have been investigated using molecular simulations for 1-*n*-butyl-3-methylimidazolium hexafluorophosphate [BMIM][PF6] and 1-*n*-butyl-3-methylimidazolium acetate [BMIM][Ac], structure shown in Figure 3.5. Simulations revealed that hydrogen bonds were disrupted by the ionic liquid at the surface of the cellulose. [BMIM][Ac] was found to have the strongest capability to disrupt cellulose hydrogen bonds (Gupta et al. 2011).

Ionic solvent combined with DMF, DMAC, or dimethylsulfoxide has been used to dissolve cellulose to provide suitable viscosity, electrical conductivity, and surface tension for electrospinning of cellulose fibers. Different dissolution power corresponded with changes in the characteristics of the fibers as observed using scanning electron microscopy. High viscosity, high surface tension, and some shear thinning were desired for fiber production by electrospinning (Härdelin et al. 2012).

BMIMAc solutions of cellulose were used to prepare all-cellulose composites from layers of Rayon and linen by solvent impregnation accompanied by heat and pressure to partially dissolve the fabrics and form a matrix phase after cellulose regeneration. Dissolution of the synthetic Rayon was greater than that of the Mercerized natural fiber, linen (Huber et al. 2012). This reflects the more porous structure of Rayon due to formation by a regeneration process.

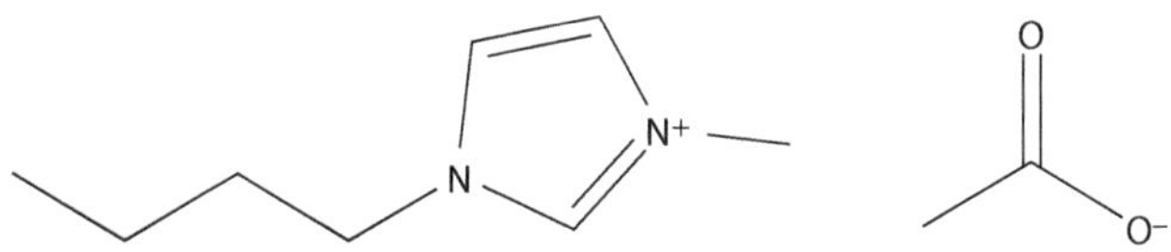

FIGURE 3.5 Ionic solvent, 1-butyl-3-methylimidazolium acetate.

3.6 CELLULOSE REGENERATION

Cellulose complexes and alkaline salts are precipitated by addition to acid, where the physical state at addition determines the cellulose morphology such as fiber, sheet, or particle. NMMO and ionic liquid solutions are regenerated by addition to water. Ionic liquids have been compared to NMMO for the Lyocell process: due to differences in rheological behavior of the solutions, the structure and physical properties of the fibers vary where 1-ethyl-3-methylimidazolium acetate and 1-ethyl-3-methylimidazolium diethyl phosphate were used (Ingildeev et al. 2012a). Blended cellulose fibers have been formed by dissolution in ionic solvents together with polyacrylamide or polyamides including m-aramide, where regeneration yielded blended fibers (Ingildeev et al. 2012b).

Regenerated cellulose films formed from solution in lithium chloride–*N,N*-dimethylacetamide was oriented by stretching to 30%. Orientation was confirmed by wide-angle X-ray scattering and the films exhibited enhanced dynamic modulus and thermal resistance (Kim et al. 2012).

The Lyocell process has been used to form cellulose fibers combined with MWCNT using different draw ratios. The Lyocell–MWCNT fibers showed enhanced mechanical properties and electrical conductivity due to alignment of MWCNT along the fiber axes (Lu et al. 2012b).

When cellulose is to be a matrix, the cellulose solution is impregnated into a fiber material and the composite is formed by immersion in water. Wood flour and bark flour derived from cedar were compression molded at 210°C with 1-butyl-3-methylimidazolium chloride (BMIC) (40%·w/w) to form composites by partial dissolution of the cellulose in BMIC. The dissolved cellulose form the matrix after regeneration in ethanol, which was used to extract BMIC. A second annealing molding step smoothed the surface and increased cellulose crystallinity by further action of residual BMIC. The annealing step increased tensile strength and thermal stability (Shibata et al. 2013).

Treatment of cellulose using the ionic liquid, 1-ethyl-3-methylimidazolium acetate, prior to enzymatic hydrolysis resulted in a loss of native cellulose crystal structure giving type II crystals. In addition to crystalline rearrangement, the ionic liquid also disrupted lignin–cellulose complexes. The resulting structural changes facilitated an increase in the rate of enzyme hydrolysis for effective use as biofuel (Cheng et al. 2011).

Ionic liquids have been used singly or in binary mixtures to dissolve cellulose to form electrospinning solutions, for example 8%·w/w cellulose in 1-ethyl-3-methylimidazolium acetate produced fibers with average diameters within 470 ± 110 nm. Surface tension of the spin solution was found to be important, so mixtures of ionic liquids, such as 1-ethyl-3-methylimidazolium acetate and 1-decyl-3-methylimidazolium chloride, were used to obtain an optimum, which gave fiber diameters of 120 ± 55 nm. Ionic liquids with water coagulation were found to provide an environmentally friendly method of fiber production, because the ionic liquids are nonvolatile and recyclable (Freire et al. 2011).

Derivatization of cellulose from eucalyptus pulp in an ionic liquid solution, 1-ethyl-3-methylimidazolium chloride, was used to prepare CA in homogeneous medium. The product CA was spun into fibers with properties dependent on the degree of substitution (Kosan et al. 2010).

All-cellulose composites were formed from ramie fibers using sodium hydroxide–urea solvent to dissolve cotton linter fibers for the matrix phase. The cellulose solution was formed by cooling mixtures to −12°C prior to rapid stirring at ambient temperature. Ramie fibers were added to the cellulose solution with stirring, and then centrifuged at 6000 rpm to remove bubbles. The composites were optically transparent, thermally stable, and exhibited high strength, due to excellent compatibility between the components (Yang et al. 2010).

3.7 ENHANCED CELLULOSE BY PARTIAL HYDROLYSIS

Selective hydrolysis with mineral acid, typically sulfuric acid, is used to remove first amorphous and defect structures, then the less-perfect crystals. More perfect crystals with their regular dense packing resist ingression of acid and are preserved. Timing is essential; after a hydrolysis period, the mixture is neutralized with alkali to terminate reaction. Separation of remaining cellulose depends on particle size. Centrifugation can be used. As particle size becomes smaller, filtration is not practical, and diffusion of water and salts through a membrane must be used.

Cellulose whiskers have found use in polyurethane composites that have thermally sensitive shape memory where they had nucleated crystallization of the hard phase. The cellulose whiskers were dispersed with some remaining aggregates at concentrations below 3.8%·w/w and rapid shape fixing was found (Han et al. 2012).

Cellulose nanowhiskers (CNW) were used to form composites with poly(ethylene oxide), where the whiskers were hierarchically structured and uniaxially oriented. The cellulose nanowhiskers were formed and oriented as fiber arrays by electrospinning, giving high tensile storage modulus along the fiber axis (Changsarn et al. 2011).

The strong interactions between cellulose nanoparticles result in difficult dispersion and cause agglomeration when they are used to form nanocomposites (Pandey et al. 2013). Surface modification, compatibilizers or surfactants, grafting to polymer, and copolymerization with the polymer are employed to give well-dispersed composites that exhibit the high-performance properties of the cellulose nanoparticles.

Cellulose nanofibrils (CNF) grafted with 3-methacryloxypropyltrimethoxysilane (MEMO) were used to render cellulose hydrophobic for composite formation with PLA. Composites were prepared by casting from DMAC solution and the composite films were transparent after DMAC evaporation at 80°C. Fourier transform infrared spectroscopy demonstrated coupling between CNF and MEMO (Qu et al. 2012).

Microfibrilated cellulose consisting of a web-array fibrils with diameters in the range 10–100 nm were used to reinforce CA by solution blending in acetone. Some microfibrils were surface treated with 3-aminopropyltriethoxysilane, which effectively doubled the composite modulus. The composites had considerably enhanced modulus and strength compared with pure CA (Lu and Drzal 2010).

A source of cellulose nanofibers is tunicate cellulose modified by oxidative treatment. The cellulose nanofibers were spun from a suspension in water into an acetone coagulation bath to give porous fibers with the component nanofibers aligned in the fiber direction. They were more porous than similar nanofiber from wood spun under the same conditions (Iwamoto et al. 2011).

CNW were used to prepare bionanocomposites based on PLA, and it was found that the CNW increased hydrolytic stability of PLA as studied in a phosphate buffer, even with concentrations of CNW as low as 1%·w/w. The stability was interpreted as due to CNW presenting a physical barrier to water absorption by PLA. The thermal stability of the PLA–CNW composites, as formed and after phosphate buffer immersion, was increased as measured by the TGA onset of mass loss. Addition of a small proportion of CNW to PLA should be effective for increasing service life (Luiz de Paula et al. 2011).

CNW were used to form bionanocomposites with a matrix of PHB with poly(ethylene glycol) plasticizer. PHB showed a wider processing window and significant increase in strain without diminishing strength. Only small fractions (0.45%·w/w) were required for the enhanced properties. It was proposed that CNW increase the chain orientation of PHB parallel to an applied load, which activated shear flow of the PHB matrix (de O. Patrício et al. 2013). CNW were used to form composites with poly(3-hydroxybutyrate-*co*-3-hydroxyvalerate) using a solution casting method. CNW were prepared from MCC using sulfuric acid partial hydrolysis. Enhanced thermal stability, mechanical, and dynamic mechanical properties were found and attributed to strong interactions between the two phases (Ten et al. 2010).

All-cellulose nanocomposites were formed from dissolved MCC forming the matrix and CNW, produced by acid hydrolysis, as the dispersed phase. The CNW were of type I crystals while the regenerated matrix phase cellulose was of type II, and each of the phases could be discriminated by Raman spectroscopy on the basis of crystal type. CNW enhanced the mechanical properties of the matrix. Deformation-dependent morphologies were characterized by Raman spectroscopy of the materials while they were under strain (Pullawan et al. 2010).

Liquefaction of wood was performed in ethylene glycol–glycerol under acidic conditions where cellulose dissolved, presumably after partial hydrolysis, leaving insoluble partially decomposed lignin and hemicellulose. The yield depended on temperature, concentration, and liquor ratio (Zhang et al. 2012). Derivatization of cellulose has been enhanced using ionic liquid solvent in a microwave-assisted reaction. Acylation, using ethanoic, butanoic, and hexanoic anhydrides, was performed in a series of ionic liquids with reaction being dependent on anhydride chain length, viscous flow, and solvatochromatic character of the ionic liquid (El Seoud et al. 2011).

Biomimetic materials based on nacre have been prepared as clay–nanofibrillated cellulose by a process based on papermaking. Hydrocolloid mixtures of clay and nanofibrillated cellulose were used on a filtration web to form the nanopapers. In addition to tough mechanical properties, the nanopapers were found to be fire retardant and resistant to oxygen permeability. Inorganic clay contents to 90%·w/w could be prepared that extended the property range of cellulosic nanopapers (Liu et al. 2011).

3.8   CONCLUSION

Cellulose is an abundant natural resource that is available through slow growing forests, crop grasses, and from agricultural water. Fibers are separated from other plant residues by a variety of techniques. Cellulose fibers can be regenerated or converted into other forms by solution processing. New solvents that are recyclable are of increasing importance. Cellulose has joined nanotechnology through preparation of nanofibers and nanocrystals that have exceptional properties and create high-value materials from the otherwise commodity resource. Cellulose, from long fibers to nanofibers, is used to prepare composites ranging from packaging products to high-performance materials. Cellulose can be used as a matrix for composites. All-cellulose composites are an addition to single material composites facilitating ease of disposal, biodegradation, or recycling.

ABBREVIATIONS

BMIMAc	1-butyl-3-methylimidazolium acetate
BMIC	1-butyl-3-methylimidazolium chloride
[BMIM][Ac]	1-*n*-butyl-3-methylimidazolium acetate
[BMIM][PF6]	1-*n*-butyl-3-methylimidazolium hexafluorophosphate
CA	cellulose acetate
CAB	cellulose acetate butyrate
CNW	cellulose nanowhisker
CNF	cellulose nanofibril
DMAC	*N,N*-dimethylacetamide
DMF	*N,N*-dimethylformamide
HPC	hydroxypropyl cellulose
LCST	lower critical solution temperature
MCC	microcrystalline cellulose
MEMO	3-methacryloxypropyltrimethoxysilane
MWCNT	multiwall carbon nanotube
NMMO	N-methylmorpholine oxide
PHBV	poly(3-hydroxybutyrate-co-3-hydroxyvalerate)
PHB	poly(hydroxybutyrate)
PLA	poly(lactic acid)
PHB	poly(hydroxybutyrate)
TGA	thermogravimetry analysis

REFERENCES

Cai, T., Yang, G., Zhang, H., Shao, H. and Hu, X. 2012. A new process for dissolution of cellulose in ionic liquids. *Polymer Engineering and Science* 52: 1708–1714.

Changsarn, S., Mendez, J.D., Shanmuganathan, K., Foster, E.J., Weder, C. and Supaphol, P. 2011. Biologically inspired hierarchical design of nanocomposites based on poly(ethylene oxide) and cellulose nanofibers. *Macromolecular Rapid Communications* 32: 1367–1372.

Cheng, G., Varanasi, P., Li, C., Liu, H., Melnichenko, Y.B., Simmons, B.A., Kent, M.S. and Singh, S. 2011. Transition of cellulose crystalline structure and surface morphology of biomass as a function of ionic liquid pretreatment and its relation to enzymatic hydrolysis. *Biomacromolecules* 12: 933–941.

Cipriani, G., Salvini, A., Baglioni, P. and Bucciarelli, E. 2010. Cellulose as a renewable resource for the synthesis of wood consolidants. *Journal of Applied Polymer Science* 118: 2939–2950.

da Silva Filho, E., Santana, S., Melo, J., Oliveira, F. and Airoldi, C. 2010. X-ray diffraction and thermogravimetry data of cellulose, chlorodeoxycellulose and aminodeoxycellulose. *Journal of Thermal Analysis and Calorimetry* 100: 315–321.

Das, B. and Choudhury, P.K. 1967. Molecular parameters of sodium cellulose xanthate in dilute solution. *Journal of Polymer Science Part A-1: Polymer Chemistry* 5: 769–777.

de O. Patrício, S.P., Pereira, F.V., dos Santos, M.C., de Souza, P.P., Roa, J.P.B. and Orefice, R.L. 2013. Increasing the elongation at break of polyhydroxybutyrate biopolymer: effect of cellulose nanowhiskers on mechanical and thermal properties. *Journal of Applied Polymer Science* 127: 3613–3621.

El Seoud, O.A., da Silva, V.C., Possidonio, S., Casarano, R., Arêas, E.P.G. and Gimenes, P. 2011. Microwave-assisted derivatization of cellulose, 2—The surprising effect of the structure of ionic liquids on the dissolution and acylation of the biopolymer. *Macromolecular Chemistry and Physics* 212: 2541–2550.

Freire, M.G., Teles, A.R.R., Ferreira, R.A.S., Carlos, L.D., Lopes-da-Silva, J.A. and Coutinho, J.A.P. 2011. Electrospun nanosized cellulose fibers using ionic liquids at room temperature. *Green Chemistry* 13: 3173–3180.

Gericke, M., Liebert, T., Seoud, O.A.E. and Heinze, T. 2011. Tailored media for homogeneous cellulose chemistry: ionic liquid/co-solvent mixtures. *Macromolecular Materials and Engineering* 296: 483–493.

Gupta, K.M., Hu, Z. and Jiang, J. 2011. Mechanistic understanding of interactions between cellulose and ionic liquids: a molecular simulation study. *Polymer* 52: 5904–5911.

Han, J., Zhu, Y., Hu, J., Luo, H., Yeung, L.Y., Li, W., Meng, Q., Ye, G., Zhang, S. and Fan, Y. 2012. Morphology, reversible phase crystallization, and thermal sensitive shape memory effect of cellulose whisker/SMPU nano-composites. *Journal of Applied Polymer Science* 123: 749–762.

Härdelin, L., Thunberg, J., Perzon, E., Westman, G., Walkenström, P. and Gatenholm, P. 2012. Electrospinning of cellulose nanofibers from ionic liquids: the effect of different cosolvents. *Journal of Applied Polymer Science* 125: 1901–1909.

Huang, Y., Zhang, L., Yang, J., Zhang, X. and Xu, M. 2013. Structure and properties of cellulose films reinforced by chitin whiskers. *Macromolecular Materials and Engineering* 298: 303–310, doi 10.1002/mame.201200011.

Huber, T., Pang, S. and Staiger, M.P. 2012. All-cellulose composite laminates. *Composites Part A: Applied Science and Manufacturing* 43: 1738–1745.

Ingildeev, D., Effenberger, F., Bredereck, K. and Hermanutz, F. 2012a. Comparison of direct solvents for regenerated cellulosic fibers via the Lyocell process and by means of ionic liquids. *Journal of Applied Polymer Science* 128(6): 4141–4150, doi 10.1002/app.38470.

Ingildeev, D., Hermanutz, F., Bredereck, K. and Effenberger, F. 2012b. Novel cellulose/ polymer blend fibers obtained using ionic liquids. *Macromolecular Materials and Engineering* 297: 585–594.

Iwamoto, S., Isogai, A. and Iwata, T. 2011. Structure and mechanical properties of wet-spun fibers made from natural cellulose nanofibers. *Biomacromolecules* 12: 831–836.

Jacquet, N., Quivy, N., Vanderghem, C., Janas, S., Blecker, C., Wathelet, B., Devaux, J. and Paquot, M. 2011. Influence of steam explosion on the thermal stability of cellulose fibres. *Polymer Degradation and Stability* 96: 1582–1588.

Kim, J.W., Park, S., Harper, D.P. and Rials, T.G. 2012. Structure and thermomechanical properties of stretched cellulose films. *Journal of Applied Polymer Science* 128(1): 181–187, doi 10.1002/app.38149.

Kim, U.J., Eom, S.H. and Wada, M. 2010. Thermal decomposition of native cellulose: influence on crystallite size. *Polymer Degradation and Stability* 95: 778–781.

Kosan, B., Dorn, S., Meister, F. and Heinze, T. 2010. Preparation and subsequent shaping of cellulose acetates using ionic liquids. *Macromolecular Materials and Engineering* 295: 676–681.

Le Moigne, N. and Navard, P. 2004. Physics of cellulose xanthate dissolution in sodium hydroxide–water mixtures: a rheo-optical study. *Cellulose Chemistry and Technology* 44: 217–221.

Liao, R., Lei, Y., Wan, J., Tang, Z., Guo, B. and Zhang, L. 2012. Dispersing graphene in hydroxypropyl cellulose by utilizing its LCST behavior. *Macromolecular Chemistry and Physics* 213: 1370–1377.

Liu, A., Walther, A., Ikkala, O., Belova, L. and Berglund, L. A. 2011. Clay nanopaper with tough cellulose nanofiber matrix for fire retardancy and gas barrier functions. *Biomacromolecules* 12(3): 633–641.

Lu, F., Cheng, B., Song, J. and Liang, Y. 2012a. Rheological characterization of concentrated cellulose solutions in 1-allyl-3-methylimidazolium chloride. *Journal of Applied Polymer Science* 124: 3419–3425.

Lu, J. and Drzal, L.T. 2010. Microfibrillated cellulose/cellulose acetate composites: effect of surface treatment. *Journal of Polymer Science Part B: Polymer Physics* 48: 153–161.

Lu, J., Zhang, H., Jian, Y., Shao, H. and Hu, X. 2012b. Properties and structure of MWNTs/cellulose composite fibers prepared by Lyocell process. *Journal of Applied Polymer Science* 123: 956–961.

Lue, A. and Zhang, L. 2010. Effects of carbon nanotubes on rheological behavior in cellulose solution dissolved at low temperature. *Polymer* 51: 2748–2754.

Luiz de Paula, E., Mano, V. and Pereira, F.V. 2011. Influence of cellulose nanowhiskers on the hydrolytic degradation behavior of poly(D,L-lactide). *Polymer Degradation and Stability* 96: 1631–1638.

Ouajai, S. and Shanks, R.A. 2006. Solvent and enzyme induced recrystallization of mechanically degraded hemp cellulose. *Cellulose* 13: 31–44.

Ouajai, S. and Shanks, R.A. 2009a. Biocomposites of cellulose acetate butyrate with modified hemp cellulose fibres. *Macromolecular Materials and Engineering* 294: 213–221.

Ouajai, S. and Shanks, R. A. 2009b. Preparation, structure and mechanical properties of all-hemp cellulose biocomposites. *Composites Science and Technology* 69: 2119–2126.

Ouajai, S., Hodzic, A. and Shanks, R.A. 2004. Morphological and grafting modification of natural cellulose fibers. *Journal of Applied Polymer Science* 94: 2456–2465.

Pandey, J.K., Nakagaito, A.N. and Takagi, H. 2013. Fabrication and applications of cellulose nanoparticle-based polymer composites. *Polymer Engineering and Science* 53: 1–8.

Poletto, M., Zattera, A.J. and Santana, R.M.C. 2012. Structural differences between wood species: evidence from chemical composition, FTIR spectroscopy, and thermogravimetric analysis. *Journal of Applied Polymer Science* 126: E337–E344.

Pullawan, T., Wilkinson, A.N. and Eichhorn, S.J. 2010. Discrimination of matrix-fibre interactions in all-cellulose nanocomposites. *Composites Science and Technology* 70: 2325–2330.

Qin, X., Lu, A. and Zhang, L. 2012. Effect of stirring conditions on cellulose dissolution in NaOH/urea aqueous solution at low temperature. *Journal of Applied Polymer Science* 126: E470–E477.

Qu, P., Zhou, Y., Zhang, X., Yao, S. and Zhang, L. 2012. Surface modification of cellulose nanofibrils for poly(lactic acid) composite application. *Journal of Applied Polymer Science* 125: 3084–3091.

Sawpan, M.A., Pickering, K.L. and Fernyhough, A. 2011. Effect of various chemical treatments on the fibre structure and tensile properties of industrial hemp fibres. *Composites Part A: Applied Science and Manufacturing* 42: 888–895.

Schweizer, E. 1857. Das Kupferoxyd-Ammoniak, ein Auflösungsmittel für die Pflanzenfaser [Copper oxide ammonia, a means of dissolution for vegetable fibers.]. *Journal für Praktische Chemie* 72: 109–111.

Shanks, R.A., Hodzic, A. and Ridderhof, D. 2006. Composites of poly(lactic acid) with flax fibers modified by interstitial polymerization. *Journal of Applied Polymer Science* 99: 2305–2313.

Shibata, M., Yamazoe, K., Kuribayashi, M. and Okuyama, Y. 2013. All-wood biocomposites by partial dissolution of wood flour in 1-butyl-3-methylimidazolium chloride. *Journal of Applied Polymer Science* 127: 4802–4808.

Tanpichai, S., Quero, F., Nogi, M., Yano, H., Young, R.J., Lindstrom, T., Sampson, W.W. and Eichhorn, S. J. 2012. Effective Young's modulus of bacterial and microfibrillated cellulose fibrils in fibrous networks. *Biomacromolecules* 13: 1340–1349.

Tatárová, I., MacNaughtan, W., Manian, A.P., Široká, B. and Bechtold, T. 2012. Steam processing of regenerated cellulose fabric in concentrated LiCl/urea solutions. *Macromolecular Materials and Engineering* 297: 540–549.

Ten, E., Turtle, J., Bahr, D., Jiang, L. and Wolcott, M. 2010. Thermal and mechanical properties of poly(3-hydroxybutyrate-*co*-3-hydroxyvalerate)/cellulose nanowhiskers composites. *Polymer* 51: 2652–2660.

Ververis, C., Georghiou, K., Christodoulakis, N., Santas, P., Santas, R. 2004. Fiber dimensions, lignin and cellulose content of various plant materials and their suitability for paper production. *Industrial Crops and Products* 19: 245–254.

Wong, S., Shanks, R.A. and Hodzic, A. 2007. Effect of additives on the interfacial strength of poly(L-lactic acid) and poly(3-hydroxy butyric acid)-flax fibre composites. *Composites Science and Technology* 67: 2478–2484.

Yang, Q., Lue, A. and Zhang, L. 2010. Reinforcement of ramie fibers on regenerated cellulose films. *Composites Science and Technology* 70: 2319–2324.

Zhang, H., Pang, H., Shi, J., Fu, T. and Liao, B. 2012. Investigation of liquefied wood residues based on cellulose, hemicellulose, and lignin. *Journal of Applied Polymer Science* 123: 850–856.

4 Hemp and Hemp-Based Composites

Hao Wang and Alan K.T. Lau

CONTENTS

4.1 INTRODUCTION

Fiber-reinforced composites are strong, stiff, and lightweight materials. They are widely used in aerospace, transportation, and construction industries. The most commonly used reinforcement fibers for composites are glass, carbon, and aramid fibers. They all have the advantages of high strength, high stiffness, and light weight; however, they are all synthetic fibers, which are not renewable and have the problem of disposal at the end of their lifetimes.

Material renewability and sustainability have become more and more important issues in materials development and applications. Government regulations and a growing environmental awareness throughout the world have encouraged the development and utilization of materials compatible with the environment. Natural fibers, especially hemp and flax, have become increasingly suitable alternatives to glass and carbon fibers, and have the potential to be used in cheaper, more sustainable, and more environmentally friendly composite materials (Bledzki and Gassan 1999; Kabir et al. 2012a; Mohanty et al. 2002; Wambua et al. 2003). Natural fibers are cheap, abundant, and renewable, and can be produced at low cost in many parts of the developing world. They are strong and stiff, and due to their low densities, have the potential to produce composites with similar specific properties to those of glass fiber. There is far less energy required for natural fiber cultivation than for synthetic fibers production. All plant-derived fibers absorb carbon dioxide when they are grown. On the other hand, synthetic fibers require the burning of fossil fuels to provide the energy needed for production, which releases CO_2 into the atmosphere.

Natural fibers, however, display large variations in fiber properties from plant to plant, such as strength, stiffness, fiber length, and cross-sectional area. These variations can ultimately lead to difficulties in composite design and performance predictions. Natural fibers are also thermally unstable compared to most synthetic fibers and are limited to processing and working temperatures below 200°C. Another major drawback when using natural fibers is the fact that they are hydrophilic (absorb water) and polar in nature, whereas common polymer matrices, thermoplastics and thermosets, are hydrophobic (do not absorb water) and nonpolar. Natural fibers used in polymer matrices display poor fiber–matrix interfacial bonding, which results in poor composite mechanical properties (Saheb and Jog 1999). It is therefore necessary to modify the fibers and the matrix, or both, to produce a composite with improved mechanical properties. Much work still needs to be done in this area to enable natural fiber–reinforced composites to compete with glass and carbon fiber composites in terms of strength and stiffness.

4.2 PLANT FIBERS AND HEMP FIBER

Natural fibers can be derived from either animal or plant sources. All plant fibers are composed of cellulose, whereas animal fibers consist of proteins. Natural cellulose fibers tend to be stronger and stiffer than their animal counterparts, and are therefore more suitable for use in composite materials. Plant-based natural fibers can be further classified as leaf, bast, fruit, seed, wood, cereal straw, and other grass fibers. In the

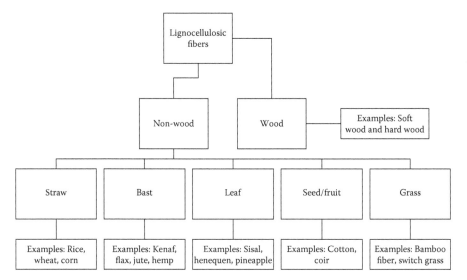

FIGURE 4.1   Classification of natural fibers.

classification of plant fibers, shown in Figure 4.1, wood-based and non-wood-based bast fibers are commonly used as reinforcements or fillers in composite materials.

Among the natural fibers, the strongest cellulose fibers are hemp, jute, and flax, with hemp and flax having the highest values for Young's modulus. Hemp and flax fibers also have high aspect ratios (length/width), which is a desirable attribute for fibers to be used as composite reinforcement. Flax is more widely accessible and slightly cheaper than hemp due to its widespread use in the textiles industry. Hemp, however, has the potential for much higher fiber yields that could result in lower cost with improvements in cultivation techniques. Hemp, compared to flax, also has the advantage of being extremely disease and pest resistant, and can be planted at high densities to prevent weeds from growing between the plants.

4.2.1 HEMP FIBER

Hemp is one of the oldest bast plants in the world. It has been widely used as an important raw material in many civilizations. It is estimated that the earliest use of hemp was over 6000 years ago. The most usual purpose of hemp cultivation is to isolate the fibers present in the bark of hemp stem surface for production of ropes, textiles, and paper. Other useful materials from hemp are the seed, which can be used for oil production, and the cannabinoids for medical, spiritual, and recreational purposes.

Hemp fibers have long been valued for their high strength and long fiber lengths, and have been extensively used in the fabrication of ropes and sails, as well as for paper and textiles. In the middle of the nineteenth century, hemp cultivation decreased with competition from other natural fibers (cotton, sisal, and jute) and synthetic fibers, as well as with the disappearance of traditional sailing ship navies.

Due to its relationship with marijuana, hemp was banned in many countries with the introduction of stricter drug laws in the nineteenth century. In 1951, the United States passed a law to prohibit hemp planting and application, as people realized that a toxic ingredient, called tetrahydrocannabinol (THC), can be extracted from hemp to make provocative drugs.

The essential difference between industrial hemp and marijuana is the amount of THC present within the plant. THC is a psychoactive chemical thought to have a range of detrimental effects to human beings. There is ongoing debate as to whether the narcotic hemp plant is actually a different variety of plant to the industrial hemp that is bred to produce a high level of THC. An industrial hemp plant will naturally attain a THC content of 0.6%, and it is generally accepted that a content of 2% is required to have a noticeable effect on the human body. Industrial hemp planting and application were reopened in 1990s throughout the world due to the acceptance of the low THC (<0.3%) hemp in "green" textile/composites industry and agriculture.

Industrial hemp is a part of the mulberry genus and is a fast growing annual plant. It is reported that hemp can grow up to 5 m in height and can reach between 6 and 60 mm in diameter in 12 weeks, depending on the plantation density (Mohanty et al. 2000). Figure 4.2 shows micrographs of hemp stalk. The hemp stem consists of wood core (xylem) with a hollow pith, surrounded by an outer layer of bark consisting of cambium, bast fiber (phloem), cortex, and epidermis (Figure 4.3). The woody core, called the hurd, is responsible for providing stiffness to the hemp stem. The bast fiber in the bark provides tensile and flexural strength. An epidermis situated on the outside of the stem gives protection from parasites (Garcia-Jaldon et al. 1998; Mediavilla et al. 2001). The bast section of the stem represents about one quarter of the total stem volume, while the hurd represents about three quarters.

Figure 4.4a shows a cross section of bast fiber bundle (Garcia-Jaldon et al. 1998). Fiber is hollow in the middle, and individual fiber sticks together and overlaps over a considerable length to form the bundle. Individual bast fiber and "knee" can be observed in Figure 4.4b. In hemp fiber, cellulose is the main constituent, which takes up to 70%–75%. The remaining constituents include hemicellulose (18%–22%), lignin (4%–6%), pectin (1%), and waxes (0.8%) (Zhang 2009). Hemp fiber possesses a density of ~1480 kg/m^3. The fiber is long, stiff, and strong, with tensile strength, elastics modulus and elongation at failure around 700 MPa, 70 GPa, and 1.7%

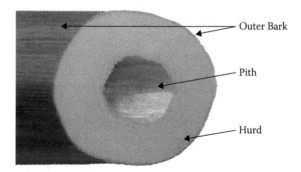

FIGURE 4.2 Cross section of a hemp stem.

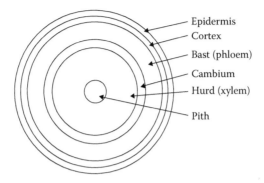

FIGURE 4.3   Schematic of cross-sectional structure of a hemp stem (not to scale).

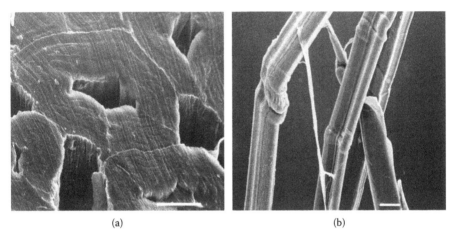

            (a)   (b)

FIGURE 4.4   Scanning electron micrographs of hemp bast fibers. (a) Bundled fibers; scale bar: 10 mm. (b) Separated fibers; scale bar: 10 mm. (From Garcia-Jaldon, C., Dupeyre, D. and Vignon, M.R. *Biomass and Bioenergy*, 14, 251–260, 1998.)

(Garcia-Jaldon et al. 1998), respectively, although these properties vary from fiber grown place, age, growth history, and the separate technique being used.

Individual fiber, also call elementary fiber, can actually be considered as composite itself as it consists of helically wound cellulose microfibrils in an amorphous matrix of lignin and hemicellulose. Each fiber consists of many microfibrils that run along the length of fiber, as shown in Figure 4.5 (Dickison 2000). The fiber wall, or cell wall, is divided into two distinct parts, namely the primary and the secondary cell wall. The primary cell wall is relatively thin, about 0.2 μm, and consists of pectin, some lignin, and cellulose. The secondary cell wall makes up most of the fiber diameter, and consists of oriented, highly crystalline microfibrils, amorphous lignin and hemicellulose. The microfibrils are packed together in a fibrillar structure, the mesofibrils, with the microfibrils oriented spirally at ~10° to the fiber axis. The mesofibrils are presumably glued together by a hemicellulose and lignin-rich phase.

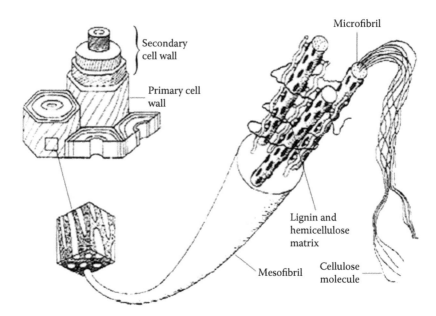

FIGURE 4.5   Fiber with primary and secondary walls. Cellulose molecules are united to form microfibrils, which in turn compose mesofibrils. (From Dickison, W.C. *Integrative Plant Anatomy.* Academic Press, Burlington, VT, 2000.)

The mechanical properties of fiber are strongly influenced by many factors, particularly chemical constituent composition and internal fiber structure, which differ between different parts of a plant as well as different plants. The most efficient cellulose fibers are those with high cellulose content coupled with a low microfibril angle in the range of 7°–12° to the fiber axis. Other factors that may affect the fiber properties are maturity, separation processes, and microscopic and molecular defectors (such as pits and nods), soil type, and weather conditions under which they were grown. The highly oriented crystalline structure of cellulose makes the fibers stiff and strong in tension, but also sensitive toward kink band formation under compressive loading. The presence of kink bands significantly reduces fiber strength in compression and in tension.

The hurd fibers are situated in the woody core of the hemp stalk, and are very thin-walled and short compared with bast fibers. They have large lumens, high lignin, and low cellulose content compared to bast fibers, and therefore weak and brittle.

4.2.2   HEMP FIBER CONSTITUENTS AND FIBER PROCESSING

4.2.2.1   Hemp Fiber Constituents

The chemical composition of hemp varies according to the variety, the area of production, and the maturation of the plant. Hemp fibers mainly composed of cellulose, hemicelluloses, lignin, and pectins. Bast fibers contain higher cellulose content and are therefore stronger than hurd fibers, whereas hurd fibers contain high level of lignin, which is undesirable for being used as reinforcement in composite materials.

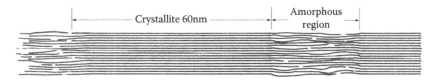

FIGURE 4.6 Schematic representation of the crystallite structure of cellulose. (From Milner, K.C., Anacker, R.L., Fukushi, K., Haskins, W.T., Landy, M., Malmgren, B. and Ribi, E. *Symposium on Relationship of Structure of Microorganisms to their Immunological Properties.* American Society for Microbiology, Cleveland, Ohio, 1963.)

4.2.2.1.1 Cellulose

The long and thin crystalline microfibrils that dominate the secondary cell walls of lignocellulose fibers are made of cellulose and are responsible for providing tensile strength to the fibers. Cellulose is a linear polymer consisting of d-anhydroglucose units joined together by β-1,4-glycosidic linkages (Eichhorn et al. 2001). Cellulose may either be crystalline or amorphous (noncrystalline), and native cellulose is usually composed of crystalline segments alternating with regions of amorphous cellulose (Milner et al. 1963) (Figure 4.6). Most plant-derived cellulose is highly crystalline and may contain as much as 80% crystalline cellulose.

The rigidity and strength of cellulose and lignocellulose-based materials is a result of hydrogen bonding, both between chains and within chains. The amorphous cellulose regions have fewer interchain hydrogen bonds, thus exposing reactive interchain hydroxyl groups (OH) for bonding with water molecules. Amorphous cellulose can therefore be considered hydrophilic due to its tendency to bond with water. On the other hand, crystalline cellulose is closely packed and very few accessible interchain OH groups are available for bonding with water. As a result, crystalline cellulose is far less hydrophilic than amorphous cellulose. Crystalline microfibrils consist of tightly packed cellulose chains with accessible hydroxyl groups present on the surface of the structure. Only the strongest acids and alkalis can penetrate and modify the crystalline lattice of cellulose.

4.2.2.1.2 Lignin

Lignin is a poorly understood hydrocarbon polymer with a highly complex structure consisting of aliphatic and aromatic constituents, which form a matrix sheath around the cellulose microfibrils and fibers. Lignin provides compressive strength to the slender microfibrils and prevents them from buckling under compressive loads. No regular structure for lignin has been demonstrated, and it is totally amorphous as opposed to the ordered structure of crystalline cellulose. Lignin is distributed throughout the primary and secondary cell walls, with the highest concentration being found in the middle lamella. The dissolution of lignin in the middle lamella using chemicals aids fiber separation, as is commonly done in the pulp and paper industries. When exposed to ultraviolet light, lignin undergoes photochemical degradation.

4.2.2.1.3 Hemicellulose

The hemicellulose fraction of the fiber contains a collection of polysaccharide polymers containing mainly the sugars D-xylopyranose, D-glocopyranose, D-galactopyranose,

L-arabinofuranose, D-mannopyranose, and D-glycopyranosyluronic acid with small quantities of other sugars (Rowell et al. 2000). Hemicelluloses appear to form a link between cellulose and lignin, thus permitting the effective transfer of shear stresses between the cellulose microfibrils and the lignin. Unlike cellulose, the hemicellulose polymer chains are rarely crystalline and are mainly responsible for the water absorption in the fiber wall. They are also heavily branched and have short side chains, whereas cellulose is a long unbranched polymer. It is suspected that no chemical bonding occurs between cellulose and hemicelluloses, but sufficient mutual adhesion is provided by hydrogen bonds and van der Waals forces. Hemicelluloses have greater solubility in solvents compared to cellulose and can be broken down in high temperature environments.

4.2.2.1.4   Pectins

Pectins are a collective name for the heteropolysaccharides found in the primary cell walls of most non-wood plant fibers, and they consist of α-1,4-linked galacturonic acid units, sugar units of various compositions, and their respective methyl esters. Pectins, along with lignin and hemicellulose, are used to connect the elementary fibers together and can easily be hydrolyzed at elevated temperatures. Pectins are the most hydrophilic compounds in plant fibers due to the presence of carboxylic acid groups.

4.2.2.2   Issues with Using Hemp Fiber in Composites

Natural fibers present many advantages compared to synthetic fibers, thus making them attractive reinforcements for composite materials. They are cheap, abundant, and renewable, and have good specific properties including tensile strength and stiffness. Unlike brittle synthetic fibers, natural fibers are flexible and are less likely to fracture during composite processing. This enables the fibers to maintain the appropriate aspect ratios to provide good composite reinforcement. Despite these fiber advantages, untreated natural fiber composites have performed well below their potential capabilities and have therefore not been used extensively in the composites industry. The main reasons for the lack of use of natural fiber composites are listed below.

4.2.2.2.1   Poor Interfacial Bonding

All plant-derived cellulose fibers are polar and hydrophilic in nature, mainly as a consequence of their chemical structure. Plant fibers contain noncellulosic components such as hemicelluloses, lignin, and pectins, of which the hemicelluloses and pectins are hydrophilic. These components contain many accessible hydroxyl (OH) and carboxylic acid groups, which are active sites for the sorption of water. The cellulose component also contains many OH groups, but little water can be accommodated within the highly ordered and highly crystalline microfibrils. As a result, only unbonded OH groups on the microfibril surfaces are available for water absorption.

Polymers, both thermoplastics and thermosets, are largely nonpolar and hydrophobic in nature. The incompatibility of polar cellulose fibers and nonpolar polymer matrix leads to poor adhesion, which then results in a composite material with poor mechanical properties. To fully utilize the mechanical properties of the reinforcing fibers and thereby improve the composite properties, it is necessary to improve the adhesion between the fibers and matrix. This can be achieved either by modifying the surface of the fibers to make them more compatible with the matrix or by modifying the matrix.

4.2.2.2.2 Low Processing Temperature

Unlike many synthetic fibers, lignocellulosic fibers are inherently thermally unstable, and thermal degradation starts to occur at temperatures of around 200°C. This results in the exclusion of some manufacturing processes. It is suggested that temperatures above 150°C can lead to permanent alterations of the physical and chemical properties of lignocellulosic fibers such as wood (Yildiz et al. 2006). It has been shown by several authors that the thermal stability of lignocellulosic fibers can be improved somewhat by means of alkali fiber treatment (Kabir et al. 2012a; Islam et al. 2005).

4.2.2.2.3 High Moisture Absorption

A further problem associated with using lignocellulosic fibers in composite materials is high moisture absorption. A moisture built up in the fiber cell can lead to fiber swelling and dimensional changes in the composite, particularly in the direction of fiber thickness. Another problem associated with fiber swelling is a reduction in the adhesion between the fiber and the matrix, leading to a reduction in mechanical properties of the composite. The debonding between fiber and matrix may be initiated by the development of osmotic pressure pockets at the surface of the fiber, which is a result of the leaching of water-soluble substances from the fiber surface.

Besides dimensional stability, the hydrophilic nature of hemp fiber also influences the processability of the composite. The tendency of lignocellulosic fibers to absorb moisture results in the release of water vapor in the composite during high-temperature compounding, leading to the formation of a highly porous material. These pores can act as stress concentration points and can lead to premature failure of the composite during loading.

4.2.2.2.4 Poor Fiber Separation and Dispersion

The incorporation of short hemp fiber into polymer is often associated with poor fiber dispersion due to the large differences in polarity between the fibers and polymer, and the strong intermolecular hydrogen bonds between the fibers. A good distribution implies that the fibers are fully separated from each other, and each fiber is fully surrounded by the matrix. Insufficient fiber dispersion can lead to clumping and agglomeration of the fibers, resulting in an inhomogeneous mixture of resin-rich and fiber-rich areas. This segregation is undesirable, as the resin-rich areas are weak, while the fiber-rich areas (clumps) are susceptible to microcracking. Microcracks contribute to inferior mechanical properties of the composite.

To ensure good distribution and dispersion of fibers within a composite matrix, it is necessary to separate the fibers from each other, modify the fibers and/or matrix to improve compatibility, and ensure that the fiber lengths are such that fiber entanglement does not occur. To separate the fibers from their fiber bundles, it is necessary to dissolve the pectins that bind the individual fibers together. Fiber separation can easily be performed by treating the fibers with a strong alkali solution. Fiber separation can also be achieved during composite compounding by means of high-energy processing techniques such as extrusion and injection molding.

4.2.2.2.5 Biodegradability of the Fiber

Natural lignocellulosic fibers degrade easily when exposed to nature. Degradation mechanisms include biological, thermal, aqueous, photochemical, chemical, and mechanical processes. In order to produce hemp fiber composites with a long service life, it is necessary to retard this natural degradation. One way of preventing or slowing down the natural degradation process is by modifying the cell wall chemistry. Undesirable natural fiber characteristics such as dimensional instability, flammability, biodegradability, and chemical degradation can be eliminated or impeded in this manner (Rowell 1997). Chemical treatments can reduce the water uptake in the fibers and can therefore reduce the amount of fiber swelling and biological degradation by blocking the available OH groups on the fiber surface.

4.2.2.3 Fiber and Matrix Treatments

To improve interfacial bonding, modifications can be made to the fibers, the matrix, or both. Matrix modifications generally involve the addition of chemical coupling agents and compatiblizers to the polymer matrix, with the purpose of improving the polymer reactivity and wetting of the reinforcing fibers. Hemp fiber treatments may be biological, physical, or chemical, and are performed to achieve one or more of the following objectives: (1) removing undesirable fiber constituents; (2) roughening of the fiber surface; (3) separating individual fibers from their fiber bundles; (4) modifying the chemical nature of the fiber surface; and (5) reducing the hydrophilicity of the hemp fiber.

4.2.2.3.1 Biological Retting

Retting is the controlled degradation of plant stems to free the bast fibers from their fiber bundles, as well as to separate them from the woody core and epidermis. During the retting process, bacteria and fungi release enzymes to degrade pectic and hemicellulosic compounds in the middle lamella between the individual fiber cells. This results in the separation of the bast fibers from the woody core, and leaves the fibers soft and clean. The retting duration is an important parameter, as under-retting can result in incomplete fiber separation, while over-retting can weaken the fibers and can lead to higher fiber mass losses during processing.

Retting may take the form of water retting where the plant stems are submerged in water, or dew retting, where plant stems are left in the field to partially degrade. Water retting requires abundant supplies of water and produces a more uniform and higher quality fiber, but is not generally practiced because of environmental pollution problems. Dew retting, on the other hand, is a much slower process and can only be performed in regions where sufficient dew is released at night.

4.2.2.3.2 Steam Explosion and Plasma Treatment

Steam explosion is an effective and a low-energy method of fiber separation that could be used as an alternative to environmentally unsound fiber separation techniques such as water retting. Steam explosion requires less time and is better controlled than traditional retting procedures (Vignon et al. 1996). For the steam explosion method, semi-retted bast fiber is removed from the woody core and impregnated with a weak solution of sodium hydroxide under a vacuum. The fibers are drained prior to loading into a steam reactor and steamed at 200°C (1.5 MPa) for 90 seconds (Vignon et al. 1996). The pressure is released suddenly from the steam reactor, resulting in explosive decompression. As the water in the fibers rapidly vaporizes and increases in volume, the fiber bundles are blown apart and separated at the middle lamella. During the steam treatment, pectins, hemicelluloses, and lignin are partially degraded and rendered soluble in the sodium hydroxide solution, which then ensures greater separation of the fibers.

Plasma treatment can modify the chemical and physical structure of the fiber surface layers. Physical modifications occur due to surface roughening of the fiber by means of the sputtering effect, which causes an enlargement of the fiber contact area and thus increases the friction between the fiber and the polymer matrix. Chemical modifications, depending on the type and nature of the plasma gases used, involve the implantation of active polar groups on the fiber surface, which reduces the fiber surface energy and promotes chemical bonding between the fiber and the polymer matrix. Plasma treatment also increases fiber surface energy, surface cross linking and the formation of reactive free radicals.

4.2.2.3.3 Alkaline Treatment

Treatment of hemp fiber by NaOH is widely used for modifying cellulosic molecular structure of the fiber. Alkali treatment, especially performed at elevated temperatures, can result in the selective degradation of lignin, pectins, and hemicelluloses in the fiber wall, while having little effect on the cellulose components. The removal of these cementing materials can result in stronger natural fiber–reinforced composites. Pectins and hemicelluloses are easily removed by alkali treatments, but lignin in hemp bast fiber can be difficult to remove. It has been reported that lignin consists of strong carbon–carbon linkages and aromatic groups that are highly resistant to chemical attack, thus limiting the degradation and fragmentation of lignin (Wang et al. 2003).

We have used different concentration alkaline solutions to treat the hemp fiber (Wang et al. 2011). Fiber composition before and after treatments is shown in Table 4.1. Alkaline treatment effectively removed lignin and hemicellulose, therefore increasing the cellulose content in the fiber. Higher strength was observed in the composites with high cellulose content hemp fiber. It is not always possible to remove all the lignin from the fiber by means of alkali treatment. The lignin in the middle lamella may be easily accessed and degraded, but access to lignin in the fiber secondary layers may be restricted due to coverage by the primary layer as well as the swelling of cellulose.

Alkali treatments generally result in a rougher fiber topography, which can further improve fiber–matrix adhesion in a composite by providing additional sites for mechanical interlocking. Figure 4.7 shows the scanning electron micrographs (SEM) of untreated and NaOH-treated hemp fibers. It can be observed that the gummy polysaccharides of lignin, pectin, and hemicellulose are localized on the surfaces of the untreated fibers. In contrast, the NaOH-treated fibers appear to have clean but rough surfaces containing large numbers of etched striations.

TABLE 4.1

Chemical Constituents of Untreated and NaOH-Treated Hemp Fibers

Fiber Treatment	Cellulose (%)	Hemicellulose (%)	Lignin (%)	Composite Bending Strength (MPa)
Untreated	80.5	6.2	7.3	190
6% NaOH	84.1	4.9	5.0	220
10% NaOH	94.2	5.0	3.5	243

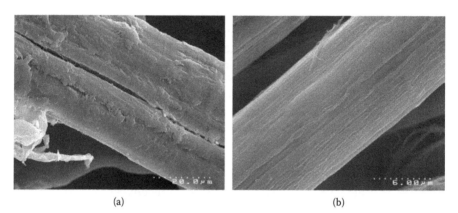

(a)   (b)

FIGURE 4.7 Scanning electron micrographs of hemp fiber: (a) untreated and (b) NaOH treated.

It has been reported that the removal of cementing materials from the fiber walls leads to better packing of the cellulose chains, thus increasing the crystallinity index of the fiber (Ouajai and Shanks 2005). In addition, treatments with NaOH can lead to a decrease in the spiral angle and an increase in the molecular orientation of the cellulose chains. These fiber changes may result in improvements in fiber strength, and hence stronger composite materials.

4.2.2.3.4 Acetylation Treatment

Acetylation treatment uses acetic anhydride to reduce the hydrophilic characteristics of cellulose fibers (Zafeiropoulos et al. 2002), as well as to improve fiber dispersion within the polymer matrix. Acetic anhydride is a compatibilizer that lowers the surface energy of the fiber to make it nonpolar and more similar to the polymer matrix (Lu et al. 2000). This is achieved by the reaction of the acetyl group ($-CH_3CO$) with the hydrophilic hydroxyl groups ($-OH$) of hemp fiber. Once the OH groups of the fiber have bonded with the acetic anhydride, they are no longer reactive and are therefore no longer free to bond with other OH groups, water, or other chemicals.

Baillie and Zafeiropoulos (Eichhorn et al. 2001) treated dew-retted and green flax fibers with acetic anhydride. It was reported that acetylation slightly improved the interfacial bonding of dew-retted flax fiber–reinforced isostatic polypropylene (PP) composites and resulted in a greater improvement for similar composites reinforced with green flax. An interfacial shear strength increase of 2.3% was observed for the acetylated dew-retted flax composites, and an interfacial shear strength increase of 83% was observed for the acetylated green flax composites.

4.2.2.3.5 Silane Treatments

Organic–inorganic silane coupling agents have been used to couple glass fibers with polymeric matrices. However, they have been used recently to couple cellulose fibers. The chemical composition of silane coupling agents allows the formation of a chemical link between the surface of the cellulose fiber and the resin through a siloxane bridge. In the presence of moisture, the hydrolyzable alkoxy group leads to the formation of silanols. Then, one end of the silanol reacts with the hydroxyl group of the fiber and the other end reacts with the functional group of the matrix. This coreactivity provides molecular continuity across the interfacial region of the composite and improves the fiber–matrix adhesion.

Silane-treated fiber composites provide better tensile strength properties than the alkali-treated fiber composites (Valadez-Gonzalez et al. 1999). Seki (2009) investigated the effect of alkali (5% NaOH for 2 hours) and silane (1% oligomeric siloxane with 96% alcohol solution for 1 hour) treatments on flexural properties of jute epoxy and jute polyester composites. For jute epoxy composites, silane over alkali treatments showed about 12% and 7% higher strength and modulus properties compared to the alkali treatment alone. Similar treatments reported around 20% and 8% improvement for jute polyester composites.

In our study, hemp fibers were pretreated with 8% NaOH solutions for 3 hours, then treated with silane with 3% oligomeric siloxane in methyl alcohol solution for 3 hours (Kabir et al. 2012b). Composites with treated fibers showed increases in both bending strength (14%) and strain (17%) compared with the untreated fiber sample. The improved interface bonding between fiber and matrix was shown in the fracture surface (Figure 4.8), where a clear gap appeared in the untreated samples while the silane-treated sample had much better interface adhesion.

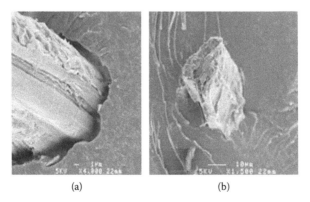

(a)   (b)

FIGURE 4.8 Fracture surface of composites with untreated hemp fiber (a) and silane-treated hemp fiber (b) shows the interface between fiber and matrix.

4.2.2.3.6 Maleic Anhydride Treatment

Maleated coupling agents are widely used to strengthen thermoplastic composites containing fillers and fiber reinforcements. The difference with other chemical treatments is that maleic anhydride (MA) is not only used to modify fiber surface but also thermoplastic matrix to achieve better interfacial bonding and mechanical properties in composites. Maleic anhydride–grafted polypropylene (MAPP) is a coupling agent that also acts as a compatibilizer. It consists of long polymer chains with an MA functional group grafted onto them. MAPP acts as a bridge between the nonpolar PP matrix and the polar fibers by chemically bonding with the cellulose fibers through the MA groups, and bonding to the matrix by means of polymer chain entanglement.

The MA functional group interacts strongly with the fiber surface through covalent and hydrogen bonding with the reactive OH groups on the surface of the cellulose and lignin (Figure 4.9). The polymer chains of MAPP then combine with the unreactive PP matrix by means of chain entanglement (Sanadi et al. 1997). These entanglements function as physical cross links that provide some mechanical integrity up to and above the glass transition temperature of the matrix. The MAPP polymer chain length is an important factor regarding the level of chain entanglement that can be achieved by the coupling agent. When the MAPP polymer chains are very short, there is little chance of entanglement between coupling agent and matrix chains as they can easily slide past one another (Figure 4.10). When the MAPP polymer chains are longer, entanglement can then occur, but the viscosity of the coupling agent also becomes high, resulting in poor fiber wetting. If the MAPP chains are extremely long, they may then entangle with the PP molecules so that the MA groups on the MAPP have difficulty migrating to the OH groups on the fiber surface. An optimum chain length or a critical molecular weight is therefore necessary to develop sufficient entanglement with the matrix polymer.

Improvements in the mechanical properties of natural fiber–reinforced thermoplastic composites as a result of the inclusion of MAPP have been shown by several researchers, including Sanadi et al. (1997), who used MAPP with kenaf fiber–reinforced PP composites. The resulting composites with 50 wt% kenaf fiber and 3% MAPP displayed a tensile strength increase of 88% and a Young's modulus increase of 350%.

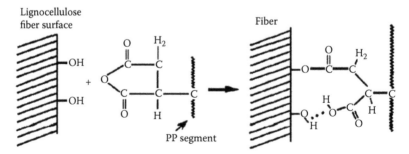

FIGURE 4.9 Reaction mechanisms of MAPP with the surface of a lignocellulosic fiber. Note the potential for both covalent and hydrogen bonding. (From Sanadi, A.R., Caulfield, D.F. and Jacobson, R.E. *Paper and Composites from Agro-Based Resources*. CRC Lewis Publishers, 1997.)

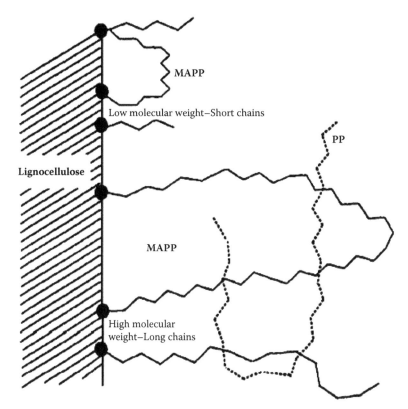

FIGURE 4.10 Schematic of possible PP molecular entanglements with longer chains of MAPP. Shorter chains of MAPP have less opportunity to entangle with the PP. (From Sanadi, A.R., Caulfield, D.F. and Jacobson, R.E. *Paper and Composites from Agro-based Resources.* CRC Lewis Publishers, 1997.)

4.3 HEMP-BASED THERMOPLASTIC COMPOSITES

In recent years, natural fiber–reinforced thermoplastic composites have gained increasing interest because of their low cost and recyclable properties. The most commonly used thermoplastics for this purpose are PP, polyethylene (PE), and polyvinyl chloride (PVC).

4.3.1 PROCESSING METHODS

The processing methods used to fabricate hemp fiber–reinforced thermoplastic composites are generally the same as those used to produce similar composites containing synthetic fibers. The common mixing techniques used for mixing fiber with thermoplastic polymer are melt mixing, extrusion compounding, and solution mixing. The common composite forming techniques used are injection molding and compression molding; in some instances extrusion is also used. No serious attempts have yet been made to develop large-scale fabrication techniques for natural fiber–reinforced thermoplastic composites with aligned fiber orientations.

4.3.1.1 Melt Mixing

Melt mixing using a radial flow (turbulent) mixer is a commonly used method for compounding short reinforcing fibers with thermoplastic polymers. The thermoplastic polymer is initially heated to its melting temperature, and then hemp fibers are added to the mix. After mixing, the composite mixture can be rolled into sheet or formed into other shape. Several mixing settings, including mixing duration, rotor speed, and melt-chamber temperature, can determine the mixing results. Joseph et al. (1999) combined sisal fiber with PP using a Haake Rheocord mixer. They showed that ineffective mixing and poor fiber dispersion occurred at short mixing times (<10 min) and low mixing speeds (<50 rpm), while low mixing temperatures (<170°C) resulted in extensive fiber breakages. Composite strength loss due to fiber breakage also occurred at high mixing times (>10 min) and high mixing speeds (>50 rpm). High mixing temperature can result in fiber degradation and poor fiber dispersion.

Melt mixing generally results in very good mixing of fiber with thermoplastic polymer, but it is a noncontinuous process, requiring stoppage to remove the mixture material from the mixer. It also only allows the processing of a limited amount of material at any given time.

4.3.1.2 Extrusion Compounding

Extrusion is one of the most effective methods of compounding natural fibers and thermoplastic polymers. A thermoplastic polymer (which is usually obtained in pellet or powder form) and short hemp fibers are drawn and combined into a heated extrusion barrel by means of a single screw or two corotating screws, depending on the type of extruder (Figure 4.11). The polymer is melted and mixed with hemp fiber to form a composite melt, which is then drawn forward through the extruder barrel and further mixed and compressed to improve the melt homogeneity. The melt then exits the barrel through a shaped die, which determines the shape of the extruded composite.

Extrusion of short fiber-reinforced thermoplastic composites using a twin-screw extruder is often carried out prior to injection molding because of the excellent fiber distribution being achieved within the polymer matrix. The extruder processing variables, such as barrel length, temperature profile, screw configuration, and screw speed, are optimized to achieve a uniform fiber distribution. Incorrect extruder setup

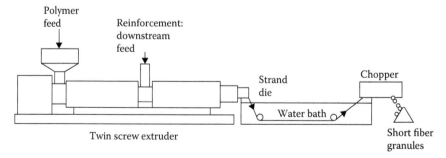

FIGURE 4.11 The extrusion compounding process for the manufacture of short fiber-reinforced thermoplastic granules for injection molding.

can result in poor dispersion and wetting of fiber in polymer matrix, as well as severe fiber damage and length reduction. The extruded composite can then be chopped into pellets that can easily be injection molded into more complex shapes.

Most fiber damage is caused by inter-fiber friction, friction between fiber and polymer, and friction between fiber and the extruder. One method of minimizing fiber breakage is to feed the polymer into the main in-feed port and to feed the fibers into a second port further down the barrel (Yam et al. 1990), as can be seen in Figure 4.11. This enables the polymer to melt before coming into contact with the fibers, thus reducing the shear forces acting on the fibers. A second method of preserving fiber integrity during compounding is to keep the compounding distance as short as possible, and also to minimize the number of kneading elements on the extruder screws.

Another variable that can affect the composite strength is processing temperature. If the temperature is too low, the polymer matrix will be too viscous to flow around the fibers, resulting in poor fiber wetting and hence a weaker composite. If the temperature is too high, degradation of the fibers and polymer will occur, also resulting in composite strength reduction.

Unlike melt mixing, extrusion is a continuous process that can accommodate high feed rates and allows the fast and efficient processing of materials. A shaped die can also be used to create a finished product with a desired profile.

4.3.1.3 Solution Mixing

Solution mixing is an alternative to physical mixing method where polymer is melted during compounding with fiber, such as melt mixing, extrusion and injection molding. This technique involves dissolving the polymer in a suitable solvent, adding the fiber to the polymer/solvent blend, and precipitating the polymer from the solvent in a vacuum oven. The polymer precipitates are weakly bound to the fibers, and final consolidation of the fiber and polymer can be achieved by extrusion or compression molding. Solution mixing avoids fiber damage that normally occurs during blending of fiber and thermoplastics by melt mixing and extrusion compounding, as it employs modest temperatures and shear stresses (polymer and fibers are mixed in a low viscosity solution).

Solvents do not produce permanent chemical changes to thermoplastic polymer matrices, but instead produce physical changes that involve the separation of individual polymeric chains. For example, xylene can be used to dissolve PP. The mechanical properties of PP will not be affected by dissolution in xylene followed by precipitation. Solvents are also known to have little damage to the structure and integrity of cellulose fibers.

4.3.1.4 Injection Molding

Injection molding is one of the most widely used processes for manufacturing molded parts from thermoplastic and reinforced thermoplastic materials. It is one of the few manufacturing processes capable of producing net shape composite parts in high volumes and at high production rates.

Short hemp fiber–reinforced composites can be processed into complex shaped components using standard thermoplastic injection molding equipment. Composite materials for use in injection molding applications must be capable of fluid-like flow during processing, and thus usually consist of short fibers with a relatively low fiber fraction (typically <50 wt% or 30 vol%). The composite fiber content needs to be carefully considered, as a low fiber content can result in insufficient reinforcement, and a high fiber content can lead to poor molding and reduced mechanical properties of the composite.

Injection molder performs the function of melting the preformed (usually by extrusion compounding) composite pellets in a heated barrel, delivering a homogeneous melt to the machine nozzle, and injecting the melt into a closed mold. Injection molding process does not induce the same level of mechanical friction on the composite melt as mixing processes such as extrusion and melt mixing, at the same time, it also does not lead to significant fiber damage. The mold unit, which comprises a fixed section and a movable section, encloses the shaped cavity into which the composite is injected and cooled, and is thus responsible for determining the final shape of the molded part.

4.3.1.5 Compression Molding

Compression molding (hot pressing) is a commonly used processing technique for producing large, relatively simple composite parts with good mechanical properties. The compression molding of glass mat–reinforced thermoplastics is widely used in the production of complex semistructural components, notably for the automotive industry. Compression molding basically involves the hot pressing of randomly orientated or aligned fiber mats, either chopped or in continuous form, with a thermoplastic material.

The compression molding operation begins with the placement of a stack of alternating fiber mat and thermoplastic sheets onto the bottom half of a preheated mold cavity. The top half of the mold is lowered at a constant rate until the desired processing pressure is reached, thus causing melting of the polymeric matrix and consolidation of the composite. Once the composite has been pressed, it is cooled and removed from the mold.

4.3.2 Hemp Polypropylene Composites

One of the main disadvantages of using lignocellulosic fibers as reinforcements in thermoplastic matrix composites is that they degrade at medium to high processing temperatures. The processing temperatures of hemp fiber–reinforced thermoplastic composites are limited to less than 200°C, although it is possible to use higher temperatures for short periods of time. Low temperature processes are desirable. PP, high-density PE, low-density PE, and PVC have melting temperatures below 200°C. These plastics are all commonly used in the plastics industry. They are not expensive, and can be recycled.

4.3.2.1 Hemp Fiber and Noil Hemp Fiber

Figure 4.12 shows the scutched hemp fiber (SHF), which is extracted from retted hemp stem without much fiber processing. The scutched fiber is regarded as the starting form for any fiber application. SHF usually has a length of 0.8–1.5 m with an average fiber diameter of about 20 μm. The fiber has a rough surface as most of the pectin and lignin are still attached on the surface (Figure 4.12d).

Currently, a large amount of hemp fiber is used for textile purposes. The scutched fiber is chemically degummed to remove the majority of pectin, lignin, and also hemicellulose, and then proceeds for combing, carding, and spinning. During these processes, fiber that is not long enough for textile purposes will be discarded, and it is called noil fiber. Figure 4.12a shows the noil hemp fiber (NHF). In the natural fiber textile industry, over half of the fiber will end up as noil fiber. At the world annual production of textile natural fiber (excluding cotton) of 5 million tons, there is about 2.5–3.0 million tons of noil fiber available every year. As the by-product, the price of NHF is only one-fifth of SHF. Because of the high-level degumming, the noil fiber has relatively high cellulose content (over 90%), which is actually beneficial to fiber strength. The chemical constituents in NHF and SHF are shown in Table 4.2. Besides its short length, the NHF is also smaller in diameter with a much cleaner surface (Figure 4.12b).

SHF was cut into short fibers with length about 15 mm. Both NHF and SHF were mixed with PP using an internal mixer. The mixing parameters including temperature, rotor speed, and mixing time were optimized to achieve a uniform dispersion of fiber in PP. The mixture was then sheeted with a roll mill and crushed

(a)

(b)

(c)

(d)

FIGURE 4.12 Photos and micrographs of noil hemp fiber (a) and (b), and unprocessed scutched hemp fiber (c) and (d).

TABLE 4.2

Constituent Composition (wt%) in NHF and SHF

	NHF	SHF
Cellulose	91.9	56.5
Hemicellulose	3.5	16.7
Lignin	1.7	7.7
Pectin	1.2	8.1
Others	1.7	11.0

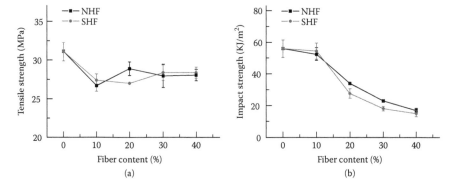

FIGURE 4.13   The effect of fiber content on mechanical properties of hemp/PP composites: (a) tensile strength and (b) impact strength.

into granules with a size of about 5–7 mm by a crusher. The granules were then fed into an injection molding machine to produce specimen for a mechanical property test (Yan et al. 2013).

4.3.2.1.1   Effect of Fiber Volume Fraction

Figure 4.13 shows the effect of hemp fiber addition on mechanical properties. Both NHF and SHF reduced tensile strength. Addition of natural fiber reduced the tensile strength from 32 to 28 MPa. For impact strength, the adverse effect of natural hemp fiber was even much substantial (Figure 4.13b). After the first 10%, further addition of hemp to 40%, impact strength reduced significantly from 55 to 20 kJ/m^2. Comparing these two hemp fibers, NHF and SHF had very similar behavior in PP.

The reduction of the mechanical properties is due to the poor bonding between hemp fiber and PP. It is well known that natural fiber is hydrophilic in nature, while plastic resin is hydrophobic. The incompatibility results in a poor interfacial bonding between them. Therefore, any applied load will not be effectively transferred to fiber. The poor bonding of hemp fiber and PP was demonstrated in the fracture surface of the tensile sample (30% hemp/PP) (Figure 4.14), where fiber pullout is the dominant fracture feature. The result indicates that resin modification or fiber treatment is needed in natural fiber composites.

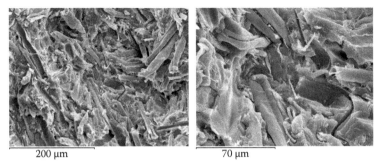

FIGURE 4.14 Fracture surface of tensile sample of 30% hemp/PP composite showing fiber pullout.

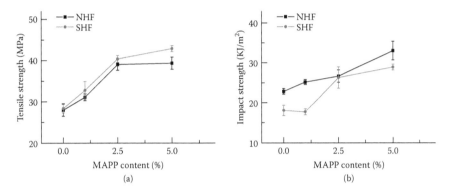

FIGURE 4.15 The effect of MAPP modification on mechanical properties of hemp/PP composites: (a) tensile strength and (b) impact strength.

4.3.2.1.2 Effect of Resin Modification

MAPP was added to modify PP. The hemp fiber content was kept at 30%. Figure 4.15 shows the effect of MAPP on mechanical properties. The addition of MAPP significantly improved composite strengths. In tensile strength (Figure 4.15a), it sharply increased from 28 to 32 MPa at 1.2% MAPP addition, and further to 40 MPa at 2.5% addition. Tensile strength stabilized at this level with further MAPP addition. Significant improvements were also found in flexural strength and impact strength (Figure 4.15b). NHF showed similar reinforcement ability as SHF in the presence of MAPP.

Figure 4.16 shows the SEM images of the fracture surface of the tensile sample with 5% MAPP addition. It was found that fiber pullout was much less compared to Figure 4.14. Plastic deformation on the polymer matrix was also visible. The results demonstrated a much better interfacial bonding between fiber and resin, which contributed to the improvement of mechanical properties.

4.3.2.1.3 Effect of Fiber Treatment

Hydrophobically modified hydroxyethylcellulose (HMHEC) was used to reduce fiber hydrophilicity and to make it more compatible with hydrophobic thermoplastics.

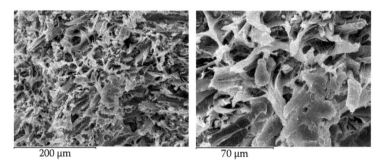

200 μm   70 μm

FIGURE 4.16   Fracture surface of tensile sample of 30% hemp/PP composite with 5% MAPP.

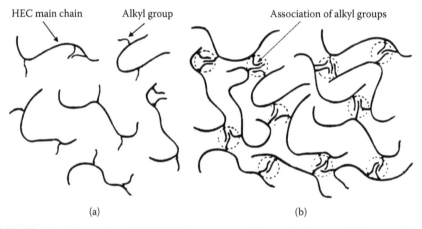

HEC main chain   Alkyl group   Association of alkyl groups

(a)   (b)

FIGURE 4.17   Illustration of HMHEC in aqueous solutions: (a) showing HEC main chain/ alkyl groups and (b) showing association of alkyl groups.

HMHEC is a kind of polymer with hydroxyl-ethylcellulose as the hydrophilic main chain and an alkyl group as the hydrophobic graft chain (Figure 4.17). When hemp fiber was impregnated in HMHEC solution, the HMHEC was coated to the fiber, so its surface has alkyl groups available to attach to the same hydrophobic PP. It should be noted that HMHEC concentration in the solution cannot be too high, otherwise the alkyl groups can group themselves instead of coating on fiber (Figure 4.17b).

The hemp fiber was immersed in HMHEC solution at different concentrations (0.25, 0.5, and 1.0 wt%) for 0.5 hours. The treated and untreated hemp fibers were then compounded in an internal mixer with PP (65 wt%), MAPP (5 wt%) and other additives. The mechanical properties of composites are shown in Figure 4.18. The treatment has different effects on the three properties. With the increase of HMHEC concentration, tensile strength increases, flexural strength decreases, and impact strength increases initially and then decreases. When the HMHEC concentration was lower than 0.5 wt%, the treatment to hemp fiber resulted in a good improvement on tensile strength and impact strength. Slight decrease in flexural strength was observed. Increasing HMHEC concentration to 1.0 wt%, tensile strength was continuously increased; however, flexural strength and impact strength decreased.

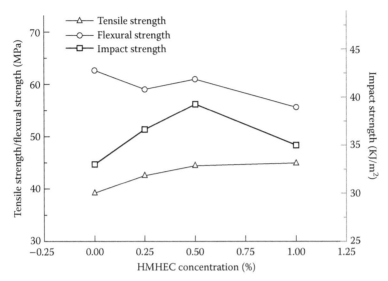

FIGURE 4.18 Mechanical properties of composites with untreated and treated noil hemp fiber.

In this study, 0.5 wt% HMHEC showed the highest mechanical properties, with tensile strength 44.5 MPa, flexural strength 60.9 MPa, and impact strength 39.3 kJ/m². Compared with the untreated and unmodified hemp/PP composites, the improvement was 59% (from 28 MPa), 45% (from 42 MPa), and 79% (from 22 kJ/m²), respectively.

4.3.3 HEMP POLYPROPYLENE PELLETS

In the plastics industry, injection molding is the most economical manufacturing process capable of producing net shape composite parts in high volumes and at high production rates (Brooks 2000). Usually, injection molding is fed with pellets. For composite materials, fiber and thermoplastic are compounded using twin screw extruder. The extruded composites were then chopped into pellets.

Pellets used in injection molding usually consist of short fibers with a relatively low fiber fraction (typically < 30 vol%) to achieve a fluid-like flow.

To produce hemp/PP composite pellets, a twin screw extruder was specifically modified to feed fiber into the extruder, see Figure 4.19. After fiber treatment, hemp fiber, if necessary, was chopped into 3–10 mm length using industry granulator. It is a technical challenge to continuously feed the chopped fiber into the extruder at a controllable fraction to PP with good quality of mixing. The extruded composite material was then pelletized.

Nova Institute in Germany has developed a method that chopped hemp fiber was first fabricated into preform hemp pellets then fed into the extruder to mix with thermoplastic (Figure 4.20). The preform hemp pellets make the accurate control of fiber/PP ratio and the quality of the final composite pellets easier.

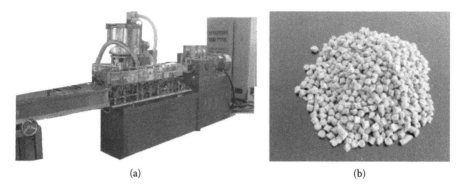

(a)   (b)

FIGURE 4.19 A modified twin screw extruder (a) and the produced hemp/PP composite pellets (b).

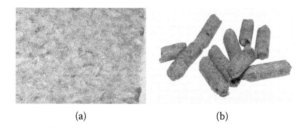

(a)   (b)

FIGURE 4.20 Chopped hemp fiber (a) and the prehemp pellets (b).

4.4   HEMP-BASED THERMOSET COMPOSITES

Thermosets are the plastics that cannot be remelted once cured. They have low viscosity, which provides good fiber wetting. Thermoset composites are used where exceptional strength, high stiffness, and low density are required. Unsaturated polyester, epoxy, phenolic, and vinyl ester are the most commonly used thermoset matrices.

4.4.1   Processing Methods

In principle, processing techniques of natural fiber thermoset composites are similar to those used in the processing of synthetic fiber composites. The basic fabrication methods are hand lay-up, vacuum-assisted resin transfer molding (VARTM), and compression molding. Other processes include pultrusion and filament winding, in which continuous fibers are used.

4.4.1.1   Hand Lay-Up

This is the simplest way of processing thermoset resin into final products. Prior to lamination, mold is cleaned carefully, and release agent is applied to the mold. Subsequently, layers of reinforcement are added to build laminate thickness. The laminating resin is applied by pouring, brushing, spraying, or using paint rollers. Paint rollers, or squeegees, are used to consolidate the laminate, thoroughly wetting the reinforcement and removing the entrapped air. The curing of a product takes usually 4–12 hours, depending on size, thickness, and complexity of the product. After curing, the product is taken out of the mold.

Vacuum bag molding can be applied to hand lay-up. Here, a flexible nylon or PE film is placed over the wet lay-up with the edges sealed and a vacuum drawn. By reducing the pressure inside the vacuum bag, external atmospheric pressure exerts force on the bag. The pressure on the laminate removes entrapped air and excess resin, and compacts the laminate.

4.4.1.2 Resin Transfer Molding

Fiber mats are placed inside a mold. The mold can be two solid parts or a single solid mold at the bottom and a vacuum bag on the top. Resin is pumped and injected into the mold. Vacuum assistance can be used to enhance resin flow in the mold cavity, then the process is called VARTM. After curing, the mold is opened and the product is removed. In VARTM, the resin impregnation quality in a composite is much better than hand lay-up, and the void content can be reduced to minimal.

4.4.1.3 Compression Molding

In this process, a weighed charge of molding compound is placed in a open mold. The two halves of the mold are closed and pressure is applied. Curing time depends on thickness, size, and shape of the part.

4.4.2 Hemp Polyester Composites

Hemp fiber–reinforced polyester composites were fabricated by VARTM. Prior to composite fabrication, hemp fibers were treated with alkali solution. The alkali-treated fiber could be further chemically treated with acetylation and silane (Wang et al. 2011).

Chemical treatments can have several effects on hemp fiber. First, chemical treatments partially remove lignin, hemicellulose, pectins, and other impurities from fiber; therefore, cellulose content in the fiber is increased. In lignocellulosic fiber, it is the cellulose content and microfibril angle that contribute to fiber strength. A high cellulose content and low microfibril angle are desirable properties in a fiber to be used as reinforcement in composites. Second, hemicellulose and pectins are the cementing constituents that glue the cellulose together, so excessive removal of these components will result in damage to the fiber and reduce the fiber strength. Third, the other effect of chemical treatments is that they reduce the hydrophilicity of hemp fiber and make it more compatible to polymer matrix. In addition, some of the treatments can include the coupling function to make the fiber more adhesive to polymer resin. To composite materials, it is the combination of these effects (the strength of the reinforcing fiber and the strength of the fiber/resin interfacial bonding) that determine the final material performance.

Figure 4.21 shows the stress–strain curves of bending testing of untreated and treated composite samples. Flexural strengths are plotted in Figure 4.22. It was observed that flexural strength of the composite samples reinforced by alkali, acetylation, and alkali + acetylation–treated fiber increased 12%, 33%, and 17%, respectively, compared to the composite with untreated fiber. The improvement in strength was attributed to (1) the increase of cellulose content after treatment; (2) the increase

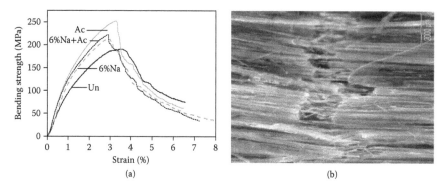

FIGURE 4.21 (a) Bending stress–strain curves of untreated, alkali, acetalytion, and alkali + acetylation–treated composites. (b) Flexural failure of the tested sample.

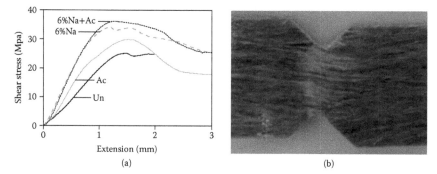

FIGURE 4.22 (a) Shear and (b) flexural strength of untreated, alkali, acetalytion, and alkali + acetylation–treated composites.

of interfacial bonding between fiber and matrix due to the reduced hydrophilicity in fiber; and (3) the increase of mechanical bonding because of fiber surface roughness (Wang et al. 2011).

The acetylation-treated samples had the highest flexural strength among the tested samples. This might be because the acetylation treatment had an additional etching function to the fiber surface. Non-uniform etching increased surface roughness and fiber/matrix contact area.

It was observed that alkali + acetylation treatment has a slightly higher flexural strength than the single alkali treatment but lower than the single acetylation treatment. The results from this study confirmed that both alkali and acetylation treatments had the effect of removing lignin and hemicellulose. Two contiguous treatments can cause excessive removal of lignin and hemicellulose, which act as cementing materials to hold the cellulose in fiber structure. As a result, the fiber itself became weaker in strength.

In the shear test, one side of V-notched specimen was fixed in the fixture and other side displaced vertically to the fiber direction. The stress transfer of the composites initiated through the matrix slides over the fiber surface. As a result, frictional stress transfer occurred along the fiber–matrix interface. From Figure 4.23, it shows that

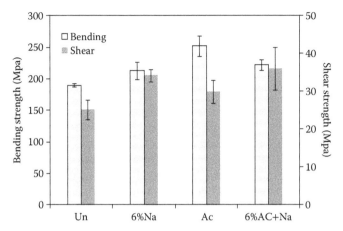

FIGURE 4.23 Shear and bending strengths of untreated, alkali-, acetalytion-, and alkali + acetylation–treated composites.

FIGURE 4.24 Nonwoven hemp mat and the hemp thermoset composite part produced using vacuum resin infusion.

alkali, acetylation, and alkali + acetylation–treated fiber composite samples showed 36.43%, 19.11%, and 43.73% higher shear strength compared to the untreated fiber sample. This was presumably due to the higher surface roughness of the treated fiber contributing good bonding with the matrix and providing greater frictional stress transfer along the interface (Kabir 2013).

4.4.3 HEMP MAT THERMOSET COMPOSITES

Using nonwoven hemp mat and vacuum resin infusion is considered one of the most viable routes for hemp thermoset composites production. To promote market development, Composites Innovation Centre in Canada has produced two sets of nonwoven hemp mat at commercial scale roll lengths (28 and 20 m) and width (1.3 m). One has an areal weight of 500 g/m² and thickness of 6.5 mm; and the other has areal weight 750 g/m² and thickness 10 mm (Figure 4.24). The mats were produced using 100% randomly orientated hemp fiber. The hemp fiber mats were freely distributed to research and manufacturing sectors all over the world to assist them in developing products, producing prototypes, and performing research (McKay 2011).

4.5 HEMP HURD COMPOSITES

Approximately three-quarters of a hemp stalk is hurd and only one-quarter is bast fiber. So development of hemp hurd applications is important to the hemp industry and hemp economy. Typically, hemp hurd has less cellulose (33%–49%) and higher hemicellulose (16%–23%) and lignin (16%–28%) compared to the hemp fiber. The content of lignin in hemp hurd is six times higher than that in hemp fiber, thus hemp hurd is significantly stiffer but less flexible and strong than the bast fiber.

4.5.1 ACTIVATED CARBON

Activated carbon is a very useful adsorbent material for purification and separation in various applications. Activated carbons can be produced from many carbonaceous precursors, and the most commonly used raw materials are wood, coal, coconut shells, and some polymers (Rosas et al. 2008). Activated carbon can also use hemp hurd and produce high apparent surface area. Rosas et al. (2008) used hemp stem (cut into 3 cm long sections) after removing hemp fibers as the starting material (Figure 4.25a). Hemp hurd was impregnated with H_3PO_4 and was activated under N_2 flow with different peak temperature ranging from 350°C–550°C. The activated carbon possessed an apparent surface area of 1500 m^2/g and a mesopore volume of 0.6 cm^3/g (Figure 4.25c). The resulted activated carbon showed a well-developed porous texture with a large quantity of micropores and some mesopores.

4.5.2 HEMP PLASTIC COMPOSITES

Wood plastic composites (WPC) use wood powder as filler in thermoplastics. WPC products have been widely used in building, furniture, and automotive industries.

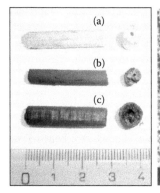
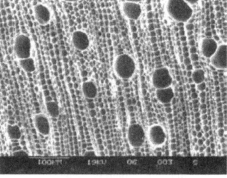

FIGURE 4.25 Photographs of hemp hurd stem (a), directly carbonized without activator (b), activated carbon (c), and the porous texture of the activated carbon from hemp hurd.

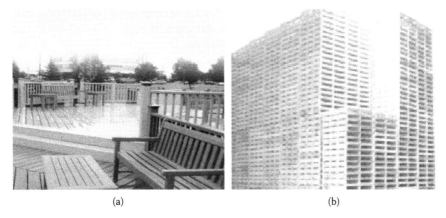

(a)   (b)

FIGURE 4.26 Extruded hemp plastic composite products: (a) outdoor furniture and (b) shipping pallets. Hemp hurd powder content is 50 wt%.

FIGURE 4.27 Injection molded hemp plastic composite products: nursery pot, shoe heel, and cloth stand.

Usually, the wood powder has a particle size ranging from 50 to 400 μm. Hemp hurd can also be the source for such wood powder and can be used to develop hemp plastic composites (Wang, H. and Warner, P., personal communication, 2012). The plastic matrix here can be PE, PP, or PVC. Coupling agent (such as MAPE and MAPP), plasticizer, and lubricant are added. Hemp plastic composites product can be used for decking, decoration, outdoor furniture, packing trays, storage shells, shipping pallet, and even railway sleepers (Figure 4.26).

Most WPC products are produced by an extrusion process as the addition of wood powder increases the viscosity of the resin, which makes injection molding not possible. For injection molding of WPC, the wood powder content is usually lower than 20 wt%. When hemp powder is used for injection molding, the hemp hurd powder content can be as high as 40 wt%. Figure 4.27 shows some of the hemp plastic composites products produced by injection molding. About 40 wt% hemp hurd powders are used as filler in PP matrix.

4.6 CONCLUSION

Hemp fibers played an important part in the technical and cultural history of mankind. Today, China, Canada, and Europe are the main hemp cultivation areas in the world.

The total cultivation area is about 80,000 ha with hemp fiber production at about 130,000 tons. In Europe, hemp fiber is currently used mainly in technical applications such as natural fiber–reinforced composites (automotive, industrial, and consumer goods), specialty paper, insulation material, and mulch. In China and Canada, hemp is mainly used in textile applications. With the trend of developing and utilizing more and more renewable and environmental friendly materials, as a proven high CO_2 capture, low energy required, and high quality fiber material, hemp will and is playing an important role in the production of innovative bio-based products such as natural fiber–reinforced composites, insulation, and construction materials.

REFERENCES

Bledzki, A.K. and Gassan, J. 1999. Composites reinforced with cellulose based fibres. *Progress in Polymer Science* 24: 221–274.

Brooks, R. 2000. Injection molding based techniques. In: *Comprehensive Composite Materials*. Eds. Kelly, A. and Zweben, C. Elsevier Science Ltd: Oxford.

Dickison, W.C. 2000. Plant growth, development, and cellular organization. In: *Integrative Plant Anatomy*, Ed. Dickison, W.C., 3–49. Burlington, VT: Academic Press.

Eichhorn, S.J., Baillie, C.A., Zafeiropoulos, N. et al. 2001. Current international research into cellulosic fibres and composites. *Journal of Materials Science* 36: 2107–2131.

Garcia-Jaldon, C., Dupeyre, D. and Vignon, M.R. 1998. Fibres from semi-retted hemp bundles by steam explosion treatment. *Biomass and Bioenergy* 14: 251–260.

Islam, M.S., Pickering, K.L. and Beckermann, G.W. 2005. The effect of fibre treatment using alkali on industrial hemp fibre/epoxy resin composites. In: *ICME*, Ed. Nuruzzaman, D.M. Dhaka, Bangladesh: Scientific & Academic Publishing.

Joseph, P.V., Joseph, K. and Thomas, S. 1999. Effect of processing variables on the mechanical properties of sisal-fiber-reinforced polypropylene composites. *Composites Science and Technology* 59: 1625–1640.

Kabir, M.M. 2013. Effects of Chemical Treatments on Hemp Fibre Reinforced Polyester Composites, PhD Thesis, University of Southern Queensland.

Kabir, M.M., Wang, H., Lau, K.T. and Cardona, F. 2012a. Chemical treatments on plant-based natural fibre reinforced polymer composites: an overview. *Composites: Part B* 43: 2883–2892.

Kabir, M.M., Wang, H., Lau, K.T. et al. 2012b. Mechanical properties of chemically-treated hemp fibre reinforced sandwich composites. *Composites: Part B* 43: 159–169.

Lu, J.Z., Wu, Q. and McNabb, H.S. 2000. Chemical coupling in wood fiber and polymer composites: A review of coupling agents and treatments. *Wood and Fiber Science* 32: 88–104.

McKay, S. 2011. Biocomposites-bridging the technology gaps. In: *The 1st International Symposium on Green Materials*. Beijing, China.

Mediavilla, V., Leupin, M. and Keller, A. 2001. Influence of the growth stage of industrial hemp on the yield formation in relation to certain fibre quality traits. *Industrial Crops and Products* 13: 49–56.

Milner, K.C., Anacker, R.L., Fukushi, K. et al. 1963. Structure and biological properties of surface antigens from gram-negative bacteria. In: *Symposium on Relationship of Structure of Microorganisms to their Immunological Properties*. American Society for Microbiology, Cleveland, Ohio.

Mohanty, A.K., Misra, M. and Drzal, L.T. 2002. Sustainable bio-composites from renewable resources: Opportunities and challenges in the green materials world. *Journal of Polymers and the Environment* 10: 19–26.

Mohanty, A.K., Misra M. and Hinrichsen, G. 2000. Biofibres, biodegradable polymers and biocomposites: An overview. *Macromolecular Materials and Engineering* 276/277: 1–24.

Ouajai, S. and Shanks, R.A. 2005. Composition, structure and thermal degradation of hemp cellulose after chemical treatments. *Polymer Degradation and Stability* 89: 327–335.

Rosas, J.M., Bedia, J., Rodrıguez-Mirasol, J. and Cordero, T. 2008. Preparation of hemp-derived activated carbon monoliths. Adsorption of water vapor. *Industrial & Engineering Chemistry Research* 47: 1288–1296.

Rowell, R.M. 1997. Chemical modification of agro-resources for property enhancement. In: *Paper and Composites from Agro-based Resources*, Eds. Rowell, R.M., Young, R.A. and Rowell, J.K. 351–375. Oxon, UK: CRC Press.

Rowell, R.M., Han, J.S. and Rowell, J.S. 2000. Characterization and factors effecting fiber properties. In: *Natural Polymers and Agrofibres Composites*, Eds. Frollini, E., Leao, A. and Mattoso, L.H.C., 115–134. Sao Carlos, Brazil: Embrapa Instrumentacao Agropecuaria.

Saheb, D.N. and Jog, J.P. 1999. Natural fibre polymer composites: A review. *Advances in Polymer Technology* 18: 351–363.

Sanadi, A.R., Caulfield, D.F. and Jacobson, R.E. 1997. Agro-fiber/thermoplastic composites. In: *Paper and Composites from Agro-based Resources*, Eds. Rowell, R.M., Young, R.A. and Rowell, J.K. 377–401. Oxon, UK: CRC Lewis Publishers.

Seki, Y. 2009. Innovative multifunctional siloxane treatment of jute fibre surface and its effect on the mechanical properties of jute/thermoset composites. *Materials Science and Engineering: A* 508: 247–252.

Valadez-Gonzalez, A., Cervantes-Uc, J.M., Olayo, R. and Herrera-Franco, P.J. 1999. Chemical modification of henequen fibers with an organosilane coupling agent. *Composites Part B: Engineering* 30: 321–331.

Vignon, M.R., Dupeyre, D. and Garcia-Jaldon, C. 1996. Morphological characterization of steam-exploded hemp fibers and their utilization in polypropylene-based composites. *Bioresource Technology* 58: 203–215.

Wambua, P., Ivens, J. and Verpoest, I. 2003. Natural fibres: Can they replace glass in fibre reinforced plastics? *Composites Science and Technology* 63: 1259–1264.

Wang, H., Kabir, M.M. and Lau, K.T. 2011. The effect of fibre chemical treatments on hemp reinforced composites. In: *18th International Conference on Composite Materials (ICCM-18)*, 21–26 Aug 2011, Jeju Island, Korea.

Wang, H.M., Postle, R., Kessler, R.W. and Kessler W. 2003. Removing pectin and lignin during chemical processing of hemp for textile applications. *Textile Research Journal* 73: 664–669.

Yam, K.L., Gogoi, B.K., Lai, C.C. et al. 1990. Composites from compounding wood fibres with recycled high density polyethylene. *Polymer Engineering and Science* 30: 693–699.

Yan, Z.L., Wang, H., Lau, K.T. et al. 2013. Reinforcement of polypropylene with hemp fibres. *Composites Part B* 46: 221–226.

Yildiz, S., Gezer, E.D. and Yildiz, U.C. 2006. Mechanical and chemical behavior of spruce wood modified by heat. *Building and Environment* 41: 1762–1766.

Zafeiropoulos, N.E., Williams, D.R., Baillie, C.A. and Matthews, F.L. 2002. Engineering and characterisation of the interface in flax fibre/polypropylene composite materials. Part I. Development and investigation of surface treatments. *Composites Part A: Applied Science and Manufacturing*. 33: 1083–1093.

Zhang, J.C. 2009. *Structure and Properties of China Hemp Fibre*. Beijing: Chemistry Industry Press.

5 Plant Fiber–Based Composites

Bessy M. Philip, Eldho Abraham, Deepa B.,
Laly A. Pothan, and Sabu Thomas

CONTENTS

5.1   INTRODUCTION

Recent interest in the utilization of greener materials has reinitiated the interest in natural fibers and fibrils as reinforcement for polymers (Blaker et al. 2010). The use of polymer composites from renewable sources has advantages over synthetic sources, particularly as a solution to the environmental problems generated by plastic waste. Green composites is today widely researched because of the need for innovations in the development of materials from biodegradable polymers, preservation of fossil-based raw materials, complete biological degradability, and reduction in the volume of carbon dioxide release into the atmosphere.

Plant fibers are found suitable to reinforce polymers. Continuous plant fiber–reinforced composites have made a huge impact in recent decades in the aerospace, transport, and oil and gas industries (Satyanarayana et al. 2009). They have relatively high strength and stiffness, low cost of acquisition, low density, nonabrasive nature, and produce low CO_2 emissions. They are also biodegradable and are annually renewable compared to other fibrous materials (Satyanarayana et al. 2009). The term natural fiber covers a broad range of vegetables and animal and mineral fibers. Availability of natural fibers and ease of manufacturing is tempting researchers to try locally available inexpensive natural fibers as reinforcement in polymer matrix (Dong et al. 2012).

Most of the natural fibers consist of long cells with relatively thick cell walls that make them stiff and strong. The fiber plant's cells are glued together into long thin fibers, the length of which is dependent on the length of the plant. The fibers may differ in coarseness, in the length of the cells, and in the strength and stiffness of the cell walls (Siqueira et al. 2010).

5.2   CHEMICAL COMPOSITION AND STRUCTURE OF PLANT FIBERS

The chemical composition as well as the structure of the plant fibers is fairly complicated. Plant fibers are essentially a biocomposite by itself in which rigid cellulose microfibrils are embedded in a matrix mainly composed of lignin and hemicelluloses (Bledzki and Gassan 1999; Satyanarayana et al. 1990). Natural fibers are basically constituted of cellulose, lignin, and hemicellulose. Pectin, pigments, and extractives can be found in lower quantities. They are basically a rigid, crystalline cellulose microfibril–reinforced amorphous lignin with or without hemicellulosic matrix. Most plant fibers, except for cotton, are composed of cellulose, hemicellulose, lignin, waxes, and some water-soluble compounds, where cellulose, hemicelluloses, and lignin are the major constituents. For this reason, natural fibers are also referred to as cellulosic or lignocellulosic fibers.

Cellulose, the most abundant material in the biosphere, is the basic structural component of all plant fibers and is the most important organic compound produced by plant. The properties of the constituents contribute to the overall properties of the fiber. The properties of cellulosic fibers are strongly influenced by many factors, for example, chemical composition, internal fiber structure, microfibril angle, cell dimensions, and defects, which differ from part to part of a plant as well as from plant to plant (Dufresne 2008). The part of the plant that grows and the composition of the soil contribute to the composition of the fiber. The mechanical properties of natural fibers also depend on their cellulose type, because each type of cellulose has its own crystalline organization, which can determine the mechanical properties. Table 5.1 shows the chemical composition of some of the important natural fibers, which varies according to their origin. In the context of both biomass valorization and nanocomposite materials development, cellulose nanofibers can be used as reinforcing filler in various polymeric matrices (Philippe et al. 2010).

The cellulose molecules consist of glucose units linked together in long chains, which in turn are linked together in bundles called microfibrils. Figure 5.1 shows the structure of a typical plant fiber showing the cellobiose repeat unit where cellulose is derived from D-glucose units, which condense through $\beta(1\rightarrow4)$-glycosidic bonds (Figure 5.2). The tensile strength of the cellulose microfibrils is enormous, being the strongest known material with a theoretically estimated tensile strength of 7.5 GPa or 1,087,500 lb/in^2.

TABLE 5.1
Chemical Composition of Some Lignocellulosic Fibers

Samples	Natural Fiber	Cellulose	Hemicellulose	Pectin	Lignin
Bast fibers	Flax	71	19	1	2
	Hemp	75	18	1	4
	Banana	72	14	>1	14
	Jute	70	14	2	18
	Ramie	75	15	2	1
Leaf fibers	Abaca	70	22	1	1
	Sisal	73	13	1	7
	PALF	85	4	<1	3
Seed hair fibers	Cotton	93	3	3	1
	Wheat straw	51	26	–	7
	Coir	40	<2	–	45

Source: Philippe, T. et al. 2010. Functional polymer nanocomposite materials from microfibrillated cellulose. In: *Nanotechnology and Nanomaterials, Advances in Nanocomposite Technology,* Ed. Hashim, A. InTech. ISBN 978-953-307-347-7.

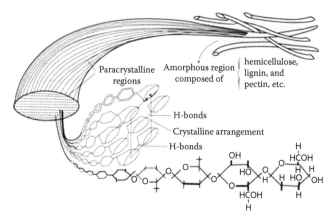

FIGURE 5.1   Structure of plant fiber showing the cellobiose repeat unit.

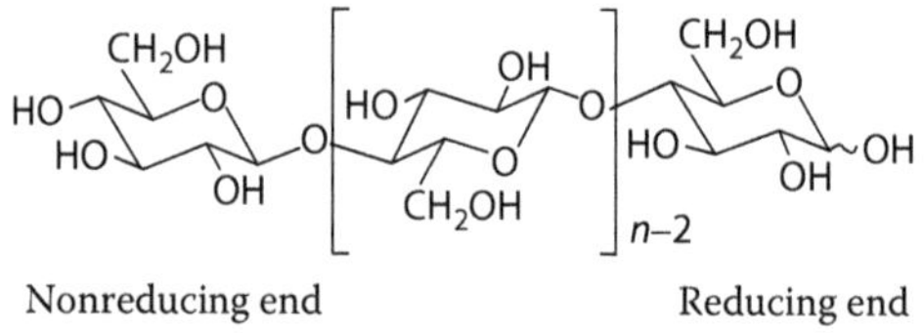

CH$_2$OH   OH   CH$_2$OH

HO   HO   O   HO
HO   O   O   OH
              OH   CH$_2$OH   OH
                                      $n-2$

Nonreducing end   Reducing end

FIGURE 5.2   Structure of cellulose.

Hemicelluloses are also found in all plant fibers. Hemicelluloses are polysaccharides bonded together in relatively short, branching chains. They are intimately associated with the cellulose microfibrils, embedding the cellulose in a matrix. Hemicelluloses are very hydrophilic (i.e., containing many sites to which water can readily bond).

Lignin is a Latin word for wood. Lignin is the compound that gives rigidity to the plant. Without lignin, plants could not attain great heights (e.g., trees) or the rigidity found in some annual crops (e.g., straw). Lignin is a three-dimensional (3D) polymer with an amorphous structure and a high molecular weight. Of the three main constituents in fibers, it is expected that lignin would be the one with the least affinity for water. Another important feature of lignin is that it is thermoplastic (i.e., at temperatures around 90°C it starts to soften and at temperatures around 170°C it starts to flow).

The combined effect of the three main constituents results in properties that are unique for plant fibers. The most important are

- Very good strength properties, especially tensile strength. In relation to its weight the best bast fibers attain strength similar to that of Kevlar.
- Very good heat, sound, and electrical insulating properties.
- Combustibility: From a waste disposal point of view, combustibility is an advantage. Products can be disposed of through burning at the end of their useful service lives, and energy can be simultaneously generated.
- Dimensional stability: As a consequence of the hygroscopicity of the fibers, products and materials based on plant fibers are not dimensionally stable under changing moisture conditions. This is the greatest disadvantage in relation to industrial use of plant fibers. However, if necessary, this may

be controlled at an extra cost by a number of known treatments (e.g., heat treatments or chemical modification procedures such as acetylation).

- Reactivity: The hydroxyl groups present in the cell wall constituents not only provide sites for water absorption but are also available for chemical modification (Dufresne 2008) (e.g., to introduce dimensional stability, durability, or improved oil/heavy metal absorption properties).
- Biodegradability: As a result of their tendency to absorb water, fibers will biodegrade under certain circumstances through the actions of fungi and/or bacteria.

Cellulose is composed of crystalline and amorphous regions. Strong intramolecular hydrogen bonds with large molecules are formed by the crystallite cellulose. Compactness of the crystalline region creates cellulose blocks that make it difficult for chemical penetration to occur. However, dyes and resins are easily absorbed by the amorphous region. In addition to this, plant fibers are highly polar and hydrophilic in character as hydroxyl groups are present in their structures. For distension of the crystalline region, elimination of hydrophilic hydroxyl groups, and removal of surface impurities (waxy substances), natural fibers need to be chemically modified. Chemical treatments such as mercerization, silane treatment, acetylation, and benzoylation with or without heat are widely being applied to modify chemistry.

5.3 DIFFERENT TYPES OF REINFORCING PLANT FIBERS

Plant fiber–reinforced composites are termed biocomposites where the matrix is any polymer system. In a fiber-reinforced polymer, the fibers serve as a reinforcement and show high tensile strength and stiffness, while the matrix holds the fibers together, transmits the shear forces, and also functions as a coating. The material behavior of matrices is usually characterized by a functional relationship of time and temperature, a considerably lower tensile strength, and a comparatively higher elongation. Therefore, the mechanical properties of the fibers determine the stiffness and tensile strength of the composite.

Depending on their origin, plant fibers may be grouped into Sections 5.3.1 through 5.3.6.

5.3.1 LEAF

The leaf fibers, also referred to as "hard" fibers, are obtained from the leaves or leaf stalks of various monocotyledonous plants. Abaca, banana, flax, pineapple, sisal, and so on are examples of plants providing leaf fibers. They can be extracted from their leaves by retting, boiling, and mechanical extraction methods. Water retting is a traditional biodegradation process involving microbial decomposition (breaking of the chemical bonds) of sisal. Boiling is another extraction method, in which leaves of sisal plant are boiled, subsequently beating is done, and after washing and sun drying we may get the usable clean leaf fibers. This method is not suitable for large-scale extraction. Mechanical extraction involves inserting leaves into a machine called a "raspador machine" and pulling the raw material out. This process does not deteriorate quality and is suitable for small-scale operations and is an efficient, versatile, cost effective, and eco-friendly process.

5.3.2 Bast (stem)

Basts (stems) are obtained from the stems of various dicotyledonous plants and are also referred to as "soft" fibers to distinguish them from leafs. Flax, hemp, jute, and kenaf come under the bast fiber category. Bast bundles are composed of elongated thick-walled ultimate cells joined together both end to end and side by side and arranged in bundles along the length of the stem. Bast bundles are removed from parent material by the decorticating process, which consists of removing them from the stem, the "cortex" comprising the bast and outer barks (Robson et al. 1993).

5.3.3 Seeds

The seeds are generally formed from a single biological cell. It is reported that more than one cell takes part in the growth of seeds. The most important of the seed fibers are cotton and Kapok.

5.3.4 Fruit

Coir has been used in India for about 3000 years. It is obtained from the fruit of the coconut palm and the fibrous tissue lies between the exocarp and the endocarp surrounding the kernel. There are three types of coconut, namely the longest and the finest called "white," a coarser one known as "brittle," and a shorter staple known as "mattress." The brittle and mattress fibers are often referred to as "brown." The retting process is traditionally used to extract coir bundles. The decorticating process can also be used to separate the bundle (Jarman and Jayasundera 1975).

5.3.5 Wood

Using polymers and reinforcing fibers from wood is another way to produce renewable and biodegradable composite materials for packaging and structural applications. Because of their low density and high mechanical properties, wood fibers are suitable for reinforcement in composite materials (Almgren 2010).

5.3.6 Grass

Many grasses have been considered as a source for reinforcing filler in polymer composites. Because of limited availability or processing difficulties, most of these grass fibers (except sugarcane bagasse) have never become widely used; however, they are often common in certain localities. These include sugarcane bagasse, bamboo, esparto, and sabai grass.

5.4 IMPORTANT PROPERTIES OF PLANT FIBERS

Plant fibers, as reinforcement, have recently attracted the attention of researchers because of their advantages over other established materials. They are environmentally friendly, fully biodegradable, abundantly available, renewable and cheap, and have low density. Plant fibers are light compared to glass, carbon, and aramid fibers. The biodegradability of plant fibers can contribute to a healthy ecosystem while their low cost and high performance fulfils the economic interest of industry. When natural fiber–reinforced plastics are subjected, at the end of their life cycle, to combustion process or landfill, the released amount of CO_2 of the fibers is neutral with respect to the assimilated amount during their growth (Wambua et al. 2003).

The abrasive nature of natural fiber–reinforced plastics is much lower leading to advantages in regard to the technical and recycling processing of the composite materials in general. Natural fiber–reinforced plastics, by using biodegradable polymers as matrices, are the most environmental friendly materials, which decompose at the end of their life cycle. Plant fiber composites are used in place of glass mostly in nonstructural applications. A number of automotive components previously made with glass composites are now being manufactured using environmentally friendly composites. Although plant fibers and their composites are environmental friendly and renewable (unlike traditional sources of energy, i.e., coal, oil, and gas), these have several bottlenecks. These have poor wettability, incompatibility with some polymeric matrices, and high moisture absorption. Composite materials made of unmodified plant fibers frequently exhibit unsatisfactory mechanical properties. To overcome this, in many cases, a surface treatment or compatibilizing agents need to be used prior to composite fabrication. The properties can be improved both by physical treatments (cold plasma treatment, corona treatment) and chemical treatments (maleic anhydride organosilanes, isocyanates, sodium hydroxide permanganate, and peroxide). Mechanical properties of plant fibers are much lower when compared to those of the most widely used competing reinforcing glass fibers. However, because of their low density, the specific properties (property-to-density ratio), strength, and stiffness of plant fibers are comparable to the values of glass fibers (Taj et al. 2007).

5.4.1 MECHANICAL PROPERTIES OF PLANT FIBERS

It is worth mentioning that fibrous materials are commonly subjected to the following deformations while in the plant source or in use: tension, compression, bending, torsion, shear, abrasion, wear, and flexing. Physical properties such as morphology, regularity, or irregularity along and across the main axis, crystalline packing order, amorphous content, and chemical composition all have an effect on the mechanical properties of plant fibers. In fact, plant fibers are composites in nature with cellulose microfibrils as the reinforcement in a matrix of lignin and hemicellulose.

5.4.2 MECHANICAL PROPERTIES OF SOME PLANT FIBERS COMPARED TO E-GLASS

The determination of mechanical properties can be predicted by using the rule of mixtures (ROM). For instance, Equation 5.1 is used to estimate the stiffness or modulus of elasticity of the plant cell wall along the axis.

$$E_f = V_c E_c \cos^2\theta + V_{nc} E_{nc} \tag{5.1}$$

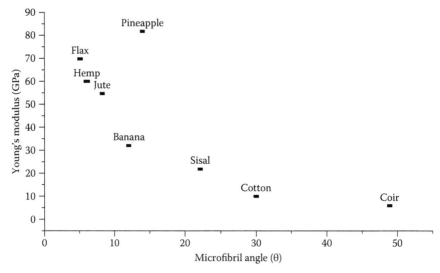

FIGURE 5.3 The relationship between the changes in E' with respect to the microfibril angle. (From Mwaikambo, L.Y., *Afr. J. Sci. Technol. (AJST) Sci. Eng. Ser.*, 7, 120, 2006.)

E_f is the effective modulus of the fiber, E_c and E_{nc} are the elastic moduli of the crystalline and noncrystalline regions, V_c and V_{nc} are the volume fractions of crystalline and noncrystalline regions, and θ is the microfibril angle (Bodig et al. 1982; McLaughlin and Tait 1980).

The relationship between the stiffness (Young's modulus) and the microfibril angle is given in Figure 5.3. The figure shows the decrease in the stiffness as the microfibril angle is increased. Plant fibers with lower microfibril angle have the highest stiffness. Pineapple leaf fiber appears to deviate from the ROM theory, which leads to a suggestion that further work is needed on the study of the fine structure of the pineapple. Plant fibers can also develop defects during harvesting and subsequent processes such as scutching and kinks can be formed during the needle punching in the mat-forming process.

Figure 5.3 shows that pineapple exhibits the highest specific strength and specific modulus followed by flax and hemp. This implies that the weight of pineapple, flax, and hemp fibers will be more suitable in composite manufacture due to the savings on weight economies over the rest of the fibers. More importantly, the same fibers exhibit high stiffness, which means that they can be used as a replacement for glass, carbon, and high-performance synthetic fibers such as Kevlar fibers in end uses where extreme stiffness is not a prerequisite. (Hughes et al. 2007). Table 5.2 shows the mechanical properties of plant fibers compared to conventional fibers.

5.4.3 MOISTURE AND DURABILITY

The major chemical constituents of natural fibers contain hydroxyl and other oxygen-containing groups that attract moisture through hydrogen bonding (Rowell 1984). The moisture content of these fibers can vary greatly depending on the fiber

TABLE 5.2

Mechanical Properties of Plant Fibers Compared to Conventional Fibers

Source	Density (g/cm³)	Tensile Strength (MPa)	Young's Modulus (GPa)	Elongation (%)
Flax	1.5	345–1035	27.6	2.7–3.2
Ramie	–	400–938	60–130	3.6–3.8
Cotton	1.5–1.6	287–587	5.5–12.6	7.0–8.0
Jute	1.3	393–773	26.5	1.5–1.8
Coir	1.2	175	4.0–6.0	30.0
Sisal	1.5	511–635	9.4–22.0	2.0–2.5
Hemp	–	690	–	1.6
Viscose (cord)	–	593	11	11.4
E-glass	2.5	2000–3500	70	2.5
Carbon	1.4	4000	230–240	1.4–1.8
Aramid	1.4	3000–3200	62–70	3.3–3.7

Source: Mwaikambo, L.Y., *Afr. J. Sci. Technol. (AJST) Sci. Eng. Ser.*, 7, 120, 2006.

type. This hygroscopicity can create challenges both in composite fabrication and in the performance of the end product. Plant fibers absorb less moisture in the final composites because they are at least partially encapsulated by the polymer matrix.

However, even small quantities of absorbed moisture can affect performance. Moisture can plasticize the composite, altering the composite's performance. In addition, volume changes associated with moisture sorption can reduce matrix adhesion and damage the matrix (Peyer and Wolcott 2000). Methods of reducing moisture sorption include adequately dispersing and encapsulating the plant fibers in the matrix during compounding, limiting content, improving fiber–matrix bonding, and chemically modifying, or simply protecting, the composite from moisture exposure.

Plant fibers undergo photochemical degradation when exposed to UV radiation. They are degraded biologically because organisms recognize the chemical constituents in the cell wall and can hydrolyze them into digestible units using specific enzyme systems (Dwivedi and Pande 2012). Because of their low thermal stability, natural plant fibers are generally processed with plastics for which high temperatures are not required (less than about 200°C). Above such temperatures, many of the polymeric constituents in natural fibers begin to decompose. Since cellulose is more thermally stable than other chemical constituents, highly pulped fibers that are nearly all cellulose have been used to extend this processing window (Sears et al. 2001).

The release of volatile gases can, before, during, and after processing, lead to odor issues in applications where the composite is in an enclosed environment, such as in many automotive applications, and especially when moisture is present (Clemons and Caulfield 2005).

5.5 SURFACE MODIFICATION OF PLANT FIBERS

5.5.1 METHODS OF SURFACE MODIFICATION

Pretreatments of the plant fiber will clean the surface, chemically modify the surface, stop the moisture absorption process, and increase the surface roughness. The surface is influenced by polymer morphology, extractive chemicals, and processing conditions. The extent of the matrix interface is significant for the application of natural and wood fibers as reinforcement for plastics. There are two types of methods used to optimize the surface, physical methods and chemical methods. These modification methods are of different efficiencies for improving the adhesion between the matrix and the filler (Dissanayake et al. 2010; Sreekala et al. 2000).

5.5.2 PHYSICAL METHODS: CORONA AND COLD PLASMA

Reinforcing plant fibers can be modified by physical methods such as stretching, calendering, thermo treatment, and production of hybrid yarns. Physical treatments change structural and surface properties of the fiber, thereby influencing the mechanical bonding to polymers. Electric discharge (corona, cold plasma) is another method of physical treatment. Corona treatment is one of the most interesting techniques for surface oxidation activation. This process changes the surface energy of the cellulose fibers, and in the case of wood, surface activation increases the amount of aldehyde groups (Belgacem et al. 1994). Plasma (glow discharge) is an ionized gas with an essentially equal density of positive and negative charges. It can exist over an extremely wide range of temperature and pressure. The solar corona, a lightning bolt, a flame, and a neon sign are all examples of plasma. For the purposes of textile modification, the low-pressure (0.01–1 mbar) plasma found in the neon sign or fluorescent light bulb is used. However, for the plasma treatment of polymeric substrates, the extremely energetic chemical environment of the plasma is used. The same effects are reached by cold plasma treatment. Depending on the type and nature of the gases used, a variety of surface modifications could be achieved: surface cross-linking could be introduced, surface energy could be increased or decreased, and reactive free radicals and groups could be produced.

Electric discharge methods are known to be very effective for non-active polymer substrates such as polystyrene (PS), polyethylene, polypropylene (PP), and so on. They are successfully used for cellulose modification, to decrease the melt viscosity of cellulose polyethylene composites, and to improve mechanical properties of cellulose polypropylene composites (Beg 2007; Dong et al. 1992).

5.5.3 CHEMICAL METHODS

Cellulose fibers that are strongly polarized are inherently incompatible with hydrophobic polymers due to their hydrophilic nature. When two materials are incompatible, it is possible in many cases to bring about compatibility by introducing a third material that has properties intermediate between those of the other two. There are several mechanisms of coupling in materials:

- Weak boundary layers: coupling agents eliminate weak boundary layers
- Deformable layers: coupling agents produce a tough, flexible layer

- Restrained layers: coupling agents develop a highly cross-linked interface region, with a modulus intermediate between that of the substrate and of the polymer
- Wettability: coupling agents improve the wetting between polymer and substrate (critical surface tension factor)
- Chemical bonding: coupling agents form covalent bonds with both materials
- Acid–base effect: coupling agents alter the acidity of the substrate surface

The development of a definite theory for the mechanism of bonding by coupling agents in composites is a complex problem. The main chemical bonding theory alone is not sufficient. So the consideration of other concepts appears to be necessary, including the morphology of the interface, the acid–base reactions at the interface, the surface energy, and the wetting phenomenon (Verma et al. 2012).

5.5.4 CHANGE OF SURFACE TENSION

The surface energy of plant fibers is closely related to the hydrophilic nature of the material. Some investigations are concerned with methods to decrease hydrophilicity. Silane-coupling agents may contribute hydrophilic properties to the interface, especially when amino functional silanes, such as epoxies and urethane silanes, are used as primers for reactive polymers. The primer may supply much more amine functionality than can possibly react with the resin at the interface. Those amines that cannot react are hydrophilic and therefore responsible for the poor water resistance of bonds. An effective way to use hydrophilic silanes is to blend them with hydrophobic silanes such as phenyltrimethoxysilane. Mixed siloxane primers also have an improved thermal stability, which is typical for aromatic silicone.

Ismail and coworkers treated oil palm empty fruit bunch and coir fibers with silane-coupling agent. They found that the addition of silane increased the scorch time and cure time and enhanced the tensile strength, tensile modulus, tear strength, fatigue life, and hardness (Bledzki et al. 2002; Khalil and Ismail 2001).

5.5.5 IMPREGNATION OF PLANT FIBERS

A better combination of plant fiber and polymer is achieved by impregnation of the reinforcing fabrics with polymer matrices compatible to the polymer. For this purpose, polymer solutions or dispersions of low viscosity are used. For a number of interesting polymers, the lack of solvents limits the use of the method of impregnation. When cellulose fibers are impregnated with a butyl benzyl phthalate-plasticized polyvinyl chloride (PVC) dispersion, excellent partitions can be achieved in PS. This significantly lowers the viscosity of the compound and the plasticator and results in cosolvent action for both PS and PVC. Valadez and coworkers found that preimpregnated henequen–high density polyethene (HDPE) composites gave better wetting (Valadez-Gonzalez et al. 1999). It is being used in automotive interiors. Jute- and flax-reinforced composites have also used for impregnation (Garkhail et al. 2000; Mishra et al. 2000; Mitra et al. 1998).

5.5.6 MERCERIZATION

An old method of cellulose modification is mercerization; it has been widely used on cotton textiles. Mercerization is an alkali treatment of cellulose fibers, which depends on the type and concentration of the solution, temperature, time of treatment, and tension of the material, as well as on the additives. At present there is a tendency to use mercerization on other natural fibers as well (Sulawan et al. 2013). Alkalization successfully modifies the structure of plant fibers (hemp, sisal, jute, and kapok) and these modifications will most likely improve the performance of natural composites by promoting better resin bonding (Mwaikambo and Ansell 1999).

Bledzki and coworkers examined alkali-treated jute fibers. Figure 5.4 shows the SEM images of raw and alkali-treated jute fiber. Shrinkage of fibers during treatment had significant effects on the structure, as well as on the mechanical properties such as tensile strength and modulus (Gassan et al. 1999). Alkali-treated jute yarns exhibited an increase in yarn tensile strength and modulus of about 120% and 150%, respectively. These changes in mechanical properties were affected by modifying the structure, basically via the crystallinity ratio, degree of polymerization, and orientation. Sisal, coir, jute, and flax have also been mercerized and their properties evaluated (George et al. 1999; Ray et al. 2001; Rout et al. 2001).

5.5.7 CHEMICAL COUPLING

An effective method of chemical modification of plant fibers is graft copolymerization, which improves the interfacial adhesion in composites. The fiber surface is treated with a compound that forms a bridge of chemical bonds between fiber and matrix. This reaction is initiated by free radicals of the cellulose molecule. The cellulose is treated with an aqueous solution containing selected ions and is exposed to high-energy radiation. Then the cellulose molecule cracks and radicals are formed. Afterward, the radical sites of the cellulose are treated with a suitable solution, compatible with the polymer matrix. The resulting copolymer possesses properties with characteristics of both fibrous cellulose and grafted polymer. After this treatment, the surface energy of the fibers is increased to a level much closer to the surface energy of the matrix. Thus, a better wettability and a higher interfacial adhesion are obtained. The graft copolymerization method is effective, but complex (Singha and Rana 2012).

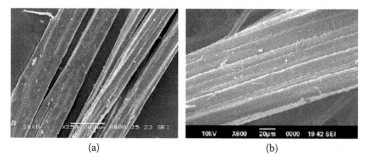

(a)   (b)

FIGURE 5.4   SEM of jute fiber. (a) Untreated and (b) 8% NaOH-treated. (From Mwaikambo, L., and Bisanda, E., *Polymer Testing*, 18, 181, 1999.)

5.5.8 Classification of Coupling Agents

Coupling agents are classified into organic, inorganic, and organic–inorganic groups. Organic coupling agents are isocyanates, anhydrides, amides, imides, acrylates, chlorotriazines, epoxides, organic acids, monomers, polymers, and copolymers. Inorganic coupling agents include silicates while organic–inorganic agents include silanes and titanates (John et al. 2000). Organic coupling agents normally have a bi- or multifunctional group in their molecular structure. These functional groups, such as (–N=C=O) of isocyanates, [–(CO)$_2$O–] of maleic anhydrides, and (Cl–) of dichloro triazine derivatives, interact with the polar groups (mainly hydroxyl groups (–OH)) of cellulose and lignin to form covalent or hydrogen bonding. Alternatively, organic coupling agents can modify the polymer matrix by graft copolymerization, thus resulting in strong adhesion, even cross-linking at the interface (Pothan et al. 2002).

Inorganic coupling agents possibly act as dispersing agents to counteract the surface polarity of cellulose and improve the compatibility between and polymer. Only a few inorganic coupling agents have been used so far in wood polymer composites. Organic–inorganic coupling agents are hybrid compounds in structure. For example, titanates usually contain a titanium center and an organic part surrounding this inorganic atom. The functionalities of the organic part in these agents determine their coupling effectiveness. Organic–inorganic coupling agents are between organic and inorganic agents in functions (Dalvag et al. 1984; Zadorecki and Flodin 1985).

The molecular chain of maleic anhydride is much shorter than that of polymer matrix and wood fibers. This discrete nature makes maleic anhydride not so effective in improving the interfacial adhesion. Therefore, long chains of high molecular weight are obtained usually by grafting maleic anhydride with polyethylene, PP, and PS to make it an ideal coupling agent for polymer composites. The formed copolymers such as maleated polyethylene (maleic anhydride-modified-polyethylene), maleated polypropylene ([MAPP] maleic anhydride-modified-polypropylene), styrene/maleic anhydride, and styrene–ethylene–butylene–styrene/maleic anhydride are used as coupling agents in polymer composites, creating both covalent bonding to the surface and extensive molecular entanglement to improve properties of the interface. Theoretically, extremely long chains may reduce the possibility of migration of the coupling agents to the fiber surface because of the short processing time (Rowell et al. 1996). If the molecular weight of the coupling agent is too high, it may entangle with the PP molecules so that the polar groups on the coupling agent have difficulty "finding" the –OH groups on the surface. The treatment of cellulose fibers with hot polypropylene–maleic anhydride copolymers provides covalent bonds across the interface (Felix and Gatenholm 1991; Kim et al. 2007; Raj and Kokta 1991). The mechanism of reaction can be divided into two steps and is shown in Scheme 5.1 (Bledzki et al. 2002).

The interactions between nonpolar thermoplastics such as PP and any coupling agents such as MAPP resulted in predominant chain entanglement between them. Stresses applied to one chain can be transmitted to other entangled chains, and stress is distributed among many chains. These entanglements function like physical cross-links, that provide some mechanical integrity up to and above the glass-transition temperature (T_g), but become ineffective at much higher temperatures (Neilsen 1974). When polymer chains are very short, there is little chance of entanglement between

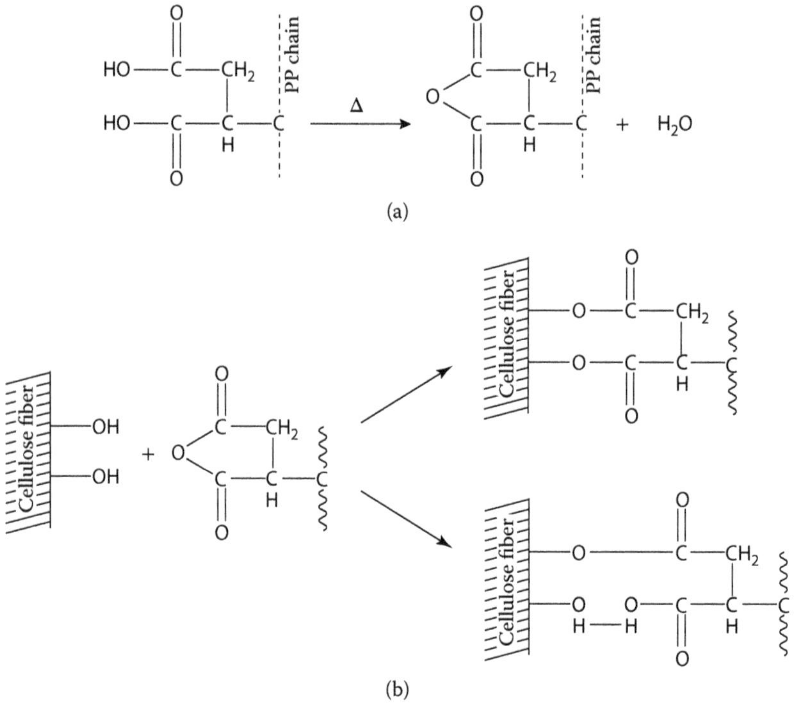

SCHEME 5.1   The reaction mechanisms involved in the treatment of cellulose fibers with hot polypropylene–maleic anhydride copolymers. (a) Activation of the copolymer by heating ($t = 170°C$) (before fiber treatment). (b) Esterification of cellulose.

chains and they can easily slide past one another. When the polymer chains are longer, entanglement between chains can occur and the viscosity of polymer becomes high. A minimum chain length or a critical molecular weight is necessary to develop these entanglements and a typical polymer has a chain length between entanglements equivalent to a molecular weight varying from 10,000 to about 40,000 (Nielsen 1977).

Silane-coupling agents usually improve the degree of cross-linking in the interface region and offer a perfect bonding. Among the various coupling agents, silane-coupling agents were found to be effective in modifying the natural matrix interface. The efficiency of silane treatment was higher for the alkali-treated fibers than for the untreated fibers because more reactive site can be generated for the silane reaction. Therefore, fibers were pretreated with NaOH for about half an hour prior to its coupling with silane. Fibers were then washed many times in distilled water and finally dried. Silane-coupling agents may reduce the number of cellulose hydroxyl groups in the matrix interface. In the presence of moisture, hydrolyzable alkoxy group leads to the formation of silanols. The silanol then reacts with the hydroxyl group of the cellulose, forming stable covalent bonds to the cell wall that are chemisorbed onto the surface (Coutinho et al. 1997). Therefore, the hydrocarbon chains obtained by the application of silane restrain the swelling of fibers by creating a cross-linked network due to covalent bonding between the matrix, and the silanes were effective in improving the

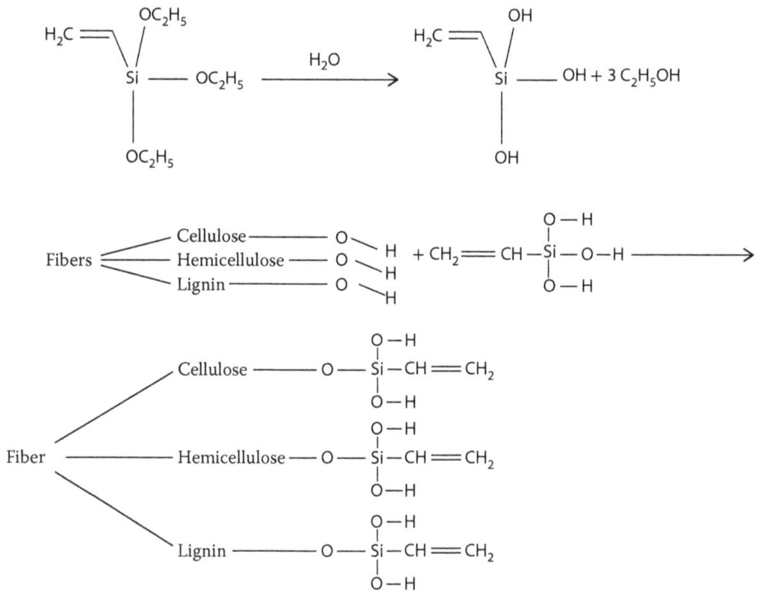

SCHEME 5.2 The reaction of cellulose fibres with silanes.

interface properties (Coutinho et al. 1997; Kalia et al. 2009). Alkoxy silanes are able to form bonds with hydroxyl groups. Treatment with toluene diisocyanate and triethoxyvinyl silane could improve the interfacial properties. Silanes after hydrolysis undergo condensation and a bond formation stage and can form polysiloxane structures by reaction with hydroxyl group of the fibers. The reactions are given in Scheme 5.2 (Kalia et al. 2009; Sreekala et al. 2000).

Dichlorotriazines and derivatives have multifunctional groups in their molecular structure. These groups have different functions in the coupling reactions. On heterocyclic ring, the reactive chlorines react with the hydroxyl group (–OH) of wood fiber and give rise to the ether linkage between cellulose and the coupling agent. The electronegative nitrogen may link the hydroxyl group through hydrogen bonding. On the alkyl chain, the carbon carbon double bonds (C=C) form covalent bonds with polymer matrix by grafting. At the same time, the electronegative nitrogen in the amino group and oxygen in the carboxylate group also link the cellulose phase through hydrogen bonding (Zadorecki and Flodin 1985).

5.6 PLANT FIBER–REINFORCED COMPOSITES

Composites are materials that comprise strong load-carrying material (known as reinforcement) imbedded in weaker material (known as matrix). Reinforcement provides strength and rigidity, helping to support structural load. The matrix or binder (organic or inorganic) maintains the position and orientation of the reinforcement. Significantly, constituents of the composites retain their individual, physical and chemical properties; yet, together they produce a combination of qualities that individual constituents would be incapable of producing alone.

The role of matrix in a reinforced composite is to transfer stress between the fibers, to provide a barrier against an adverse environment, and to protect the surface of the fibers from mechanical abrasion. The matrix plays a major role in the tensile load-carrying capacity of a composite structure. The binding agent or matrix in the composite is of critical importance. Four major types of matrices have been reported: polymeric, metallic, ceramic, and carbon. Most of the composites used in the industry today are based on polymer matrices (Hull and Clyne 1996). Polymer resins have been divided broadly into two categories: thermosets and thermoplastics.

5.6.1 Thermosets and Thermoplastics

Thermoset is a hard and stiff cross-linked material that does not soften or become moldable when heated. Thermosets are stiff and do not stretch the way that elastomers and thermoplastics do (Coutinho et al. 1997). Several types of polymers have been used as matrices for natural composites. The most commonly used thermoset polymers are epoxy resins and other resins (unsaturated polyester resins [as in glass], vinyl ester, phenolic epoxy, novolac, and polyamide) (Bledzki et al. 1998; Chawla 1987; Hull and Clyne 1996).

Unsaturated polyesters are extremely versatile in properties and applications and have been a popular thermoset used as the polymer matrix in composites. They are widely produced industrially as they possess many advantages compared to other thermosets including room temperature cure capability, good mechanical properties, and transparency. The reinforcement of polyesters with cellulosic fibers have been widely reported. Polyester–jute, Polyester–sisal, polyester–coir, polyester–banana–cotton, polyester–straw, polyester–pineapple leaf, and polyester–cotton–kapok (Mwaikambo et al. 1999) are some of the promising systems. Thermoplastics are polymers that require heat to make them processable. After cooling, such materials retain their shape. In addition, these polymers may be reheated and reformed, often without significant changes in their properties. Thermoplastics that have been used as matrix for natural reinforced composites are as follows: HDPE, low-density polyethene, chlorinated polyethylene, PP, normal PS, PVC, mixtures of polymers, and recycled thermoplastics. Only those thermoplastics are useable for natural reinforced composites, whose processing temperature (temperature at which it is incorporated into polymer matrix) does not exceed 230°C. These are, most of all, polyolefins, such as polyethylene and PP. Technical thermoplastics, such as polyamides, polyesters, and polycarbonates require processing temperatures >250°C and are therefore not useable for such composite processing without degradation (Taj et al. 2007).

5.6.2 Rubber-Based Polymer Composites

The primary effects of bioreinforcement on the mechanical properties of natural rubber (NR) composites include increased modulus, increased strength with good bonding at high concentrations, decreased elongation at failure, greatly improved creep resistance over particulate-filled rubber, increased hardness, and a substantial improvement in cut, tear, and puncture resistance. Biodegradation of vulcanized rubber material is possible, although it is difficult due to the interlinkages of the poly(*cis*-1,4-isoprene) chains, which result in reduced water absorption and gas permeability of the material (Seal and Morton 1996).

Researchers have also investigated the reinforcement effects of a leaf (sisal) in NR. An interesting report on the reinforcement effect of grass (bagasse) in NR was presented by Nassar et al. (1996). Aging experiments revealed tensile strength retention of 97%. Scientists have also developed composites comprising kenaf and NR (El Sabbagh et al. 2000). An increase in rheometric and mechanical properties was observed. Pineapple and jute have also found their way as a potential reinforcement in NR (John and Thomas 2007).

The incorporation of sisal and coir in NR was seen to increase the dielectric constant of the composites. These hybrid biocomposites were found to have enormous applications as antistatic agents. In another interesting study, the preparation of composites comprising waste paper in NR along with boron carbide and paraffin wax, for radiation shielding applications, was investigated (Madani et al. 2004). In an innovative study, a unique combination of sisal and oil palm fibers in NR has been used to design hybrid biocomposites. It was seen that the incorporation of fibers resulted in increased modulus. Chemical modification of both sisal and oil palm fibers was imperative for increased interfacial adhesion and resulted in enhanced properties. The viscoelastic, water sorption, dielectric, and stress relaxation characteristics were also studied (Jacob et al. 2006).

Efforts are on the anvil for the development of "advanced green composites" made out of high-strength protein fibers (spider silk) and biodegradable matrices. Biotechnology is being used to increase the yield of specific triglycerides and oils in beans for producing resins. These resins will be inexpensive compared to those available today and if suitably modified, could be biodegradable. Research is also being conducted to develop new pathways to synthesize inexpensive biodegradable resins with better mechanical properties and thermal stability using nanotechnology.

5.7 PROCESSING METHODS

Plant fiber composites are prepared using various composite-manufacturing methods such as compression molding, injection molding, resin transfer molding (RTM), and vacuum bagging. The preforms are mostly fibers, fabrics, or nonwovens. Prepregs are also widely used to prepare composites (Njuguna et al. 2011). Equation 5.2 is commonly used in the preparation of composites.

$$V_f = \frac{W_f/\rho_f}{(W_f/\rho_f) + (W_m/\rho_m)} \tag{5.2}$$

where V_f is the volume fraction of the fiber, W_f is the weight fraction of fiber, and W_m is the weight of matrix. ρ_f and ρ_m are the densities of the fiber and matrix, respectively. The production of the composites is optimized in relation to temperature, pressure, and molding time. It is often necessary to preheat the natural fibers to reduce the moisture before processing the composites. High temperatures degrade the cellulose, thus negatively affecting the mechanical properties of the composites. Inefficient dispersion in the matrix causes agglomeration, which decreases the tensile strength. Most of the previous research on natural composites has focused on reinforcements such as flax, hemp, sisal, and jute, and matrices such as thermoplastics and thermosets. Some of these composites have been produced using matrices made of derivatives from cellulose, starch, and lactic acid to develop fully biodegradable composites or biocomposites. The emerging diversity of applications of natural composites has seen the production of sandwich structures based on natural composite skins. In some cases, these sandwich composites have been produced from paper honeycomb and natural reinforced thermoplastic or thermoset skins, depending on the applications (Njuguna et al. 2011; Mishra et al. 2004).

The main criteria for the selection of the appropriate process technology for natural composite manufacture include the desired product geometry, the performance needed, and the cost and the ease of manufacture. The fabrication methods for natural composites are similar to those used for glass fibers. The most commonly used manufacturing processes are introduced in the following. Although many variants on these techniques exist, this overview gives a good indication of the production possibilities.

5.7.1 HAND LAMINATING

The fibers are placed in a mold and the resin is later applied by rollers. One option is to cure using a vacuum bag; excess air is removed and then the atmospheric pressure exerts pressure to compact the part. The simplicity, low cost of tooling, and flexibility of design are the main advantages of the procedure. On the other end, the long production time, intensive labor, and low automation potential are some of the disadvantages.

5.7.2 RESIN TRANSFER MOLDING

The resin transfer molding (RTM) technique requires the fibers to be placed inside a mold consisting of two solid parts (close-mold technique). A tube connects the mold with a supply of liquid resin, which is injected at low pressure through the mold, impregnating the fibers. The resulting part is cured at room temperature or above until the end of the curing reaction, when the mold is opened and the product is removed. Parameters such as injection pressure, content, and mold temperature have great influence on the development of the temperature profiles and the thermal boundary layers, especially for thin cavities. This technique has the advantage of rapid manufacturing of large, complex, and high-performance parts. Several types of resins (epoxy, polyester, phenolic, and acrylic) can be used for RTM as long as their viscosity is low enough to ensure proper wetting of the plant fibers. Parameters such as injection pressure, content, and mold temperature have a great influence on the development of the temperature profiles and the thermal boundary layers, especially for thin cavities. Good knowledge of all the operating steps is very important to obtain high-quality parts (Njuguna et al. 2011).

An alternative variant of this process is vacuum injection or vacuum-assisted RTM (VARTM), where a single solid mold and a foil (polymeric film) are used. The VARTM process is a very clean and low-cost manufacturing method: resin is processed into a dry reinforcement on a vacuum-bagged tool, using only the partial vacuum to drive the resin. As one of the tool faces is flexible, the molded laminate thickness depends partially on the compressibility of the resin composite before curing and the vacuum negative pressure.

5.7.3 COMPRESSION MOLDING

Compression molding is another major technique for the construction of reinforced polymers, which involves a semifinished composite sheet widely known as sheet molding compound (SMC) that is later molded into the final parts by compression. For the SMC, the process consists of a rolling film of resin on which fibers are added. A second film of resin is then added, so as to later be compressed in a composite sheet that may be stored for few days. To get the final product, the reinforced sheet is then placed into a press to take its desired shape.

Advantages of compression molding are the very high volume production ability, the excellent part reproducibility, and the short cycle times. Processing times of <2 min are reached during the compression molding of 3D components with a high forming degree. It has also been shown that the adhesion of plant fibers and matrix resin is important to obtain good mechanical properties of plant fiber composites, and the mechanical properties were improved by the molding condition, the molding pressure, and temperature. A big concern with compression molding that always needs to be considered is the maximum pressure before the fibers and the structure are damaged.

5.7.4 INJECTION MOLDING

The injection molding process is suitable to form complex shapes and fine details with excellent surface finish and good dimensional accuracy for high production rate and low labor cost. In the injection molding resin, granules and short fibers are mixed into a heated barrel and transported to the mold cavity by a spindle. Injection molding is one of the most important processes for the manufacturing of plastics/composites and can produce from very small products such as bottle tops to very large car body parts.

5.7.5 PULTRUSION

Pultrusion is a continuous process to manufacture composite profiles at any length. The impregnated fibers are pulled through a die, which is shaped according to the desired cross section of the product. The resulting profile is shaped until the resin is dry. Advantages of this process are the ability to build thin-wall structures, the large variety of cross-sectional shapes, and the possibility for high degree of automation (Kalia et al. 2011; Njuguna et al. 2011).

5.8 PROPERTIES OF THE PLANT FIBER COMPOSITES

5.8.1 THERMAL STABILITY OF PLANT FIBER COMPOSITES

The plant fibers start degrading at about 240°C. Structural constituents of the fibers (cellulose, hemicelluloses, lignin, etc.) are sensitive to the different range of temperatures. It was reported that lignin starts degrading at a temperature around 200°C and hemicelluloses and cellulosic constituents degrade at higher temperatures (Joseph 2001). Thermal stability can be enhanced by removing a certain proportion of hemicelluloses and lignin constituents from the plant fiber by different chemical treatments. The degradation of plant fiber is an important issue in the development of natural fiber composites in both composite manufacturing and materials in service (Sgriccia et al. 2008).

5.8.2 LENGTH, LOADING, AND ORIENTATION OF THE FIBER IN MATRIX

The properties of plant fiber–reinforced polymer composites depend on the length, percentage of fiber volume/loading, distribution, and orientation into the matrix. When load is applied to the matrix, stress transfer occurs by shear at both the interface along the length and at the ends of the fiber. The extent of load transmitted to the fiber is a function of critical length (aspect ratio), direction, and orientation of the fibers relative to each other and the compatibility between matrix interfaces. Depending on the orientation of the fiber at the matrix, three types of reinforcement can be obtained. First, longitudinally aligned fibrous composites generally have higher tensile strength but lower compressive strength due to the buckling. Second, the transverse direction supports very low tensile stress, which is lower than the strength of the matrix. Finally, for the randomly oriented short composites, prediction of mechanical properties is far more difficult, because of the dispersion, orientation, and complexities for the load distribution along the fiber–matrix interface. By controlling factors such as aspect ratio, dispersion, and orientation of fibers, considerable improvements in composite property can be accomplished (Fakirov and Bhattacharya 2007; Mwaikambo et al. 1999).

5.8.3 PRESENCE OF VOIDS

During the introduction of the plant fiber into the matrix, air or other volatile substances may be trapped in the material. After curing, microvoids may form in the composite along the individuals, due to the spacing between the laminate and the resin-rich regions, which has an adverse effect on the mechanical properties of the composites. Curing and cooling rate of the resin can also be responsible for void formation (Joseph 2001). Higher void content (over 20% by volume) is responsible for lower fatigue resistance, greater affinity to water diffusion, and increased variation (scatter) in mechanical properties. Composites with higher fiber content usually display more void formation (Kenneth 1992).

5.8.4 MOISTURE ABSORPTION OF PLANT FIBER COMPOSITES

The lignocellulosic fibers are hydrophilic and absorb moisture. Large amounts of hydrogen bonds (hydroxyl groups −OH) are present between the macromolecules in the plant cell wall. When moisture from the atmosphere comes in contact with the fiber surface, the hydrogen bond breaks and hydroxyl groups form new hydrogen bonds with water molecules. The cross section of the fiber becomes the main access to the penetrating water. As a result, when hydrophilic cellulosic fiber is reinforced with hydrophobic resin, swelling within the matrix occurs. This causes weak bonding between fiber and matrix, dimensional instability, matrix cracking, and poor mechanical properties of the composites (Vaxman et al. 2004).

Therefore, the removal of moisture from fibers is an essential step before the preparation of composites. The moisture absorption of natural fibers can be reduced by different chemical treatments such as alkali treatment, silane treatment, acetylation, benzoylation, and peroxide treatment on the fiber surface to remove hydrophilic hydroxyl bonds (Wang et al. 2007).

5.9 FACTORS AFFECTING COMPOSITE PROPERTIES

Plant fibers have some problems that cause adverse effects on composite properties, as these fibers consist of different chemical constituents in their structure and undergo different reactivity with the environment and matrix during processing. The following are the major issues that need to be addressed during fabrication of composite materials (Kabir et al. 2011). The fiber–matrix interface has great influence on the mechanical properties of composites. Surface modification of natural fibres prior to their use in composite materials is also needed to facilitate fiber–matrix interfacial addition and an increase in mechanical properties. The fiber–matrix interface does not only influence strength in natural fiber composites. Tuning of the interfacial properties can also be used to control other macroscopic composite properties. Some examples of interfacial influences on engineering properties are given in Sections 5.9.1 through 5.9.7.

5.9.1 MECHANICAL STRENGTH

For wood–plastic composites with wood particles used as fillers, the composite strength is frequently used as a measure of the interfacial strength (Oksman 1996) because the wood particles act as stress raisers. For particle composites, it can be assumed that the strength improves with stronger fiber–matrix interfaces, provided that all other material parameters remain unchanged. For fiber composites, a stronger interface does not necessarily mean a stronger composite, because a too strong interface can result in a brittle and flaw-sensitive composite material. However, for most polyolefin thermoplastics, the interface should be as strong as possible if a strong composite is the aim, because the interaction between the polar natural fiber and the nonpolar aliphatic polymer matrix is very small.

5.9.2 FRACTURE TOUGHNESS

In crack propagation in fiber composites, the energy dissipation is affected by pullout of fibers in the wake of the crack. The interfacial properties may also affect to which degree the crack diverts from planar growth as it encounters fibers as obstacles at the matrix crack tip. Pullout is where the fiber–matrix interface comes into play, especially for composites with fibers primarily ordered in the load direction. The interfacial shear strength will affect the cohesive tractions of bridging fibers, which in turn will affect the fracture toughness or critical energy release rate as shown by McCartney (1989). The two extreme cases of a vanishing interfacial strength and an infinitely strong interface can be envisaged; the first case would lead to effortless pullout of fibers, whereas the second case would lead to brittle scission of fibers. The most energy dissipating mechanisms would be in an intermediate region, where a sufficiently strong interface would demand a high level of work to pull out fibers in the crack wake without breaking the fibers in a brittle manner. Thus, an optimum in fracture toughness can be expected for intermediately strong interfaces. These issues have been investigated experimentally and discussed by Feih et al. (2005).

5.9.3 STIFFNESS

Generally, the elastic properties are improved when the interface is stronger, that is, the stress transfer ability is improved. Bogren et al. (2006) measured the complex viscoelastic properties during dynamic cyclic loading, and observed a difference between the measured values and the predicted moduli based on a model with a perfect interface. This difference may be attributable, at least partly, to an imperfect interface with reduced stress transfer. Most models to predict elastic properties of composites based on the elastic properties of the constituents assume an ideal interface, for example, Neagu et al. (2006) for wood fiber composites, although there are models with imperfect interfaces (Hashin 1990), which can be applied if there is sufficient data for estimation of the parameter describing the interface imperfection. It should be borne in mind that for each envisaged unit cell or representative volume element, the crack-opening displacement for a suitable load level should be large enough to contribute to a measurable difference in stiffness compared to the case of a closed crack (Varna et al. 2001). For dispersed, dilute, and small interfacial cracks, there might not be any measurable difference in composite stiffness.

5.9.4 IMPACT TOUGHNESS

The damage process during notched impact testing is essentially similar to that in testing of fracture toughness. The main difference is that the process is several orders of magnitude faster during impact than under quasistatic conditions. Since both the polymer matrix and plant fibers are viscoelastic in nature, the behavior will be more brittle in impact. The polymer matrix has generally a more viscous behavior than natural fibers, and the transition from slow and ductile yielding to brittle fracture will be more apparent in the matrix than in the fibers, when the strain rate is increased. However, the rank in impact toughness of different composites can be assumed to be very similar to the corresponding rank in fracture toughness for the same set of composites. Park and Balatinecz (1996) found a correlation between impact toughness and the work of fracture defined by the area under the stress–strain curve for wood fiber–reinforced PP. The interface has probably the same role in impact testing as for quasistatic fracture toughness or strength testing, that is, as an energy-absorbing mechanism that contributes positively to the toughness property for a suitably strong interface. The interface should not be too weak to be pulled out without effort, and not too strong to induce brittle failure by fiber breakage in the same plane as the matrix crack. An example of the former situation has been presented by Bengtsson et al. (2006), who showed an increase in impact toughness for interface-modified wood fiber–reinforced polyethylene. This is an example where the initial interaction is very weak between the nonpolar polyethylene matrix and the polar cellulosic fibers. By silane cross-linking, the interface is made stronger resulting in an increased impact toughness.

5.9.5 MOISTURE UPTAKE

Joseph et al. (2002) showed a reduction in water uptake for sisal fiber–reinforced PP with improved interfacial adhesion. This result indicates that capillary water sorption mechanisms at the interface may be subdued with improved fiber–matrix bonding. Marcovich et al. (2005) showed that interfacial modification in lignocellulosic fiber–reinforced unsaturated polyester led to improved fiber wetting during processing, which in turn led to less moisture uptake since capillary absorption through narrow channels along the interfaces are suppressed. Baillie et al. (2000) discuss pathways of moisture absorption found in wood, while maintaining high strength and toughness. By studying solutions in nature, ideas can be conceived of how to improve engineered composites while simultaneously optimizing several properties.

5.9.6 DIMENSIONAL STABILITY

For hydrophilic materials, such as natural fiber composites, hygroelastic swelling goes hand in hand with moisture uptake: the more moisture the material absorbs, the more it will swell. If thermal deformation is ignored, the dimensional stability is typically characterized by the hygroexpansion coefficient, which describes the strain induced by a change in moisture content or in surrounding relative humidity. Increased hydrophilicity thus generally means increased dimensional instability. Typically, the cellulosic natural fibers swell more than polymer matrix materials (except, maybe, for certain thermoplastic starches). Higher fiber content leads to a higher degree of hygroexpansion. Grigoriou (2003) found less moisture-induced swelling for wood and waste paper composites containing higher degrees of resin. As for moisture uptake, the fiber–matrix bond will also affect the eventual hygroexpansion.

Naik and Mishra (2006) showed decreased moisture-induced swelling in maleic anhydride–treated natural fiber–reinforced PS, where the fiber–matrix adhesion has been improved.

5.9.7 TIME-DEPENDENT PROPERTIES

Creep and fatigue usually involve the same set of underlying damage mechanisms as in quasistatic failure, although at a different rate and to different degrees. These damage mechanisms include fiber fracture, fiber pullout, matrix yielding, interfacial debonding, crack coalescence, and so on. Sain et al. (2000) showed only a marginal improvement in creep properties by maleic and maleimide interfacial modification in wood fiber PP composites. A stronger effect of the interface is expected in cyclic loading under fatigue conditions, where an imperfect interface can lead to repeated frictional sliding, which induces temperature increase and energy dissipation. Gassan and Bledzki (1997) measured a significant decrease in energy dissipation during cyclic loading for jute fiber–reinforced PP with improved interface from maleic anhydride-grafted PP.

5.10   APPLICATIONS OF PLANT FIBER COMPOSITES

More interest is now shown in the investigation of the suitability of plant fiber composites in structural and infrastructure applications where moderate strength, lower cost, and environmental friendly features are required. The application of plant fiber composites has started in automotive industries and production of nonstructural elements. In 1986, a study was published where it was reported that coir/polyester composites have been used to produce mirror casing, paper weight, projector cover, voltage stabilizer cover, mailbox, helmet, and roof. In structural applications and infrastructure applications, plant fiber composites have been used to develop load-bearing elements such as beams, roofs, multipurpose panels, water tanks, and pedestrian bridges (Ticoalu et al. 2010).

Roof materials made of plant fiber composites that have been developed are, for example, woven mat sisal/cashew nut shell liquid and recycled paper–reinforced acrylated epoxidized soybean oil (AESO) with a foam core (Dweib et al. 2006). The recycled paper–reinforced AESO composites are in the form of sandwich panel and were used to construct a monolithic roof for a single-story A-frame house. The composite panel is usually a flat surfaced element on both sides, although corrugated panels are also possible. It is usually of uniform thickness and can be produced with different dimension. A panel can be made from homogenous material, composite laminate, or sandwich panel. LOC Composites Pty Ltd has developed and produced a composite sandwich panel that uses plant-based polymers that can be used in several applications such as balcony construction, walls, roofs, floors, and fire doors (Erp and Rogers 2008). In the area of structural rehabilitation, jute mat–reinforced composites have been used for trenchless rehabilitation of underground drain pipes and water pipes (Yu et al. 2008). The previously mentioned studies have highlighted that natural materials can be used effectively to develop load-bearing materials such as for roofs, beams, and panels. Furthermore, for infrastructure applications where the use of synthetic fibers is not suitable, natural fibers can be a suitable substitute.

The main and current drawbacks of natural composites in structural and infrastructure applications are mostly related to the large variation in the properties of the plant fibers, treatments, and manufacturing optimization. These concerns need to be addressed to develop and produce improved structural elements that can be used both for structural and infrastructure applications (Ticoalu et al. 2010).

In Europe, plant composites are mainly used by the automotive industry (Ellison and McNaught 2000). The applied semiproducts are raw fibers and nonwoven mats, and therefore, the composites possess moderate mechanical properties. This makes them nevertheless well qualified to be used as nonstructural components. Because of the high sensitivity of plant fibers toward water exposure, mostly interior components (Figure 5.5) are made of plant fiber composites, for example, door liners, boot liners, and parcel shelves. Recently, plant fibers have also been used in exterior composite components; the engine and transmission covers of a Mercedes-Benz Travego coach (Karus et al. 2002). In 2002, the total consumption of plant fibers was about

FIGURE 5.5   A side-door liner of an Audi A2 car based on a plant fiber composite material. (Courtesy of N-FibreBase.)

17,000 tons, and the average amount of plant fiber per vehicle was 10–15 kg. These numbers are anticipated to increase in the remaining western world, with the use of nonstructural composite components based on plant fibers becoming considerably more widespread, and wood fibers being by far the preferred type. In the United States, wood fibers account for about 7% of the total amount of fillers and fibers used in composite materials, and the market is fast growing. The main applications are building components, such as deckings, windows profiles, and floorings, but also products such as pallets, flowerpots, and office accessories are increasingly being made of plant-based composite materials. This development in the use of plant fibers is likely to be mirrored by Europe in the coming years (Ashori 2008).

In most of the current applications of plant composites, the reinforcement effect is of minor importance. The fibers are not aligned and the good tensile properties are therefore not efficiently used. Only one example has been found of strong plant composites used in load-bearing applications. In India, houses, grain silos, and fishing boats have been fabricated with the use of aligned jute composites (Mohanty and Misra 1995).

The industrial use of plant fibers is generally driven by reductions in cost, but also the issue of environmental awareness has become important. In Europe, the European Union directive "end-of-life vehicle" imposes that 85% by weight of all vehicle components should be recyclable by 2006, and 95% by 2015. In relation to plant composites, the term "recyclable" is, however, somewhat unresolved (Ticoalu et al. 2010). Plant fibers are fully recyclable by combustion, as well as being fully biodegradable, but the same cannot be implied for the remaining synthetic polymeric matrix and chemical additives. Moreover, as a material type, the recyclability (or reusability) of composites is a field where only little practical experience has been accumulated, and as such, it is inferior to other types of materials; examples are metals and pure plastics. Glass composites with a thermoplastic matrix can be recycled by remolding used parts, but in the case of plant composites, this has been shown to severely deteriorate the mechanical properties because of the repeated thermal expo-sure. Even though the current industrially applied plant composites are not strictly recyclable, they form nevertheless a foundation for a future increasing use of envi-ronmental friendly materials (Madsen and Lilholt 2003; Reussman et al 1999).

REFERENCES

Almgren, K.M. 2010. *Wood-Fibre Composites: Stress Transfer and Hygroexpansion*, Doctoral Thesis No. 9 KTH Fibre and Polymer Technology. Stockholm, Sweden.

Ashori, A. 2008. Wood–plastic composites as promising green-composites for automotive industries. *Bioresource Technology* 99: 4661–4667.

Baillie, C., Tual, D. and Terraillon, J.C. 2000. Interfacial pathways in wood. *Advanced Composites Letters* 9: 45–57.

Beg, M.D.H. 2007. *The Improvement of Interfacial Bonding, Weathering and Recycling of Wood Fibre Reinforced Polypropylene Composites*. The University of Waikato, Hamilton, New Zealand.

Belgacem, M.N., Bataille, P. and Sapieha, S. 1994. Effect of corona modification on the mechanical properties of polypropylene/cellulose composites. *Journal of Applied Polymer Science* 53: 379–385.

Bengtsson, M., Oksman, K. and Stark, N.M. 2006. Profile extrusion and mechanical proper-ties of crosslinked wood-thermoplastic composites. *Polymer Composites* 27: 184–194.

Blaker, J.J., Lee, K.Y. and Bismarck, A. 2010. Hierarchical composites made entirely from renewable resources. *Journal of Biobased Materials and Bioenergy* 5: 1–16.

Bledzki, A.K. and Gassan, J. 1999. Composites reinforced with cellulose based fibres. *Progress in Polymer Science* 24: 221–274.

Bledzki, A.K., Reinhmane, S. and Gassan, J. 1998. Thermoplastics reinforced with wood fill-ers. *Journal of Polymer Plastic Technology* 37: 451–468.

Bledzki, A.K., Sperber, V.E. and Faruk, O. 2002. Natural and Wood Fibre Reinforcement in Polymers. Ed. Humphreys, S. Rapra Review Reports, Report 152, 13(8). iSmithers Rapra Publishing. ISSN 0889-3144.

Bodig, J., Jayne, B.A. and Nostrand, V. 1982. *Mechanics of Wood and Wood Composites*. Von Nostrand Reinhold Company: New York, Cincinnati, Toronto.

Bogren, K.M., Gamstedt, E.K., Neagu, R.C., Åkerholm, M., Lindström, M. 2006. Dynamic-mechanical properties of wood-fibre reinforced polylactide: Experimental characteriza-tion and micromechanical modelling. *Journal of Thermoplastic Composite Materials* 19: 613–637.

Chawla, K.K. 1987. *Composite Materials Science and Engineering*. New York: Springer-Verlag.

Clemons, C.M. and Caulfield, D.F. 2005. Natural fibers. In: *Functional Fillers for Plastics*. Ed. Xanthos, M. Wiley-VCH Verlag GmbH & Co KGaA: Weinheim. ISBN 3-527-31054-1.

Coutinho, F.M.B., Costa, T.H.S. and Carvalho, D.L. 1997. Polypropylene-wood fibre com-posites: effect of treatment and mixing conditions on mechanical properties. *Journal of Applied Polymer Science* 65: 1227–1235.

Dalvag, H., Klason, C. and Stromvall, H.E. 1984. The efficiency of cellulosic fillers in common thermoplastics-part II. Filling with processing aids and coupling agents. *International Journal of Polymer Materials* 11: 9–38.

Dissanayake, N., Virk, A. and Hall, W. 2010. A review of bast fibres and their composites. Part 20 composites. *Composites Part A* 41: 1336–1344.

Dong, H., Strawheckera, K.E., Snydera, J.F., Orlickia, J.A., Reinerb, R.S. and Rudieb, A.W. 2012. Cellulose nanocrystals as a reinforcing material for electrospun poly (methyl methacrylate) fibres: Formation, properties and nanomechanical characterization. *Carbohydrate Polymers* 87: 2488–2495.

Dong, S., Sapieha, S. and Schreiber, H.P. 1992. Rheological properties of corona modified cellulose/polyethylene composites. *Polymer Engineering and Science* 32: 1734–1739.

Dufresne, A. 2008. Cellulose-based composites and nanocomposites. In: *Monomers, Polymers and Composites from Renewable Resources*, 1st edition. Eds. Gandini, A. and Belgacem, M.N., 401–418. Elsevier: Oxford, UK.

Dweib, M.A., Hu, B., Wool, R.P. and Shenton, H.W. 2006. Bio-based composite roof structure. Manufacturing and processing issues. *Composite Structures* 74: 379–388.

Dwivedi, A.H. and Pande, U.C. 2012. Photochemical degradation of halogenated compounds: A review. *Scientific Reviews and Chemical Communication* 2: 41–65.

El Sabbagh, S.H., El Hariri, D.M. and Abd. El Ghaffar, M.A. 2000. In: *Proceedings from the Third International Symposium on Natural Polymers & Composites*. May 14–17, 469–483.

Ellison, G.C. and McNaught, R. 2000. *The Use of Natural Fibres in Nonwoven Structures for Applications as Automotive Component Substrates*. Research and Development Report. Reference NF0309. Ministry of Agriculture Fisheries and Food Agri–Industrial Materials: London, UK, pp. 1–91.

Erp, G.V. and Rogers, D. 2008. A highly sustainable fibre composite building panel. Presented at the Fibre Composites in Civil Infrastructure—Past, Present and Future. Toowoomba: Australia.

Fakirov, S. and Bhattacharya, D. 2007. *Engineering Biopolymers: Homopolymers, Blends and Composites*. Auckland, New Zealand: Hanser Gardner Publications.

Feih, S., Wei, J., Kingshott, P. and Sørensen, B.F. 2005. The influence of fibre sizing on the strength and fracture toughness of glass fibre composites. *Composites Part A* 36: 245–255.

Felix, M. and Gatenholm, P. 1991. The nature of adhesion in composites of modified cellulose fibres and polypropylene. *Journal of Applied Polymer Science* 42: 609–620.

Garkhail, S.K., Heijenrath, R. and Eindhoven, P.T. 2000. Mechanical properties of natural fibre-mat-reinforced thermoplastics based on flax fibres. *Applied Composite Materials* 7: 351–372.

Gassan, J. and Bledzki, A.K. 1999. Possibilities for improving the mechanical properties of jute/epoxy composites by alkali treatment of fibres. *Composites Science & Technology* 59: 1303–1309.

Gassan, J. and Bledzki, A.K. 1997. The influence of fibre-surface treatment on the mechanical properties of jute-polypropylene composites. Composites Part A 28: 1001–1005.

George, J., Ivens, J. and Verpoest, I.L. 1999. Mechanical properties of flax fibre reinforced epoxy composites. *Macromolecular Chemistry* 272: 41–45.

Grigoriou, A.H. 2003. Waste paper-wood composites bonded with isocyanate. *Wood Science and Technology* 37: 79–89.

Hashin, Z. 1990. Thermoelastic properties of fibre composites with imperfect interface. *Mechanics of Materials* 8: 333–348.

Hughes, M., Carpenter J. and Hill, C. 2007. Deformation and fracture behaviour of flax fibre reinforced thermosetting polymer matrix composites. *Journal of Materials Science* 42: 2499–2511.

Hull, D. and Clyne, T.W. 1996. *An Introduction to Composite Materials*. Cambridge: Cambridge University Press.

Jacob, M., Francis, B., Varughese, K.T. and Thomas, S. 2006. Dynamical mechanical analysis of sisal/oil palm hybrid fibre-reinforced natural rubber composites. *Polymer Composites* 27: 671–680.

Jarman, C.G. and Jayasundera, D.S. 1975. The extraction and processing of coconut fibre. Tropical Products Institute, Ministry of Overseas Development, 25. London

John, M.J. and Thomas, S. 2007. Review: Biofibres and biocomposites. *Carbohydrate Polymers* 71: 343–364.

John, Z.L., Qingli, W. and Harold, S. 2000. Chemical coupling in wood fibre and polymer Composites: A review of coupling agents and treatments. *Wood and Fibre Science* 32: 88–104.

Joseph, P.V. 2001. *Studies on Short Sisal Fibre Reinforced Isotactic Polypropylene Composites*, PhD Thesis, Mahatma Gandhi University, India.

Joseph, P.V., Rabello, M.S., Mattoso, L.H.C., Joseph, K. and Thomas, S. 2002. Environmental effects on the degradation behaviour of sisal fibre reinforced polypropylene composites. *Composites Science and Technology* 62: 1357–1372.

Kabir, M.M., Wang, H., Aravinthan, T., Cardona, F. and Lau, K.T. 2011. *Effects of Natural Fibre Surface on Composite Properties: A Review*, Centre of Excellence in Engineered Fibre Composite (CEEFC). Faculty of Engineering and Surveying, University of Southern Queensland: Toowoomba, Australia, 95–96.

Kalia, S., Dufresne, A., Cherian, B.M., Kaith, B.S., Averous, L., Njuguna, J. and Nassiopoulos, E. 2011. Review Article Cellulose-Based Bio- and Nanocomposites. *International Journal of Polymer Science* article ID 837875.

Kalia, S., Kaith, B. S. and Kaur, I. 2009. Pretreatments of natural fibres and their application as reinforcing material in polymer composites-a review. *Polymer Engineering and Science* 49: 1253–1272.

Karus, M., Kaup, M. and Ortmann, S. 2002. *Market Survey Use of Natural Fibres in the German and Austrian Automotive Industry Status, Analysis and Trends*. Nova-Institute GmbH: Hürth, Germany.

Kenneth, J.B. 1992. Void effects on the interlaminar shear strength of unidirectional graphite fibre-reinforced composites. *Composite Material* 26: 1487–1992.

Khalil, H.P.S.A. and Ismail, H. 2001. Effect of acetylation and coupling agent treatments upon biological degradation of plant fibre reinforced polyester composites. *Polymer Testing* 20: 65–75.

Kim, H.S., Lee, B.H. and Choi, S.W. 2007. The effect of types of maleic anhydride-grafted polypropylene (MAPP) on the interfacial adhesion properties of bio-flour-filled polypropylene composites. *Composites Part A: Applied Science and Manufacturing* 38: 1473–1482.

Madani, M., Basta, A.H., Abdo, A.S. and El-Saied, H. 2004. Utilization of waste paper in the manufacture of natural rubber composite for radiation shielding progress in rubbers. *Plastics and Recycling Technology* 20: 210–287.

Madsen, B. and Lilholt, H. 2003. Physical and mechanical properties of unidirectional plant fibre composites—an evaluation of the influence of porosity. *Composites Science and Technology*, 63(9): 1265–1272.

Marcovich, N.E., Reboredo, M.M. and Aranguren, M.I. 2005. Lignocellulosic materials and unsaturated polyester matrix composites: Interfacial modifications. *Composite Interfaces* 12: 3–24.

McCartney, L.N. 1989. New theoretical model of stress transfer between fibre and matrix in a uniaxially fibre-reinforced composite. *Proceedings of the Royal Society of London Series A* 425: 215–244.

McLaughlin, E.C. and Tait, R.H. 1980. Fracture mechanism of plant fibres. *Journal of Material Science* 15: 89–95.

Mishra, H.K., Dash, B.N., Tripathy, S.S. and Padhi, P.N. 2000. Study on mechanical performance of jute-epoxy composites. *Polymer Plastics Technology and Engineering* 39: 187–198.

Mishra, S., Mohanty, A.K., Drzal, L.T., Misra, M. and Hinrichsen, G. 2004. A review on pineapple leaf fibres, sisal fibres and their biocomposites. *Macromolecular Materials and Engineering* 289(11): 955–974.

Mitra, B.C., Basak, R. K. and Sarkar, M. 1998. Studies on jute-reinforced composites, its limitations, and some solutions through chemical modifications of fibres, Indian jute industries. *Journal of Applied Polymer Science* 67: 1093–1100.

Mohanty, A.K. and Misra, M. 1995. Studies on jute composites. A literature review. *Polymer Plastics Technology and Engineering* 34: 729–792.

Mwaikambo, L. and Bisanda, E. 1999. The performance of cotton/kapok fabric-polyester composites. *Polymer Testing* 18: 181–198.

Mwaikambo, L.Y. 2006. Review of the history, properties and application of plant fibres. *African Journal of Science and Technology (AJST) Science and Engineering Series* 7: 120–133.

Mwaikambo, L.Y. and Ansell, M.P. 1999. The effect of chemical treatment on the properties of hemp, sisal, jute and kapok for composite reinforcement. *Macromolecular Chemistry* 272: 108–116.

Naik, J.B. and Mishra, S. 2006. The compatibilizing effect of maleic anhydride on swelling properties of plant-fibre-reinforced polystyrene composites. *Polymer Plastic Technology Engineering* 45: 923–927.

Nassar, M.M., Ashour, E.A. and Washid, S.S. 1996. Thermal characteristics of bagasse. *Journal of Applied Polymer Science* 61: 885–890.

Neagu, R.C., Gamstedt, E.K. and Berthold, F. 2006. Stiffness contribution of various wood fibres to composite materials. *Journal of Composite Materials* 40: 663–699.

Nielsen, L.E. 1974. *Mechanical Properties of Polymers and Composites*. Dekker: New York.

Nielsen, L.E. 1977. *Polymer Rheology*. M. Dekker: New York.

Njuguna, J., Wambua, P., Pielichowski, K. and Kayvantash, K. 2011. Natural fibre-reinforced polymer composites and nanocomposites for automotive applications. In: *Cellulose Fibres: Bio- and Nano-Polymer Composites*. Eds. Kalia, S., Kaith, B.S. and Kaur, I. Springer-Verlag: Berlin, Germany. ISBN 978 3 642 17369 1.

Oksman, K. 1996. Improved interaction between wood and synthetic polymers in wood/polymer composites. *Wood Science and Technology* 30: 197–205.

Park, B.D. and Balatinecz, J.J. 1996. Effects of impact modification on the mechanical properties of wood-fibre thermoplastic composites with high impact polypropylene (HIPP). *Journal of Thermoplastic Composite Materials* 9: 342–364.

Peyer, S. and Wolcott, M. 2000. *Engineered Wood Composites for Naval Waterfront Facilities*. Yearly report to office of Naval research, wood materials and engineering laboratory. Washington State University, Pullman, Washington.

Philippe, T., Christian, E. and Zimmermann, T. 2010. Functional polymer nanocomposite materials from microfibrillated cellulose. In: *Nanotechnology and Nanomaterials, Advances in Nanocomposite Technology,* Ed. Hashim, A. InTech. ISBN 978-953-307-347-7.

Pothan, L.A., George, J. and Thomas, S. 2002. Effect of fibre surface treatments on the fibre–matrix interaction in banana fibre reinforced polyester composites. *Composite Interfaces* 9: 335–353.

Raj, R.G. and Kokta, B.V. 1991. Improving the mechanical properties of HDPE-wood fibre composites with additives/coupling agents. In: *49th Annual Technical Conference*. Montreal, Canada, Society of Plastics Engineers, 1883–1885.

Ray, D., Sarkar, B.K., Rana, A.K. and Bose, N.R. 2001. Mechanical properties of vinyl ester resin matrix composites reinforced with alkali-treated jute fibres. *Composites Part A: Applied Science and Manufacturing* 32: 119–127.

Reussman, T., Mieck, P., Grützner, R. and Bayer, R. 1999. The recycling of polypropylene reinforced with natural fibres. *Kunststoffe Plast Europe* 89: 80–84.

Robson, D., Hague, J., Newman, G., Jeronomidis, G. and Ansell, M.P. 1993. *Survey of Natural Materials for Use in Structural Composites as Reinforcement and Matrices, Natural Materials for Composites*. Bangor: The Biocomposites Centre, University of Wales, 1–71.

Rout, J., Tripathy, S.S., Nayak, S.K., Misra, M. and Mohanty, A.K. 2001. Scanning electron microscopy study of chemically modified coir fibres. *Journal of Applied Polymer Science* 79: 1169–1177.

Rowell, R.M. 1984. The chemistry of solid wood. In: *Penetration and Reactivity of Cell Wall Components*. Rowell, R.M., ed. Washington, DC: American Chemical Society, 176.

Rowell, R.M., Young, R.A. and Rowell, J.K. 1996. Agro-fiber thermoplastic composites. In: *Paper and Composites from Agro-based Resources*. Eds. Rowell, R.M., Young, R.A. and Rowell, J.K. Lewis Publishers: New Jersey. pp. 377–401.

Sain, M.M., Balatinecz, J. and Law, S. 2000. Creep fatigue in engineered wood fibre and plastic compositions. *Journal of Applied Polymer Science* 77: 260–268.

Satyanarayana, K.G., Arizaga, G.G.C. and Wypych, F. 2009. Biodegradable composites based on lignocellulosic fibres—An overview. *Progress in Polymer Science* 34: 982–1021.

Satyanarayana, K.G., Sukumaran, K., Mukherjee, P.S., Pavithran, C. and Pillai, S.G. 1990. Natural fibre-polymer composites. *Cement & Concrete Composites* 12: 117–136.

Seal, K.J. and Morton, L.H.G. 1996. Chemical materials. *Biotechnology* 8: 583–590.

Sears, K., Jacobson, R., Caulfield, D. and Underwood, J. 2001. Reinforcement of engineering thermoplastics with high purity wood cellulose fibers. In: *Proceedings of the 6th International Conference on Woodfibre-Plastic Composites, Forest Products Society*. Madison, WI, 27–34.

Sgriccia,N., Hawley, M. and Misra, M. 2008. Characterization of natural fibre surfaces and natural fibre composites. *Composites Part A-Applied Science & Manufacturing* 39: 514–522.

Singha, A.S. and Rana, R.K. 2012. Chemically induced graft copolymerization of acrylonitrile onto lignocellulosic fibres. *Journal of Applied Polymer Science* 124: 1891–1898.

Siqueira, G., Bras, J. and Dufresne, A. 2010. The potential of flax fibres as reinforcement for composite material. *Properties and Applications Polymers* 2: 728–765.

Sreekala, M.S., Kumaran, M.G., Joseph, S., Jacob, M. and Thomas, S. 2000. Oil palm fibre reinforced phenol formaldehyde composites: influence of fibre surface modifications on the mechanical performance. *Applied Composite Materials* 7: 295–329.

Sulawan, K, Sutapun, W. and Jarukumjorn, K. 2013. Effects of interfacial modification and fibre content on physical properties of sisal fibre/polypropylene composites. *Composites Part B: Engineering* 45: 1544–1549.

Taj, S., Munawar, M.A. and Khan, S.U. 2007. Review natural fibre-reinforced polymer composites. Proc. Pakistan Acad. Sci. 44(2): 19–14.

Ticoalu, A., Aravinthan, T. and Cardona, F. 2010. A review of current development in natural fibre composites for structural and infrastructure applications. In: *Southern Region Engineering Conference*. University of Southern Queensland Toowoomba, Australia, 11–12.

Valadez-Gonzalez, A., Cervantes-Uc, J.M., Olayo, R. and Herrera-Franco, P.J.Y. 1999. Effect of fibre surface treatment on the fibre-matrix bond strength of natural fibre reinforced composites. *Composites Part B: Engineering* 30: 309–320.

Varna, J., Joffe, R. and Talreja, R. 2001. A synergistic damage-mechanics analysis of transverse cracking in cross ply laminates. *Composites Science and Technology* 61: 657–667.

Vaxman, A., Narkis, M., Siegmann, A. and Kenig, S. 2004. Void formation in short-fibre thermoplastic composites. *Polymer Composite* 10: 449–453.

Verma, D., Gope, P.C., Maheshwari, M.K. and Sharma, R.K. 2012. Bagasse fibre composites-a review. *Journal of Materials and Environmental Science* 3: 1079–1092.

Wambua, P., Ivens, U. and Verpoest, I. 2003. Natural fibres: can they replace glass in fibre-reinforced plastics? *Composites Science and Technology* 63: 1259–1264.

Wang, B., Panigrahi, S., Tabil, L. and Crerar, W. 2007. Pretreatment of flax fibres for use in rotationally molded biocomposites. *Reinforced Plastics and Composites* 26(5): 447–463.

Yu, H.N., Kim, S.S., Hwang, I.U. and Lee, D.G. 2008. Application of natural fibre reinforced composites to trenchless rehabilitation of underground pipes. *Composite Structures* 86: 285–290.

Zadorecki, P. and Flodin, P. 1985. Surface modification of cellulose fibres, the effect of cellulose fibre treatment on the performance of cellulose-polyester composites. *Journal of Applied Polymer Science* 30: 3971–3983.

6 *Eulaliopsis binata* Utilization of Waste Biomass in Green Composites

Vijay K. Thakur, Manju K. Thakur, and Raju K. Gupta

CONTENTS

6.1 INTRODUCTION

Green materials particularly from renewable waste biomass have attracted an increasing amount of concentration during the last few years (Averous 2004). Research on lignocellulosic waste biomass materials is increasing because of the prominent use of plant biomass in numerous applications ranging from traditional use of waste biomass and agricultural residues to some promising value-added products (Bledzki and Gassan 1999; Bledzki et al. 1996, 1998, 2010; Singha et al. 2009). The use of waste biomass as the most suitable alternative to traditional synthetic polymers for use in various fields (Debapriya and Adhikari 2004; Hagstrand and Oksman 2001; Singha et al. 2008; Singha and Thakur 2008) provides a number of advantages. Green polymer composites, which have been developed in recent years, have been among the most successful green materials, due to their high performance, which includes higher strength, homogenization, and machining behavior as well as ultimate disposability and the best utilization of raw material (Kabir et al. 2012a,b; Singha and Thakur 2009a,b). A number of studies are presently ongoing to use waste biomass such as cellulosic fibers in place of synthetic fibers in various applications (Panthapulakkal et al. 2006; Thakur and Singha 2010a–c).

The significant advantages of waste biomass such as cellulosic fibers over traditional synthetic fibers such as aramid and carbon fibers are eco-friendliness, biodegradability, low cost, high toughness, and low-density enhanced energy recovery, recyclability (Singha and Thakur 2010a,b; Thakur et al. 2012a–c; Wambua et al. 2003). Waste biomass such as wood fiber and other plant fibers along with cellulosic fibers composed of a large percentage of cellulose, making them the most abundant natural polymer (Thakur and Singha 2010a–c; Thakur et al. 2010; Zain et al. 2011). Cellulose has been found to be predominantly responsible for the overall properties of the fibers because of its specific properties such as a high degree of polymerization and linear orientation (Thakur and Singha 2011a,b; Thakur et al. 2011).

Among various types of natural waste biomass, *Eulaliopsis binata*, a perennial plant, belonging to the family Poaceae, which is abundantly grown in countries such as India, China, and Nepal, has high potential as a reinforcing material in green composites. The cellulosic fibers of *E. binata* have been found in leaves and hence belong to leaf fibers. *E. binata* fibers are eco-friendly, low cost, and possess good specific strength, light weight, and lower density compared to synthetic fibers. The literature review has revealed scant information on the applications of *E. binata* as reinforcement in polymer composites. Traditionally, these fibers have not been developed to full potential, because these fibers have been mainly used by local inhabitants for household applications. In this chapter, we have reported the preparation of *E. binata*-reinforced thermosetting polymer-based green biocomposites. The biocomposites thus prepared have their mechanical and thermal behavior evaluated. The prime aim of this chapter has been to assess the use of *E. binata* in polymer composites and assess its properties for industrial applications (mainly mechanical and thermal properties).

6.2 GREEN COMPOSITES REINFORCED WITH *E. binata* FIBERS

6.2.1 MATERIALS AND METHODS

E. binata fibers were collected from local resources of the Himalayan region in India. After proper purification, *E. binata* fibers were dried completely in a hot air oven. These were then subsequently grinded into particle form of dimension 250 μm and then used as reinforcing material in the thermosetting polymer matrix. For the thermosetting polymer matrix, phenol, formaldehyde solution, and sodium hydroxide of Qualigens make were used. Phenol–formaldehyde (PF) resin was used as novel polymer matrix for preparing green composites. The polymer resin was synthesized by the standard method reported elsewhere (Singha and Thakur 2009a, 2010b).

6.2.2 GREEN POLYMER COMPOSITES PROCESSING

Dried *E. binata* fibers in particle form were mixed thoroughly with the synthesized polymer resin with the help of a mechanical stirrer with diverse loadings ranging from 10% to 40% in terms of weight. The above thoroughly mixed composition was then poured into molds and spread equally. Composite sheets of size 150 mm × 150 mm × 5.0 mm were prepared by published compression molding techniques (Singha and Thakur 2009a, 2010b; Thakur and Singha 2010b,c).

6.2.3 CHARACTERIZATION OF GREEN POLYMER COMPOSITES

Mechanical properties such as wear resistance and tensile, compresive, and flexural strength of the *E. binata* fiber–reinforced green polymer composites were tested in accordance with ASTM D 3039, ASTM D 3410, ASTM D 790, and ASTM D 3702 methods, respectively (Thakur and Singha 2010b,c). Thermogravimetric properties of green polymer composite were carried out in nitrogen atmosphere on a thermal analyzer (Perkin-Elmer) at a heating rate of 10°C/min. The outer surface morphology of the optimized polymer composite sample was observed by scanning electron microscopy (SEM). For this, the excitation energy used was 5 keV. In an effort to achieve good electric conductivity, all samples were first carbon sputtered followed by sputtering a gold platinum mixture before examination. The SEM micrographs of the optimized sample thus show the morphology of the green polymer composites prepared (Thakur and Singha 2010b,c).

6.3 RESULTS AND DISCUSSION

Thermosetting polymer resin matrix was synthesized by the condensation reaction of phenol with formaldehyde (Singha and Thakur 2009a, 2010b). The synthesized thermosetting polymer contains hydroxyl groups present in the resulting matrix (Singha and Thakur 2010b). In addition to the polymer matrix, the reinforcing materials, such as *E. binata* fibers, contain hydroxyl groups that provide the active sites for cross linking with the PF during precuring and curing processes. Hence, the resulting mechanical properties of *E. binata* fibers/thermosetting polymers composites depend on (1) the extent of *E. binata* fibers–matrix bonding and (2) the load transfer from thermosetting polymer matrix to reinforcement (Singha and Thakur 2009a, 2010b; Thakur and Singha 2010b,c). A higher magnitude of bonding between the matrix and *E. binata* reinforcement facilitates the load transfer resulting in higher mechanical properties. However, *E. binata* fiber loading beyond 30% results in decreased mechanical properties due to the agglomeration of fibers at higher loading (Singha and Thakur 2009a, 2010b).

6.3.1 TENSILE STRENGTH

Figure 6.1a shows the tensile strength results of the green composites reinforced with *E. binata* fibers in particle form. It is clear from Figure 6.1a that green composites reinforced with *E. binata* fibers in particle form show enhanced tensile strength in comparison with the parent thermosetting polymer matrix.

It has been observed that green composites with 30% loading exhibit maximum tensile strength followed by the 40%, 20%, and 10% loadings. During tensile strength testing, the failure of green composites at meticulous loading can be attributed to the breaking of reinforcement at the weaker point followed by further propagation under the applied load, which is transferred to the remaining intact reinforcement, leading to complete rupture of the composites (Singha and Thakur 2009a, 2010b; Thakur and Singha 2010b,c).

6.3.2 COMPRESSIVE STRENGTH

The compressive strength results of the *E. binata* fiber–reinforced green composites have been found to follow the tensile strength (Figure 6.1b).

The failure of fiber-reinforced composite under compression has been found to occur when the reinforcement exhibits sudden and dramatic buckling (Thakur and Singha 2010b,c). It has been concluded that the prime mode of failure in the composites with different loading under compressive load can be due to the buckling of columns or microbuckling, which was preceded by debonding and microcracking of matrix (Singha and Thakur 2009a, 2010b).

6.3.3 FLEXURAL STRENGTH

The flexural strength of the *E. binata* fiber–reinforced green composites (Figure 6.1c) has been found to increase up to 30% loading and then decreases upon further higher loading, following the same trends obtained in the tensile strength and compressive strength tests (Singha and Thakur 2009a, 2010b; Thakur and Singha 2010b,c).

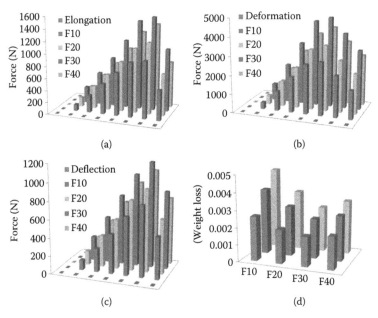

FIGURE 6.1 (a) Load elongation curve of cellulosic fiber-reinforced green composites. (b) Load deformation curve of cellulosic fiber-reinforced green composites. (c) Load deflection curve of cellulosic fiber-reinforced green composites. (d) Wear resistance curve of cellulosic fiber-reinforced green composites.

6.3.4 WEAR TEST

Wear resistance test results demonstrate that reinforcement of the polymer matrix with *E. binata* fibers significantly improved the wear resistance as compared with pristine polymer (Figure 6.1d).

Maximum wear resistance behavior is shown by green composites with 30% loading followed by 40%, 20%, and 10% loading (Thakur and Singha 2010b,c).

6.3.5 MORPHOLOGICAL AND THERMAL STUDY OF THE GREEN COMPOSITES

The morphological result (Figure 6.2) shows that the proper intimate mixing of the optimized fiber loading with the polymer resin in the green composites.

The micrograph clearly demonstrates that when thermosetting polymer matrix is reinforced with the optimum loading of fiber, morphological changes take place depending on the interfacial interaction between the fiber and the polymer matrix (Thakur and Singha 2010b,c; 2011a). The thermal stability of the *E. binata* fiber–reinforced composite was studied using thermogravimetry analysis (TGA). Table 6.1 shows the TGA results of *E. binata* fiber, polymer matrix, and the fiber-reinforced composite with 30% loading. The green composites show intermediate behavior between the fiber and the matrix and was consistent with earlier reported results (Singha and Thakur 2009a, 2010b; Thakur & Singha 2010b,c).

6.4 CONCLUSIONS

Green polymer composites were prepared using *E. binata* fibers as potential reinforcement in thermosetting polymer matrix. The variation of tensile, compressive, flexural, and wear properties along with optimization of reinforcement loading for *E. binata* fiber–reinforced phenolic composites for 0%–40% fiber content are studied. The fibers are assessed in particle form to improve the mechanical properties by increasing the reinforcement content. The overall results of the above experiment

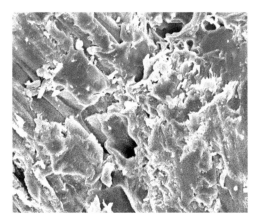

FIGURE 6.2 Scanning electron micrograph of cellulosic fiber-reinforced green composites with optimum (30%) loading at 1 KX.

TABLE 6.1
Thermogravimetric Analysis of *Eulaliopsis binata* Fibers, Polymer Resin, and Green Composites (30% Loading)

No.	Sample Code	IDT (°C)	% Wt. Loss	FDT (°C)	% Wt. Loss	Final Residue (%)
1	*Eulaliopsis binata* fibers	190	5.00	491	85.4	14.5
2	Phenol–formaldehyde resin	397	14.5	1188	46.5	53.4
3	Phenol–reinforced composites	315	22.8	975	55.1	44.8

suggest that the *E. binata* reinforcement has a beneficial effect in the adhesion of the polymer resin into the green composites.

ACKNOWLEDGMENT

The authors thank their parental institutes for providing the necessary facilities to accomplish the present research work.

REFERENCES

Averous, L. 2004. Biodegradable multiphase systems based on plasticized starch: A review. *Journal of Macromolecular Science. Polymer Reviews C* 44(3): 231–274.

Bledzki, A.K. and Gassan, J. 1999. Composites reinforced with cellulose based fibers. *Progress in Polymer Science* 24: 221–274.

Bledzki, A.K., Mamun, A. and Volk, J. 2010. Barley husk and coconut shell reinforced polypropylene composites: The effect of fibre physical, chemical and surface properties. *Composites Science and Technology* 70: 840–846.

Bledzki, A.K., Reihmane, S. and Gassan, J. 1996. Properties and modification methods for vegetable fibers for natural fiber composites. *Journal of Applied Polymer Science* 59(8): 1329–1336.

Bledzki, A.K., Reihmane, S. and Gassan, J. 1998. Thermoplastics reinforced with wood fillers: a literature review. *Polymer-Plastics Technology and Engineering* 37(4): 451–468.

Debapriya, De. and Adhikari, B. 2004. The effect of grass fiber filler on curing characteristics and mechanical properties of natural rubber. *Polymers for Advanced Technologies* 15(12): 708–715.

Hagstrand, P.O. and Oksman, K. (2001). Mechanical properties and morphology of flax fiber reinforced melamine–formaldehyde composites. *Polymer Composites* 22(4): 568–578.

Kabir, M.M., Wang, H., Lau, K.T. and Cardona, F. 2012a. Chemical treatments on plant-based natural fibre reinforced polymer composites: an overview. *Composites: Part B*, 43B: 2883–2892.

Kabir, M.M., Wang, H., Lau K.T., Cardona, F. and Aravinthan, T. 2012b. Mechanical properties of chemically-treated hemp fibre reinforced sandwich composites, *Composites: Part B*, 43: 159–169.

Panthapulakkal, S., Zereshkian, A. and Sain, M. 2006. Preparation and characterization of wheat straw fibers for reinforcing application in injection molded thermoplastic composites. *Bioresource Technology* 97: 265–272.

Singha, A.S., Shama, A. and Thakur, V.K. 2008. Pressure induced graft co-polymerization of acrylonitrile onto *Saccharum cilliare* fiber and evaluation of some properties of grafted fibers. *Bulletin of Material Science* 31(1): 1–7.

Singha, A.S., Shama, A. and Thakur, V.K. 2009. Graft copolymerization of acrylonitrile onto *Saccaharum cilliare* fiber. *E-Polymers* 105:1–12.

Singha, A.S. and Thakur, V.K. 2008. Mechanical properties of natural fiber reinforced polymer composites. *Bulletin of Material Science*. 31(5):791–799.

Singha, A.S. and Thakur, V.K. 2009a. Chemical resistance, mechanical and physical properties of biofiber based polymer composites. *Polymer-Plastics Technology and Engineering* 48(7): 736–744.

Singha, A.S. and Thakur, V.K. 2009b. Fabrication and characterization of *S. cilliare* fiber reinforced polymer composites. *Bulletin of Materials Science* 32(1): 49–58.

Singha, A.S. and Thakur, V.K. 2010a. Renewable resources based green polymer composites: Analysis and characterization. *International Journal of Polymer Analysis and Characterization* 15(3): 127–146.

Singha, A.S. and Thakur, V.K. 2010b. Synthesis, characterization and study of pine needles reinforced polymer matrix based composites. *Journal of Reinforced Plastics and Composites* 29(5): 700–709.

Thakur, V.K. and Singha, A.S. 2010a. KPS-initiated graft copolymerization onto modified cellulosic biofibers. *International Journal of Polymer Analysis and Characterization* 15(8): 471–485.

Thakur, V.K. and Singha, A.S. 2010b. Mechanical and water absorption properties of natural fibers/polymer biocomposites. *Polymer-Plastics Technology and Engineering* 49(7): 694–700.

Thakur, V.K. and Singha, A.S. 2010c. Natural fibres-based polymers: Part I—Mechanical analysis of pine needles reinforced biocomposites. *Bulletin of Materials Science* 33(3): 257–264.

Thakur, V.K., Singha, A.S., Kaur, I., Nagarajarao, R.P. and Yang, L.P. 2010. Surface modified *Hibiscus sabdariffa* fibers: physico-chemical, thermal and morphological properties evaluation. *International Journal of Polymer Analysis and Characterization* 15(7): 397–414.

Thakur, V.K. and Singha, A.S. 2011a. Physico-chemical and mechanical behavior of cellulosic pine needles based biocomposites. *International Journal of Polymer Analysis and Characterization* 16(6): 390–398.

Thakur, V.K. and Singha, A.S. 2011b. Rapid synthesis, characterization, and physicochemical analysis of biopolymer-based graft copolymers. *International Journal of Polymer Analysis and Characterization* 16(3): 153–164.

Thakur, V.K., Singha, A.S. and Misra, B.N. 2011. Graft copolymerization of methyl methacrylate onto cellulosic biofibers. *Journal of Applied Polymer Science* 122(1): 532–544.

Thakur, V.K., Singha, A.S. and Thakur, M.K. 2012a. Graft copolymerization of methyl acrylate onto cellulosic biofibers: synthesis, characterization and applications. *Journal of Polymers and the Environment*. 20(1):164–174.

Thakur, V.K., Singha, A.S. and Thakur, M.K. 2012b. Surface modification of natural polymers to impart low water absorbency. *International Journal of Polymer Analysis and Characterization* 17(2): 133–143.

Thakur, V.K., Singha, A.S. and Thakur, M.K. 2012c. In-air graft copolymerization of ethyl acrylate onto natural cellulosic polymers. *International Journal of Polymer Analysis and Characterization* 17(1): 48–60.

Wambua, P., Ivens, J. and Verpoest, I. 2003. Natural fibres: Can they replace glass in fibre reinforced plastics? *Composites Science and Technology* 63: 1259–1264.

Zain, M.F.M., Islam, M.N., Mahmud, F. and Jamil M. 2011. Production of rice husk ash for use in concrete as a supplementary cementious material. *Construction and Building Materials* 25: 798–805.

7 Bast Fibers Composites for Engineering Structural Applications
Myth or the Future Trend

Bartosz T. Weclawski and Mizi Fan

CONTENTS

7.1 INTRODUCTION

Natural fiber composites (NFCs) should be compared with relevant substitute civil engineering materials in order to analyze the capability of application in this sector. Usually, this group of materials is compared with glass fiber reinforced–composites (GFRC). Thygesen et al. (2006) in their investigation into hemp-reinforced composites analyzed mechanical properties for application in wind turbine fan blades. They concluded that at this stage material is not suitable for use in fan construction due to insufficient strength and relatively high cost.

The production of NFCs creates significantly less environmental impact due to low-energy fiber processing and carbon storage during plant growth, when compared with the high-energy melting and forming process of glass fiber. Energy efficiency of NFCs is further demonstrated when considering material transportation or uses because of its light weight. High volume fractions (Vfs) of fibrous reinforcement in NFCs reduce resin consumption. At the end of NFC life, energy recovery from the incineration or biodegradation of the NFC is possible (Joshi et al. 2004).

On the other side, the disadvantages of NFCs can be listed. These include lower mechanical properties such as stiffness and strength, and higher sensitivity to weathering and degradation. Additionally, the consistency of the material properties and price are considered inferior to GFRP at this point. High specific properties of NFCs are included as one of the main driving forces for automotive applications. These are related to the lowered fuel consumption, but are less significant in civil engineering applications. Safer NFC shrapnel during collision is another advantage related specifically only to the automotive industry. Weathering and degradation resistance together with fire properties are of significant importance in civil engineering applications.

This chapter focuses on hemp and flax, both of which are the most researched and the most popularly cultivated bast fiber crops in temperate climate zones. Both plant fibers will be highlighted with global production analysis based on statistical data. In the following part, the influence of the reinforcement structure will be connected with tensile and flexural properties of hemp and flax composites. This chapter then discusses NFC weathering properties and potential civil engineering applications. In the last part of this chapter, the prospects of NFC for civil engineering applications, such as rods, tubes, panels, and I-beams, are presented.

Hemp (*Cannabis sativa* L.) is one of the first plants to be domesticated and it is native to India and Iran. There are around 2000 varieties of plants in the mulberry family. Hemp is well known for its bast fibers (Lawrence 1969). Hemp was an important crop for millennia. Ropes were produced from hemp in China as early as 2800 BC (Joseph 1986). Hemp was used to produce the first paper in 100 BC in China during the Western Han Dynasty. In Europe, hemp was an important crop from the Middle Ages and was used for the production of ropes, sacks, threads, water hoses, sails, and textiles. For centuries, hemp was one of the most important and strategic crops (Renewable-Resources 2010). In 1937, hemp was named "the most profitable and desirable crop that can be grown" (Lower 1937) and the "new billion dollar crop" (Popular Mechanics 1938). Flax (*Linum usitatissimum*) is a plant that originated from Caucasus territories. Recent discovery in Upper Paleolithic layers at Dzudzuana Cave of a spun, weaved, and dyed wild flax linen dates it back to 30,000 BC. This was a significant discovery, proving that fabrics were produced 25,000 years earlier than the previous findings suggested (Kvavadze et al. 2009). In Egypt, almost transparent flax linen cloth with 500 threads per inch from 2500 BC was found. In Western Europe, flax linen was discovered on Switzerland Neolithic lake dwellings from 2000 BC. Paper from flax was found in China produced in 105 AD (Hollen and Saddler 1964; Joseph 1986). Flax is grown in temperate climate zones. In Europe, there were a couple of main flax cultivation and processing centers, such as Russia, Denmark, England, and

France. Russia was producing around 90% of flax before World War I (Johanson 1927). Flax was used throughout the ages for textiles, threads, paper, and nets.

7.2 HEMP AND FLAX PRODUCTION

When considering potential applications and the price of NFC, material properties and production capacities together with environmental effects need to be evaluated. Production costs and the nature of processing are directly related to the commercial viability and NFC, the key factors for successful development of NFCs supply chain.

Hemp and flax have relatively high tolerance for a variety of climate and soil and they can be cultivated in temperate zones. There are three main centers for cultivation, namely Canada, China, and Europe. Europe was one of the biggest hemp and flax producers in the world (Figures 7.1 and 7.2), and only recently in the last two decades has the production of hemp spiked in Asia. Flax production in Europe declined in the 1990s, giving opportunity to producers from Asia, but this trend has changed in recent years and since 2007 flax production in Europe has increased significantly.

World production of hemp started to fall in 1966, from almost 370,000 tons/ year to minimum in 1994 of 51,500 tons/year. When analyzing world production in terms of total area dedicated, a 91% decrease is observed between 1966 and 2009 (FAO 2012a). This decline was the outcome of revolution in the field of synthetic fiber production, and hemp regulatory laws were introduced in the United States as the Marihuana Tax Act of 1937, which later influenced global hemp production and trade. Nowadays, European hemp fiber is mainly used in the production of paper,

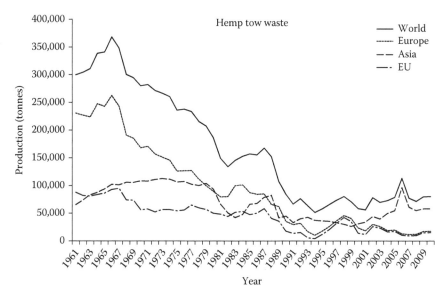

FIGURE 7.1 Production of hemp in the European Union, Europe, Asia, and the world. (From FAO. *Hemp Tow and Waste Production Data.* Food and Agriculture Organisation of the UN, 2012.)

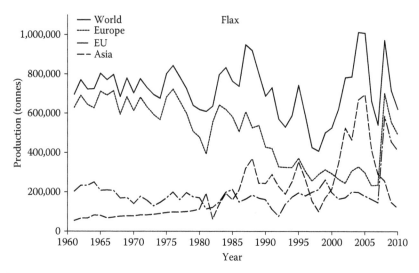

FIGURE 7.2 Production of flax in the European Union, Europe, Asia, and the world. (From FAO. *Flax Fibre and Tow Production Data*. Food and Agriculture Organisation of the UN, 2012.)

specialty paper, technical filters, cigarette paper, NFC, insulation material, cultivation fleeces, animal bedding, and mulch.

The statistical data, collected from 1961 to 2010 by the Food and Agriculture Organization of the United Nations on the production of flax fiber and tow in the world, reveals an interesting trend. Despite an almost 90% decrease in the area dedicated for flax cultivation from 2 Mha to 231 kha during the last 50 years, on average global production rates fluctuate around 716 ktonnes/year. This is a result of the increased production efficiency due to the developments of cultivation techniques, fertilization, processing machinery, and pest control. The highest improvement in efficiency is observed in the European Union (EU) with 734% more flax harvested from the same area in 2009 when compared with those in 1960s (FAO 2012b).

Hemp grows for 3–4 months and is harvested before flowering, which may impair the quality of the fiber. Planting is done during spring. Up to 4 tonnes of retted and dried hemp stems are used to produce 1 tonne of hemp fiber, giving above 2 tonnes of hemp shives material as a by-product (Kraus 2004). Similar to other high-production crops, it requires fertilization with nitrogen, phosphorus, and potassium. Since it does not require herbicides or excessive pesticides, its life cycle analysis distinguishes it as carbon negative raw material (Saskatchewan Ministry of Agriculture 2003). These properties makes hemp a relatively easy crop to grow. It does not require special treatments. The use of moderate amounts of fertilizer will give high harvest yields. It is possible to yield up to 25 tonnes/ha of hemp, which makes it highly efficient in biomass production (Struik et al. 2000). The highest increase in yield happened in the early 1990s in EU, where producers are amongst the most efficient, with the peak average production yield being 5.9 tonnes/ha in 2002. Average production in the world in the last decade was at the level of 1.5 tonnes/ha, and at the same time in EU countries, it was at the level of 3.2 tonnes/ha (FAO 2012a). Yield values increased due to process improvement, research, and development of optimal growth conditions. This is due to easy access to farming technologies, machines, climate, and economic drive related to land prices, which forced farmers to optimize their production.

For most applications, cellulose bast fibers need to be separated from other stem material, such as shives, lignin, and pectin, for the product of reinforcement. The traditional method of separation is retting and is based on soaking hemp in water tanks, where bacteria enzymes degrade lignin to free fibers. In more recent approaches, a mechanical separation process called decortication is used. Hemp stems are taken through a series of rollers that squeeze the material and separate fibers from shives. The biggest advantage of mechanical decortication process is minimization of time and area consumed by classic process. However, it was found to have a small deteriorative effect on fiber-reinforcing properties (Hepworth et al. 2000). In the next stage in order to reduce lignin and pectin level, fibers can be bleached with hydrogen peroxide, sodium hydroxide, sodium sulfite, or other alkaline solution (Wang and Postle 2003). Raw or treated fibers are then used to process mats and yarns used as composite reinforcement.

7.3 TENSILE AND FLEXURAL PROPERTIES OF NATURAL FIBER COMPOSITES

When considering NFC for civil engineering applications, there are properties that are first considered, namely tensile and flexural strength, durability, and resistance to weathering conditions. This section will focus on the analysis of the main mechanical properties of hemp and flax composites. Tensile and flexural properties from the results of the Brunel research program are presented and discussed.

7.3.1 INFLUENCE OF THE REINFORCEMENT TYPE

NFCs have the highest properties in the direction parallel to the fibers in tensile mode, what is the case for most fibrous composites. Thus, elements designed from the aligned fibrous NFC laminates should be loaded in tensile and flexural modes.

Reported properties of the flax- and hemp-extracted bast fibers vary, which is inherent for natural based materials. Hemp and flax bast fibers have similar densities that range between 1.35–1.50 g/cm^3 and 1.38–1.52 g/cm^3, respectively. The length of fibers range from 5.6 to 110 mm. Young's modulus ranges between 5.5 and 70 GPa for hemp fibers and between 12 and 100 GPa for flax fibers. The values of tensile strengths range between 690 and 1040 MPa and between 345 and 1100 MPa for hemp and flax, respectively. High discrepancies in the reported fiber properties are an outcome of the hierarchical nature of material, test methodologies used, and an inherent feature of natural material (Hagstrand and Oksman 2001; Pickering, 2008; Shazad 2011). The Young's modulus and tensile strength of E-glass are between 70 and 72 GPa and between 2000 and 3500 MPa, respectively, which are higher and more consistent when compared with those of hemp and flax fibers. However, the density of E-glass is about 50 times higher (2.5 g/cm^3) when compared with those of hemp and flax fibers (Saheb and Jog 1999). This reduces the gap between specific mechanical properties of both reinforcing fibers, hence NFCs envisioned applications where weight is one of the design factors. The above values are the first points of reference when considering the tensile properties of NFCs. When the perfect interphase bond, lack of defects, misalignments, and porosities are achieved, this difference would be decreased in regard to the Vf. Substantial work was done on the tensile strength of different types of flax and hemp composites, and this is used to evaluate changes in composition, treatments, or processing improvements.

Laminates reinforced with fibers or mats are processed by compression molding, vacuum-assisted techniques, pultrusion, and extrusion. The most popular processing technique for randomly oriented mats and fabric is compression molding. Pultrusion or filament winding gives the most repeatable results with continuous yarn reinforcement. All techniques have the same limitations as for conventional composites. In addition, the processing temperature is limited by fiber degradation. Chemical bonds start to break above 100°C, and above 200°C, celluloses start to decompose (Arno 1989), which has implications in matrix systems used and processing times. In processing techniques involving tensioning of the continuous reinforcement, such as pultrusion or filament winding, the forces used are limited by natural fiber yarn breaking load.

7.3.2 MAT AND FABRIC

A series of laboratory tests were conducted to evaluate hemp and flax mat and fabric-reinforced composites. Two unsaturated polyester resins were used: (1) ECO resin was based on 65% renewable raw materials and (2) crystic resin was entirely fossil fuel based. NFC laminates reinforced with mats and fabrics were processed using the compression molding technique with vacuum-assisted wetting. Continuous reinforcement in the form of various types of hemp or flax yarns was aligned with filament winding before compression molding.

Tensile tests conducted on the laminates, reinforced with hemp randomly oriented mat in ECO matrix, gave results of 50 MPa and 6.8 GPa tensile strength and tensile modulus, respectively. Flexural strength and modulus was at 67 MPa and 4.5 GPa, respectively. These results are in the same range as those presented by researchers working with similar types of reinforcement and thermoset resin systems. Reported laminates, reinforced with randomly oriented hemp or flax mat with Vf of fibers between 10% and 45%, gave ultimate tensile strength (UTS) range between 19 and 65 MPa. One of the highest properties were reported by Hepworth et al. (2000), but the higher result was an outcome of partial alignment of the fibers with combing. Reported values for tensile stiffness in composites reinforced with mats in thermoset resin systems range between 0.6 and 10 GPa for hemp and between 2.9 and 9.8 GPa for flax (Dhakal et al. 2007; Mehta et al. 2006; Mwaikambo et al. 2007; Rouison et al. 2006; Van Den Oever et al. 2000; Yuanjian and Isaac 2007).

Cellulosic bast fibers have to be converted into yarns and fabrics to use them in continuous industrial process. This allows for precise alignment and achievement of increased Vfs, resulting in higher mechanical properties. Yarns can be processed by conventional ring spinning process used in textile industry or by wrapping aligned short fibers with polymeric wire. Reinforcement produced in this way can be handled like most conventional filaments for winding or pultrusion processes. Conversion into fabrics or prepregs allows for molding or hot pressing. Figure 7.3 presents example of nontwisted (left) and twisted (right) yarn geometries.

Two types of flax fabric reinforcement, Twill and Hopsack, supplied by Biotex, were used in experimental work with the ECO resin system. Fabrics were made out of weaved nontwisted yarns. In summary, flax laminates with 45% Vf had 6.7 GPa tensile modulus and 63.3 MPa tensile strength. Tested flexural stiffness ranged between 5.1 and 6.3 GPa. Flexural strength ranged between 140 and 213.8 MPa. Glass fiber composite processed with the same matrix system and Twill glass fabric had 19.6 GPa flexural stiffness and 450 MPa flexural strength.

Talreja and Manson (2000) presented the results for GFRP fabric laminate with 40%–55% Vf, showing a range of tensile stiffness between 15 and 28 GPa and tensile strength between 138 and 241 MPa. Muralidhar et al. (2012) reported a flexural strength of 31–106 MPa and flexural stiffness of 0.8–2.9 GPa for flax fabric laminate with the Vf ranging from 18% to 34%. Low stiffness values usually indicate low interface properties. This can be changed by the mercerization process, where fibers are bleached using silane coupling agents. Various types of mercerization processes can increase surface adhesion and reduce lignin content (Bledzki and Gassan 1999). Both alkali treatments and use of coupling agents need to be optimized depending on the type of reinforcement and resin system. Furthermore, the treatments may increase fiber cost and environmental impact. Hence, the improvement from additional treatments or incurring extra cost should be compared with any value added to the modified materials or products to ensure a successful market.

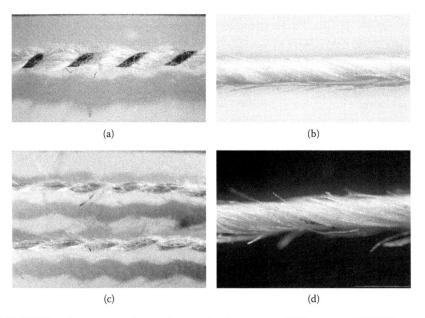

(a) (b)

(c) (d)

FIGURE 7.3 Comparison of nontwisted and twisted yarns. (a) Hemp–wool 1000 Tex nontwisted yarn, (b) hemp-twisted 39 Tex yarn, (c) flax 250 Tex nontwisted yarns, and (d) hemp 130 Tex twisted yarn.

7.3.3 Axially Aligned Laminates

In order to find maximum tensile properties, fibers are arranged in the testing direction in our study. The samples with various hemp and flax aligned yarns were used for the analysis of tensile properties. The influence of Vf, tex value, yarn type, and yarn twist on tensile properties and fracture pattern was investigated.

Four types of yarns were used. Both hemp and flax were geometrically twisted and nontwisted. From each type, six yarn sizes were available. In order to investigate the influence of yarn size on tensile and flexural properties, six grades of hemp yarns from 25 to 130 Tex were used. Coupons were tested using BS EN 527-5:1997.

Figure 7.4 presents an example of fracture area in a unidirectional sample after a tensile test. In this example, the sample was reinforced with 130 Tex twisted hemp yarn. Cracks appeared in normal to the longitudinal direction. Moreover, during material breaking, longitudinal cracks appeared along the yarns. This is caused by a combination of the shearing forces induced by uneven individual yarn tensioning as well as elastic energy release when sudden rupture occurs.

Twisted hemp fiber-reinforced composites had the highest tensile stress and stiffness properties among all tested samples. The composite directly influences material tensile properties. In order to find optimal Vf, samples were prepared with fiber content from 23% to 75% with two grades of hemp twisted yarn. Above the composite practical saturation point with fiber (specific for the fiber and matrix used), further increase in Vf leads to a decrease in stiffness properties. This is observed for ECO 39 Tex samples at 50% Vf (Figure 7.5a). This is caused by deterioration in wet-out properties, which further leads to the creation of dry spots at the interface. The highest modulus measured for composite with 39 Tex was at the level of 12 GPa, and for 130 Tex composite type, it was equal to 14 GPa. Tensile strength of the tested composites grew almost linearly throughout the whole range of Vf. For 39 Tex fine sample, tensile strength increases from 180 MPa at 23% Vf to 411 MPa for 75% Vf (Figure 7.5b). In twisted yarn composites, load is transferred by the interphase from matrix to fibers. If fiber is not wetted-out completely, the load is transferred by shear forces between the fibers.

One of the most important factors in the strength build up for the fibrous composites is an adhesion between fibers and matrix. There are three main theories of adhesion, namely mechanical coupling, molecular bonding, and thermodynamic adhesion (Awaja et al. 2009). Interface between fiber and matrix transfers the loads

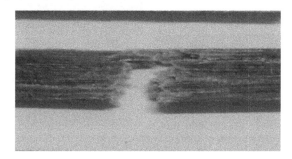

FIGURE 7.4 Example of fractured area on the 130 Tex sample fracturing in the normal to the longitudinal direction.

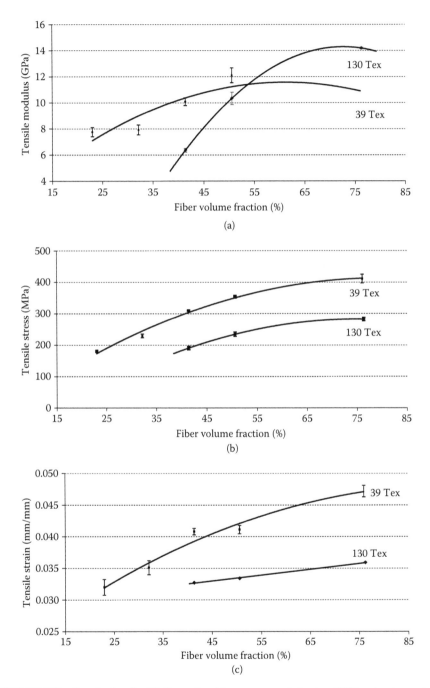

FIGURE 7.5   Results for the 39 Tex twisted hemp yarn reinforcing ECO polyester resin. Graphs illustrate relationship of property and volume fraction of fiber's (a) tensile modulus, (b) tensile strain, and (c) tensile stress.

from the matrix onto fibers. The mode of composite failure is directly linked to the fiber–matrix interface and the relationship of fiber surface area and the cross-section area. Since natural fiber yarns are composed of short fibers, the same relationships apply. In other words, total load transferring capacity by the fiber interface should be greater than the fiber strength. Then, the composite fails in fiber fracture mode as opposed to fiber pull-out mode. This can be quantified by measuring the critical fiber length for a particular fiber–matrix system (Miwa and Endo 1994).

When both yarns are compared, the composites processed with 130 Tex yarns exhibit lower stiffness within the range from 23% to 50% Vf (Figure 7.5a). The tensile strength of 39 Tex composite ranges from 32% to 39%, which is the highest across all tested Vfs. Together with the increase in fiber Vf, the elongation at break also increases for both types of materials. Additionally, the composites reinforced with coarser 130 Tex yarn exhibit higher deformation before fracture (Figure 7.5c).

Flexural modulus was tested for the same composites. Finer 39 Tex yarn produces 28% higher flexural stiffness for all Vfs in comparison with 130 Tex. The maximum flexural stiffness of 32 GPa was achieved. Ultimate flexural strength of 39 Tex composite has 26% higher UTS until 50% Vf than those of 130 Tex. Above this value, UTS stops rising with increasing Vf. Finally, maximum values for UTS were 320 MPa for 130 Tex yarn and up to 330 MPa for 39 Tex yarn for the composite made with fibers of 76% Vf.

Tex value is calculated by weighing yarn and expressing the results in terms of grams per 1000 meters (ISO 2060:1994). This value is commonly used in the textile industry. Figure 7.6 presents bar charts of the influence of yarn size on tensile properties. Each yarn type is represented with its Tex value, starting from fine 25 to 130 Tex yarn. All composites were processed with 41% Vf. The tensile modulus first increases with the decrease in yarn size to reach a maximum of 14 GPa for 60 Tex and 51 Tex laminates (Figure 7.6a). After that, further yarn size reduction results in a decrease in tensile stiffness. Hence, the use of 60 Tex yarn improves by 58% tensile stiffness when compared with 130 Tex composites. This indicates that there is a close to linear relationship between Tex value and tensile stress. By using finer yarns, tensile strength of the composite is improved, increasing from 190 to 337 MPa for the finest 25 Tex type composite (Figure 7.6a). This is a 44% improvement. Tensile strain to failure increases with yarn size reduction, 0.032–0.045 mm/mm (Figure 7.6c).

Figure 7.7 presents the relationship between the yarn Tex value and flexural properties for the hemp twisted yarn-reinforced composites. A linear decline was observed for the stiffness and UTS results with increasing Tex. Flexural stiffness ranges from 15.6 to 22.4 GPa for 25 Tex yarn. Ultimate flexural stress ranges from 226.8 to 304 MPa for the tested Tex range. Use of fine grades of yarns allows for 44% increase in flexural modulus and 34% increase in ultimate flexural strength.

Reported values for tensile strength for unidirectional composites range from 43 to 277 MPa for composites up to 50% Vf. Flax unidirectional laminates had tensile strength between 65 and 328 MPa. Tensile stiffness for unidirectional aligned NFCs ranges from 5.9 to 27 GPa for hemp-reinforced laminates (Bledzki et al. 2004; Madsen et al. 2007a; Thygesen et al. 2007) and from 15 to 29 GPa for flax-reinforced laminates (Charlet et al. 2007; Hughes et al. 2007; Oksman 2001; Van De Weyenberg

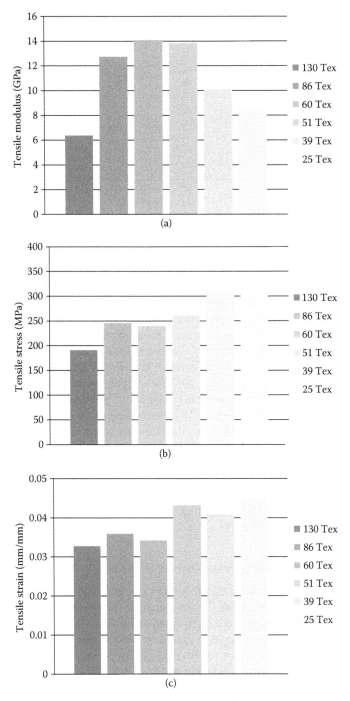

FIGURE 7.6 Influence of the hemp yarn type on the tensile properties. (a) Modulus, (b) ultimate tensile strength, and (c) strain for the unidirectional composite. All samples measured at the 41% fiber volume fraction.

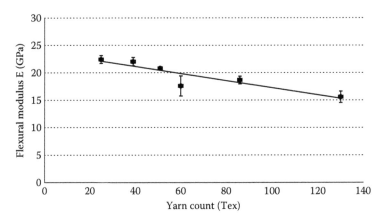

FIGURE 7.7 Relationship between flexural modulus and Tex grade for aligned composite laminate reinforced with twisted hemp yarns. Samples with 41% Vf.

et al. 2003). Those values are closer to the stiffness of glass composites, from 35 to 45 GPa (Talreja and Manson 2000). GFRP laminate with epoxy resin system at Vf ranging from 40% to 60% has stiffness ranging from 35 to 45 GPa and tensile strength from 300 to 1000 MPa (Talreja and Manson 2000).

As reported in the literature, unidirectional hemp and flax bast fiber composites were processed with Vfs from 10% to 70%. Measured flexural modulus ranged between 0.8 and 25 GPa and flexural strength was between 31 and 219 MPa (Bledzki et al. 2004; Goutianos et al. 2006; Muralidhar et al. 2012; Ochi 2006; Sawpan et al. 2012).

Flax nontwisted 250 Tex yarn laminates were investigated for the influence of fiber content on flexural properties. Figure 7.8 presents flexural modulus and Vf relationship. Composites were measured in longitudinal direction for the flexural properties. There was an apparent linear relationship between stiffness and Vf with values from 7 GPa of flexural stiffness at 22% Vf to 24 GPa at 90% Vf. Modulus was increasing linearly with the increase of fiber Vf ratio.

When examining ultimate flexural stress, Vf saturation point was observed. UTS increases until 55% Vf. After exceeding this point, further increase in Vf reduces flexural strength (Figure 7.9).

7.3.4 FIBER MISALIGNMENT IN THE NONTWISTED YARN

It is evident that the yarns with nontwisted constituent fibers and yarns with twisted constituent fibers have different average fiber orientation. The comparison between twisted and nontwisted yarn arrangement is represented in Figure 7.3. In Figure 7.3a and c, the sinusoidal pattern is visible. It might be caused by the difference in strain of polymeric yarn wrapping unidirectional flax fibers. This can be an outcome of tension induced during fiber alignment, wrapping wire shrinkage, or processing settings. The main reason for processing of nontwisted yarns is to maximize alignment of the fibers in the desired direction. Misalignments can deteriorate the properties of processed material.

Tensioned polyester wrapping wire induces bending stresses in natural fibers, causing fibers to slide and bend against one another to accommodate the stresses.

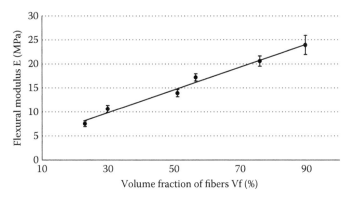

FIGURE 7.8   Flexural modulus of composites reinforced with nontwisted 250 Tex flax yarn.

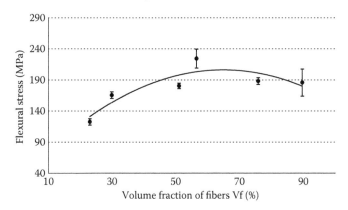

FIGURE 7.9   Flexural stress of composites reinforced with nontwisted 250 Tex flax yarn.

This may result in significant misalignments. In Figure 7.10, a nontwisted yarn deformation schematic is presented, where A is yarn that is not deformed with ideal alignment and B is the yarn subjected to tensioning with misalignment. The facts indicated
in Figure 7.10 may give rise to stress concentrations within composite reinforced
with nontwisted yarns. Additionally, these results can impair resin impregnation.

These also explain the specific angular fracture pattern exhibited by unidirectional composites reinforced with nontwisted yarns. The fibers in the twisted yarns
are held together by the means of shear through a self-locking mechanism, thus having more uniform stress distribution along the entire length and resulting in rupture
normal to the loading direction (Figure 7.11a). For the composites reinforced with
nontwisted yarns, the rupture path oscillates around 45° angles (Figure 7.11b).

Not all yarns exhibit the same magnitude of waviness along the length. In order
to quantify misalignment value, the calculation based on outer fiber angle and fiber
diameter can be performed (Madsen et al. 2007b). Fibers within ring spun twisted
yarn change their orientation with radial position inside the yarn. The angle of the surface single fibers is not equal to the fiber angle inside the yarn. Mean value of the yarn
twist angle can be calculated by measuring the outer fiber angle diameter and single
fiber diameter. Nontwisted yarns were designed in order to fully utilize the potential

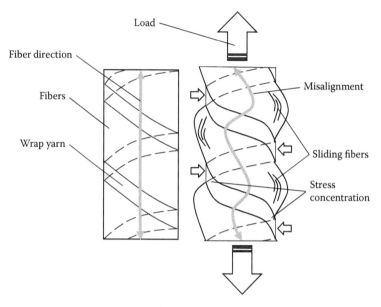

FIGURE 7.10 Diagram of nontwisted yarn with deformation morphology under load. Gray arrows indicate fiber direction before and after deformation.

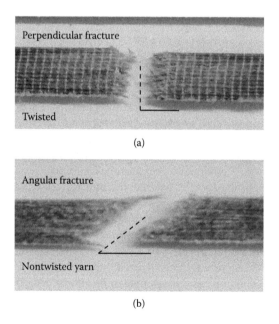

FIGURE 7.11 Comparison between twisted flax F12 yarn-reinforced unidirectional composite (UD) sample (a) and nontwisted flax yarn-reinforced UD fracture modes (b).

of the fiber longitudinal properties. Therefore, it is important to accurately apply non-twisted reinforcement in all three stages: during production of the yarn, processing of the reinforcing fabrics, and finally processing of the laminates. The control over aforementioned processes will prevent deterioration of laminate properties.

7.3.5 Hybrid Hemp–Wool Yarn 1000 Tex Reinforced Aligned Composites

Yarns can be hybridized in order to improve or include new properties. Another set of flexural tests was conducted for the composites reinforced with 1000 Tex non-twisted hemp wool yarns (Figure 7.12). The ratio between wool and hemp was 0.25. Flexural modulus increases with Vf from 11 to 16 GPa, which is significantly lower than the results of flax or hemp composites.

An interesting result was obtained for the ultimate flexural strength of the hybrid samples (Figure 7.13). There is an inverse relationship between Vf and ultimate strength. This is explained by increasing wool percentage, which has lower adhesion

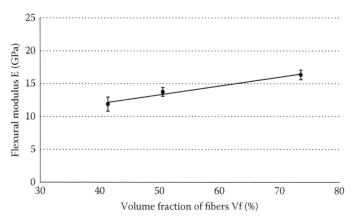

FIGURE 7.12 Relationship between volume fraction of fibers and flexural modulus for composites reinforced with hybrid hemp–wool 1000 Tex nontwisted yarns.

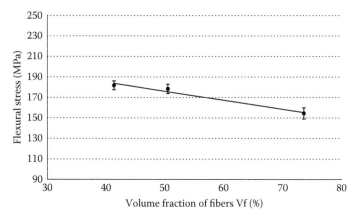

FIGURE 7.13 Relationship between volume fraction of fibers and flexural stress for composites reinforced with hybrid hemp–wool 1000 Tex nontwisted yarns.

properties in comparison with hemp fiber. Moreover, yarns of 1000 Tex thickness are more difficult to impregnate and contribute to linear drop in ultimate flexural stress.

7.4   NATURAL FIBER COMPOSITE WEATHERING

Many organisms and microorganisms use cellulose as a source of energy by degrading it to glucose in enzymatic processes (Beguin 1990). The natural degradation of cellulose allows biomass circulation in the system and at the same time degrades the properties of cellulose-based man-made materials. In order to prevent property deterioration, treatments, additives, and inhibitors are used or the material is sealed off from the environment. The latter is more difficult to achieve for construction materials and is possible only for indoor applications.

In order to assess the combined environmental effect of sunlight and moisture, a set of composites were tested for mechanical properties deterioration due to outdoor weather exposure. Sets of samples $90 \times 90mm^2$ were subjected to the temperate U.K. climate conditions. The composites were placed on the testing racks at the 45° facing south and were tested in the intervals of 1 month to observe the change of the flexural properties due to weather conditions. In 15-day intervals, composites were turned, and in 30-day intervals, composites were exposed to an elevated temperature of 110°C for 30 minutes in an air circulated oven, and then the cycle was completed. The exposure was based on, and performed in accordance with, the BS PL 4:2005 standard for glass fiber–reinforced polyester composites (BSI 2005) after which samples were tested for flexural properties (BSI 1998).

Figures 7.14 and 7.15 present the results from weathering investigation of the flax composite laminate reinforced with Hopsack fabric weaved from 250 Tex nontwisted yarns. The composites were degraded in three types of environments, namely saline solution, water, and environmental weathering. After the first month of the exposure, the composites started to lose mechanical properties. The most severe influence on

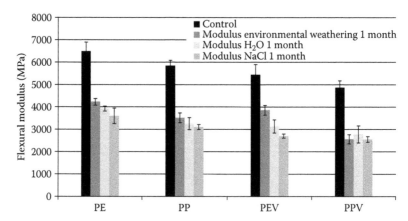

FIGURE 7.14   Influence of environmental weathering, immersion in water, and saline solution on flexural modulus of four types of flax composites.

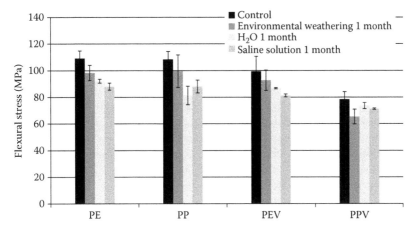

FIGURE 7.15 Influence of environmental weathering, immersion in water, and saline solution on flexural stress of four types of flax composites.

the mechanical properties of the materials happened during the immersion in saline solution. The lowest effect was observed when exposed to environmental weathering.

7.5 APPLICATIONS OF NATURAL FIBER COMPOSITES IN CIVIL ENGINEERING

Plants during production of biofibres by means of the photosynthesis process capture and store carbon dioxide, which creates a good starting point toward carbon neutral or negative materials for engineering. Renewable construction materials can bring benefits throughout the life cycle of buildings, such as reducing embodied energy, increasing energy efficiency during use, reducing waste, and adding properties including insulation and breathability (Wool and Sun 2005).

Bast natural fiber-based composite materials have recently been seen as a competitor for wood and glass fiber composites in a number of applications. To utilize natural fibers, as reinforcement for the thermoset matrix, a number of developments have been made. Composites reinforced with natural fibers have a set of preferable properties in terms of specific tensile properties, impact properties, and natural appearance, together with less favorable properties such as poor resistance to ultraviolet (UV) light and moisture, low stiffness, compression strength, and relative low price. Even though natural fiber composites have not been considered as materials for compressive load-bearing applications, most of the elements were designed for tensile or flexural modes, which is the main application for natural fiber composites. In reality, the elements will be exposed to compressive loads, at least in part. The composites are under tensile-compressive coupling reaction when subjected to flexure, and residual compressive forces existing within the composites may be caused by unbalanced stacking sequences of laminate or residual processing stresses. This part will consider application of NFC materials as nonstructural and structural components. There are many areas where natural bast fibers are considered as reinforcement at the moment.

7.5.1 Rods

NATCOM, a U.K. government funded project led by Professor Mizi Fan at Brunel University, United Kingdom, have developed circular and square rods made from NFC to work under tensile loads (Figure 7.16). This has been further commercialized at industrial scale by a pultrusion process. Peng et al. (2011) described a series of mechanical tests used to evaluate and optimize NFC reinforced with coarse hemp nontwisted yarns. Aligned fibers are tensioned along the axis of the rod, which allows for higher Vf of reinforcement and increased mechanical properties. This type of element can work as an element of indoor and outdoor furniture.

7.5.2 Panels

Flat laminated panels reinforced with fabrics or rovings are already being used in the automotive industry (Kraus 2004). In civil engineering flat laminate panels can be used as wall partitions, roofing elements, flooring, or elements of furniture. Properties can be controlled by the use of fabric reinforcements. Panels can be molded into various shapes (Figure 7.17). Natural appearance and fibrous texture are appealing and perfectly fit into the sustainable design trend.

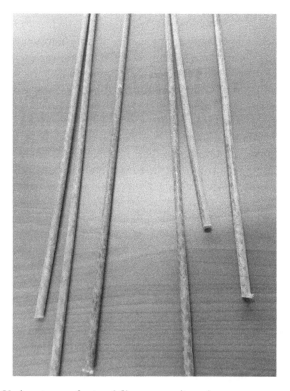

FIGURE 7.16   Various types of natural fiber composite rods.

FIGURE 7.17 Lightweight panel reinforced with hemp mat (left) and flax hopsack fabric (right).

FIGURE 7.18 Natural fiber composite tubes.

7.5.3 Tubes

An extensive program has been carried out led by Professor Mizi Fan at Brunel University to produce square or circular NFC tubes by filament winding or by pull-winding techniques (Figure 7.18). Many important outcomes have been achieved. Tensile and shear are controlled by winding or the stacking sequence of the composite.

A set of thin-walled tubular NFC elements were processed by filament winding and subsequent curing under vacuum conditions. Buckling initiation stress in the axial direction for 0.3 Vf reinforced thin-walled tubes ranged from 30 to 76 MPa. Tube properties can be controlled, as in conventional fiber composites, by control over reinforcement

FIGURE 7.19   Natural fiber composite I-beam during three-point bending test.

direction, arrangement, tube size, and wall thickness. It was found that thin-walled tubes go through similar collapse mechanisms as glass fiber–reinforced tubes.

Tubes can be loaded in tensile and flexural modes. With specific design, tubes can be applied to load-bearing application such as household staircases or furniture.

7.5.4   I-BEAMS

I-section and T-section beams have also been developed at Brunel University led by Professor Mizi Fan (Figure 7.19). The products have been formulated for load-bearing applications out of NFC, by compression molding or pultrusion processes. Reinforcement can be arranged with use for fabrics and single yarns in order to maximize mechanical properties and the ensure correct fracture mode. Optimized design with regard to shear properties within the web gives the best result. Those elements can be applied as elements of roofing, door frames, windows elements, and others.

7.6   CONCLUSIONS: FUTURE OF NATURAL FIBER CIVIL ENGINEERING APPLICATIONS

NFCs have been intensively developed over past decades and continue to be a popular research topic. Recent improvements in fiber extraction, treatments, and resin systems, and successful application in the automotive industry have created a positive future trend. Critics have recalled a long history of development, the high price of material, inconsistencies in properties and weathering performance, and competitive substitute materials, but most of these challenges have already been answered by the research community.

NFC has the highest properties in tensile or flexural modes. With control over type and arrangement of reinforcement, composite properties could be optimized for specific element requirements. NFC rods, tubes, and I-beams have been processed to meet specific needs. One of the aesthetic key points in favor of NFC application in civil engineering sustainable design is, besides specific mechanical properties, their "natural" look. Use of pigments to reduce weathering deterioration caused by UV radiation could overshadow this effect.

It was shown that the nontwisted yarn composites underwent different fracture modes that might be due to the polymeric yarn holding the fibers together. Unmodified flax NFCs were susceptible to high mechanical property degradation when exposed to UV radiation and moist conditions.

The fracture mode of unidirectional composites had a specific angular shape, which was explained by the fracture propagation along the points between polymeric wrapping yarns and natural fibers. This was due to the stress concentration induced by fiber misalignment, reduced adhesion between matrix and polymeric wrap fiber, and the existence of lowered adhesion spots. Measurement of the fiber misalignment in nontwisted yarn was proposed.

Twisted hemp fiber yarns aligned composites had the highest tensile properties among all the composites tested, with 411 MPa tensile strength and up to 14 GPa tensile stiffness. There was a linear relationship between Vf of yarns and tensile stress and strain. Tensile stiffness was found to increase with increasing Vf of fibers until the breaking point and then slightly decreases upon further increase of fiber concentration. Tensile strain was increasing linearly for the twisted yarn composites, but this was less significant for the untwisted fiber yarn composites.

There was a clear linear correlation between ultimate tensile stress and Tex value. The tensile stress was the highest for the fine 39 Tex hemp fiber yarn over other composites developed. Tensile modulus was found to be the highest for the composites reinforced with 60 and 51 Tex, showing the increased fiber quality due to the higher processing grade and increased wettability caused by small size of the yarn.

Minimum and maximum values for the main flexural properties were recorded. The highest value for flexural modulus, 32.6 GPa, was observed for the aligned composite laminate reinforced with hemp fibers arranged in helical shape in the twisted yarns. This value was measured for the composites with 76% Vf of reinforcement and 1.36 g/cm^3 density. Maximum flexural strength measured during these trials was equal to 330.7 MPa and was measured for the same aforementioned laminate with 39 Tex hemp twisted yarn reinforcement.

Composites with pure flax and hemp fibers as well as hybrid composites, such as hemp blended with wool yarns were tested. This type of composite reinforcement led to inferior mechanical flexural properties of the laminates in comparison with yarns composed of only one type of fiber. This was due to the high Tex count (1000) of the yarns used as well as the reduced interface mechanical properties between wool and matrix.

REFERENCES

Arno, P. 1989. *Concise Encyclopedia of Wood and Wood-Based Materials*, 1st ed. Elmsford, NY: Pergamon Press, 271–273.

Awaja, F., Gilbert, M., Kelly, G., Fox, B. and Pigram, P. 2009. Adhesion of polymers. *Progress in Polymer Science* 34: 948–968.

Beguin, P. 1990. Molecular biology of cellulose degradation. *Annual Review of Microbiology* 44: 219–248.

Bledzki, A.K., Fink, H.P. and Specht, K. 2004. Unidirectional hemp and flax EP- and PP-composites: Influence of defined fiber treatments. *Journal of Applied Polymer Science* 93: 2150–2156.

Bledzki, A.K. and Gassan, J. 1999. Composites reinforced with cellulose based fibres. *Progress in Polymer Science* 24: 221–274.

British Standards Institute (BSI). 1997. *BS EN 527-5:1997 BS 2782-3: Method 326G: 1997: Plastics—Determination of Tensile Properties. Part 5: Test Conditions for Unidirectional Fibre-Reinforced Plastic Composites.* London: British Standards Institute.

British Standards Institute (BSI). 1998. *BS EN ISO 14125:1998+A1:2011: Fibre-Reinforced Plastic Composites—Determination of Flexural Properties.* London: British Standards Institute.

British Standards Institute (BSI). 2005. *BS PL ISO 4:2005: Properties of Unsaturated Polyester Resins for Low Pressure Laminating of High Strength Fibre Reinforced Composites— Specification.* London: British Standards Institute.

Charlet, K., Baley, C., Morvan, C., Jernot, J.P., Gomina, M. and Breard, J. 2007. Characteristics of Hermès flax fibres as a function of their location in the stem and properties of the derived unidirectional composites. *Composites Part A: Applied Science and Manufacturing* 38: 1912–1921.

Dhakal, H.N., Zhang, Z.Y. and Richardson, M.O.W. 2007. Effect of water absorption on the mechanical properties of hemp fibre reinforced unsaturated polyester composites. *Composites Science and Technology* 67: 1674–1683.

FAO. 2012a. *Hemp Tow and Waste Production Data.* Food and Agriculture Organisation of the UN, Available from: http://faostat3.fao.org/home/index.html#SEARCH_DATA, accessed 13 November 2012.

FAO. 2012b. *Flax Fibre and Tow Production Data.* Food and Agriculture Organisation of the UN, Available from: http://faostat3.fao.org/home/index.html#SEARCH_DATA, accessed 13 November 2012.

Goutianos, S., Peijs, T., Nystrom, B. and Skrifvars, M. 2006. Development of flax fibre based textile reinforcements for composite applications. *Applied Composite Materials* 13: 199–215.

Hagstrand, P.O. and Oksman, K. 2001. Mechanical properties and morphology of flax fiber reinforced melamine-formaldehyde composites. *Polymer Composites* 22: 568–578.

Hepworth, D.G., Hobson, R.N., Bruce, D.M., Farrent, J.W. 2000. The use of unretted hemp fibre in composite manufacture. *Composites Part A: Applied Science and Manufacturing* 31: 1279–1283.

Hollen, N. and Saddler, J. 1964. *Textiles.* New York: The MacMillan Company.

Hughes, M., Carpenter, J. and Hill, C. 2007. Deformation and fracture behaviour of flax fibre reinforced thermosetting polymer matrix composites. *Journal of Materials Science* 42: 2499–2511.

Johanson, G.H. 1927. *Textile Fabrics.* New York: Harper and Brothers.

Joseph, M.L. 1986. *Introductory Textile Science.* New York: Holt Rinehart and Winston.

Joshi, S.V., Drzal, L.T., Mohanty, A.K. and Arora, S. (2004) Are natural fiber composites environmentally superior to glass fiber reinforced composites? *Composites Part A: Applied Science and Manufacturing* 35: 371–376.

Kraus, M. 2004. European hemp industry 2002: Cultivation, processing and product lines. *Journal of Industrial Hemp* 9: 93–101.

Kvavadze, E., Bar-Yosef, O., Belfer-Cohen, A., Boaretto, E., Jakeli, N., Matskevich, Z., and Meshveliani, T. 2009. 30,000-year-old wild flax fibres *Science* 325: 1359.

Lawrence, G.H.M. 1969. *Taxonomy of Vascular Plants.* New York: The MacMillan Company.

Lower, G.A. 1937. Flax and hemp: From the seed to the loom. In: Ed. Herer, J. *Emperor Wears No Clothes.* Van Nuys, CA: Hemp Publishing.

Madsen, B., Hoffmeyer, P. and Lilholt, H. 2007a. Hemp yarn reinforced composites— II. Tensile properties. *Composites Part A: Applied Science and Manufacturing* 38: 2204–2215.

Madsen, B., Hoffmeyer, P., Thomsen, A.B. and Lilholt, H. 2007b. Hemp yarn reinforced composites—I. Yarn characteristics. *Composites Part A: Applied Science and Manufacturing*
38: 2194–2203.

Mehta, G., Drzal, L.T., Mohanty, A.K. and Misra, M. 2006. Effect of fiber surface treatment on
the properties of biocomposites from nonwoven industrial hemp fiber mats and unsaturated polyester resin. *Journal of Applied Polymer Science* 99: 1055–1068.

Miwa, M. and Endo, I. 1994. Critical fibre length and tensile strength for carbon fibre-epoxy
composites. *Journal of Materials Science* 29: 1174–1178.

Muralidhar, B.A., Giridev, V.R. and Raghunathan, K. 2012. Flexural and impact properties of
flax woven, knitted and sequentially stacked knitted/woven perform reinforced epoxy
composites. *Journal of Reinforced Plastics and Composites* 31(6): 379–388.

Mwaikambo, L.Y., Tucker, N. and Clark, A.J. 2007. Mechanical properties of hemp-fibre-
reinforced euphorbia composites. *Macromolecular Materials and Engineering* 292:
993–1000.

Ochi, S. 2006. Development of high strength biodegradable composites using Manila hemp
fiber and starch-based biodegradable resin. *Composites Part A* 37: 1879–1883.

Oksman, K. 2001. High quality flax fibre composites manufactured by the resin transfer
moulding process. *Journal of Reinforced Plastics and Composites* 20: 621–627.

Peng, X., Fan, M., Hartley, J. and Al-Zubaidy, M. 2011. Properties of natural fibre composites
made by pultrusion process. *Journal of Composite Materials* 44(2): 237–246.

Pickering, K. 2008. *Properties and Performance of Natural Fibre Composites*. Cambridge:
Woodhead Publishing in Materials.

Popular Mechanics. 1938. New billion-dollar crop. *Popular Mechanics* 69(2): 238–239.

Renewable-Resources. 2010. *Biowerkstoff, Report on Bio-based Plastics and Composites*. 7th
edn. Hürth: Nova Institut.

Rouison, D., Sain, M. and Couturier, M. 2006. Resin transfer molding of hemp fiber composites: optimization of the process and mechanical properties of the materials. *Composites
Science and Technology* 66: 895–906.

Saheb, D.N. and Jog, J.P. 1999. Natural fiber polymer composites: A review. *Advances in
Polymer Technology* 18: 351–363.

Saskatchewan Ministry of Agriculture. 2003. *Hemp Production in Saskatchewan*. Government
of Saskatchewan. Available from: http://www.agriculture.gov.sk.ca/Default.aspx?DN=
e60e706d-c852-4206-9959-e4b134782175, accessed 13 November 2012.

Sawpan, M.A., Pickering, K.L. and Fernyhough, A. 2012. Flexural properties of hemp fibre
reinforced polylactide and unsaturated polyester composites. *Composites: Part A* 43:
519–526.

Shazad, A. 2011. Hemp fibre and its composites—a review. *Journal of Composite Materials*
46(8): 973–986.

Struik, P.C., Amaducci, S., Bullard, M.J., Stutterheim, N.C., Venturi, G. and Cromack, H.T.
2000. Agronomy of fibre hemp (*Cannabis sativa* L.) in Europe. *Journal of Industrial
Crops and Products* 11: 107–118.

Talreja, R. and Manson, J. 2000. *Comprehensive Composite Materials, Volume 2: Polymer
Matrix Composites*. Amsterdam: Elsevier.

Thygesen, A., Thomsen, A.B., Daniel, G. and Lilholt, H. Comparison of hemp fibres with
glass fibres for wind power turbines: effect of composite density. In: *Proceedings of
the 27th Riso International Symposium on Materials Science, 2006*. Roskilde: Riso
National Laboratory.

Thygesen, A., Thomsen, A.B., Daniel, G. and Lilholt, H. 2007. Comparison of composites
made from fungal defibrilated hemp with composites of traditional hemp yarn. *Industrial
Crops and Products* 25: 147–159.

Wang, H.M. and Postle, R. 2003. Removing pectin and lignin during chemical processing of
hemp for textile applications. *Textile Research Journal* 73: 664–669.

Van De Weyenberg, I., Ivens, J., De Coster, A., Kino, B., Baetens, E. and Verpoest, I. 2003. Influence of processing and chemical treatment of flax fibres on their composites. *Composites Science and Technology* 63: 1241–1246.

Van Den Oever, M.J.A., Bos, H.L. and Van Kemenade, M.J.J.M. 2000. Influence of the physical structure of flax fibres on the mechanical properties of flax fibre reinforced polypropylene composites. *Applied Composite Materials* 7: 387–402.

Wool, R.P. and Sun, X.S. 2005. *Bio-Based Polymers and Composites*. London: Elsevier Inc.

Yuanjian, T. and Isaac, D.H. 2007. Impact and fatigue behaviour of hemp fibre composites. *Composites Science and Technology* 67: 3300–3307.

8 Life Cycle Assessment for Natural Fiber Composites

Nilmini P.J. Dissanayake and John Summerscales

CONTENTS

8.1 INTRODUCTION

Concern for the environment is not a new phenomenon. In the fifth century BC, Plato wrote about the effects of unsustainable practice regarding forests, referring to the deforestation of the hills around Athens as a result of logging for shipbuilding and to clear agricultural land (Moll et al. 2005). von Carlowitz (1713) explained that if forest resources were not used with caution (i.e., planned on a sustainable basis to achieve continuity between increment and felling), then humanity would plunge into poverty and destitution. Malthus (1798) presented his *Essay on the Principles of Population* in which he proposed that population tends to increase faster than the means of subsistence, and its growth could only be checked by moral restraint or disease and war. Carson (1962) published *Silent Spring*, which warned against profligate use of synthetic chemical pesticides, challenged the agricultural practices of the time, and suggested that there were methods of pest control that were less damaging to the natural world. Her work is often suggested to be the primer for the rise of the environmental movement in the 1960s. The World Commission on Environment and Development (1987) suggested that Sustainable Development should be defined as "Meeting the needs of the present without compromising the ability of future generations to meet their own needs." There is a far more extensive literature on concern for the environment but that topic is beyond the scope of this review.

The composites industry has grown rapidly over the past 75 years since the introduction of commercial continuous fiber reinforcements (glass in 1937, carbon in the 1960s, and aramids in 1971). Natural fibers were used in resin matrix composites in the early years of the industry including flax and hemp fibers for the bodywork of a Henry Ford car in 1941 (Lewington 2003), but then fell from favor. The current focus on sustainable materials has renewed interest in bio-based fibers and resins. The fibers most likely to be adopted as reinforcements are bast (stem) cellulose fibers from plants including flax and hemp (in temperate zones) or jute and kenaf (in tropical zones). Bast fibers and the composites have been reviewed recently by numerous authors as compiled by Summerscales (2013).

8.2 MATERIALS

Natural bast/cellulose fibers and man-made glass fibers have similar specific elastic modulus, with the lower moduli (cellulose $\approx$ 50 GPa, E-glass $\approx$ 70 GPa) compensated by lower density (cellulose $\approx$ 1500 kg/m^3, glass $\approx$ 2500 kg/m^3). Bast fibers have finite length (i.e., not continuous filaments unless spun into that form), lower strengths, higher variability in many properties relative to synthetic fibers and more rapid degradation in hot and/or wet environments. The issue of cost is currently complicated due to the established production of glass fibers at industrial scale while the natural fiber market is still in its infancy.

Fibers such as flax/linseed and hemp are currently grown commercially in the United Kingdom and Europe and such natural fibers as composites in a polymer matrix are used for a wide range of automotive applications such as the interior panels of passenger cars and truck cabins, door panels, and cabin linings as substitutes for glass fiber composites.

8.2.1 PRODUCTION OF NATURAL FIBERS

The typical U.K. production cycle for flax fibers (Turner 1987) is described below. A similar route is followed for the other bast fibers.

Tillage: the preparation of land for cropping by ploughing and similar operations. Conservation tillage is one part of conservation agriculture, which has three basic principles: (1) reduction in tillage, (2) retention of crop residues at the soil surface, and (3) use of crop rotations (Govaerts et al. 2009).

Drilling (planting) the seed: this usually occurs between the end of February and early April in Belgium, France, and the Netherlands or in early April in Northern Ireland (NI). For flax in NI, the suggested levels of fertilizer are 20 kg N/ha, 20 kg P_2O_5/ha, and 80 kg K_2O/ha (Henfaes Research Centre 2004).

Weed control: it is essential to minimize weeds to avoid contamination of the scutched (i.e., decorticated) flax fibers.

Plant growth: the life cycle of the flax plant consists of a 45- to 60-day vegetative period, a 15- to 25-day flowering period, and a maturation period of 30–40 days.

Desiccation: glyphosate is typically applied 10–14 days after full flower, at about mid-July in NI. Glyphosate is only used where stand retting is adopted followed by direct combine-harvesting of the crop.

Harvest: by either combining or pulling, in August/September.

Rippling: the removal of flax seed capsules by drawing pulled stems through a coarse steel comb.

Retting is defined for flax as the "subjection of crop or deseeded straw to chemical or biological treatment to make the fibre bundles more easily separable from the woody part of the stem. Flax is described as water-retted, dew-retted or chemically–retted, etc., according to the process employed" (Farnfield and Alvey 1975). Enzymes may be used to assist the retting process, but termination of the retting process may be a problem and failure to achieve this can result in reduced fiber properties. Preharvest retting of flax with glyphosate (Sharma et al. 1989) applied at the midpoint of flowering depends on uniform desiccation of the entire stem and is difficult to achieve during a dry season. As in dew retting, stand retting of the desiccated flax in the field relies on microorganisms and is dependent on the vagaries of the weather. Sharma et al. (1989) presented and discussed the role of microbial enzymes, screening of flax cultivars, fiber quality, and upgrading of coarse fibers in this context. Windrowing of flax is part of the field or dew-retting process.

Decortication is the mechanical removal of nonfibrous material from retted stalks or from ribbons or strips of stem or leaf fibers to extract the fibers. For flax, the process is usually referred to as "scutching." This can be achieved by a manual operation, hammer mill, inclined plane/fluted rollers, or willower. Harwood et al. (2008) described a novel environmentally friendly and cost-effective method using shock waves generated by high-voltage pulsed electrical discharges on fibers immersed in water.

Hackling is the combing of line flax in order to remove short fibers, parallelize the remaining long (line) fibers, and also remove any extraneous matter (shive).

Carding is defined as "the disentanglement of fibers by working them between two closely spaced, relatively moving surfaces clothed with pointed wire, pins, spikes or saw teeth" (Farnfield and Alvey 1975).

Spinning is the drafting (decreasing the mass per unit length) and twisting of natural (or man-made) fibers for the production of yarns or filaments. In the bast- and leaf-fiber industries, the terms "wet spinning" and "dry spinning" refer to the spinning of fibers in the wet and dry state, respectively.

Subsequent treatment of natural fiber textiles may include the following:

Acetylation (Farnfield and Alvey 1975) is defined as the "process of introducing an acetyl radical into an organic molecule" and the term "is used to describe the process of combining cellulose with acetic acid."

Grafting is the incorporation of monomers (e.g., the cyanoethylation reaction with acrylonitrile) or oligomers (short chain polymers) by chemical reaction at the fiber surface.

Mercerization (Farnfield and Alvey 1975) is the "treatment of cellulosic textiles in yarn or fabric form with a concentrated solution of caustic alkali (soda), whereby the fibers are swollen, the strength and dye affinity of the materials are increased, and their handle is modified. The process takes its name from its discoverer, John Mercer (1844)."

Plasma treatment may be used to modify the surface chemistry.

For a more complete description of fiber chemical treatments and interface engineering, there are a number of good reviews (George et al. 2001, Li et al. 2007, Kalia et al. 2009, Zafeiropoulos 2011, Kabir et al. 2012).

8.2.2 PRODUCTION OF GLASS FIBERS

Glass fibers come in a variety of forms although >95% of all reinforcements are E-glass. The formulation of these fibers includes several minerals (see e.g., Net Composites undated):

- Sand—particles of minerals including quartz (silica: SiO_2), mica (complex silicates usually with K, Na, Li, H, and Mg), and feldspar (aluminum silicates with varying amounts of K, Na, Ca, and Ba). "Pure sand is white in colour and consists of silica" (Tottle 1984).
- Kaolin—hydrated aluminum silicate ($Al_2O_3 \cdot 2SiO_2 \cdot 2H_2O$) also known as china clay (Tottle 1984).
- Limestone—mostly calcium carbonate ($CaCO_3$) with other oxides (Si, Al, Fe), carbonates (Fe, Mg), and calcium phosphate (Net Composites undated).
- Colemanite—hydrated calcium borate ($2CaO \cdot 3B_2O_3 \cdot 5H_2O$) (Tottle 1984).

These materials are melted at ~1600°C and spun through microfine bushings to produce filaments of 5–24 μm in diameter. The cooled filaments are drawn together into a strand (closely associated) or roving (loosely associated), and coated with a "size" (Net Composites undated). Surface treatments on fibers exist for a variety of reasons: filament cohesion, antistatic agents, lubricants for textile processes, and coupling agents to promote good adhesion between the fibers and the resin. The most common coupling agents are organofunctional silanes.

Manufacture of glass fibers may involve significant transport distances both from the raw materials source to the industrial-scale fiber factory and from the factory to the customer.

8.2.3 END-OF-LIFE COMPOSITES

Pickering (2006) reviewed the technologies for recycling thermoset composite materials. A key issue here is the grinding technique used to reduce the scrap material to handleable particles or to recyclate products suitable for mechanical recycling as fillers or partial reinforcement in new composite material. There are a range of thermal recycling processes for energy (incineration) and material recovery (pyrolysis).

At the end of life, there is potential for controlled degradation of natural fibers and plant-based resins by composting or other methods. A biodegradable material is expected to reach a defined extent of degradation by biological activity under specific environmental conditions within a given time under standard test conditions (Murphy and Bartle 2004). The EU Directive on Packaging and Packaging Waste criteria for biodegradability are set out in BS EN 13432:2000 (2000), while the criteria in North America are set out in ASTM D6400-04 (2004). The requirements of the standard include the following:

1. Biodegradation: over 90% relative to the standard (cellulose) in 180 days under conditions of controlled composting using respirometric methods (ISO14855:-1:2005) (2005)
2. Disintegration: over 90% in 3 months
3. Ecotoxicity: test results for aquatic and terrestrial organisms (*Daphnia magna*, worm test, germination test) as for reference compost
4. Absence of hazardous chemicals (included in a reference list)

The biodegradation of polymeric materials under controlled composting conditions can be determined using standard methods including ASTM D 5338-98 (2003) or ISO 14852:1999 (1999). There are essentially two options, as follows: (1) aerobic, carried out either in open air windrows or in enclosed vessels or (2) anaerobic, required when animal by-products or catering wastes are included. A demonstration-scale anaerobic digestion plant is operating at Dufferin (Toronto, Canada) solid waste transfer station with a mass balance (based on 100 tonnes/day) of 50% biogas and effluent, 25% digestate, and 25% residue (Goldstein 2005). The biogas varies due to the batch operation but is typically 110 m^3/tonne with an average of 56% methane (ranges from 45% to 73%) by volume. Jana et al. (2001) suggested that the biogas is typically "60%–65% methane, 35% carbon dioxide and a small amount of other impurities." Similarly, "pure landfill gas can contain up to 35% carbon dioxide, 65% methane and no oxygen" (Greenham and Walsh 2004).

Organisms that possess cellulase (the enzyme that cleaves sugar from the cellulose molecule) include bacteria, some flagellate and ciliate protozoa, and fungi (Turner 1987). If an animal is to digest cellulose, it must enter into an alliance with such an organism. For example, termites have a symbiotic relationship with fungi that provides the symbionts with a rich source of cellulose for food in return for access to glucose cleaved from the cellulose and additionally to protein, vitamins, and essential amino acids produced by the fungi. Termites hatch without this essential intestinal flora and are inoculated with it by being fed feces and regurgitant that contain the symbionts.

8.3   LIFE CYCLE ASSESSMENT

Life Cycle Assessment (LCA) is an environmental assessment method which, according to the international standard Environmental Management—Life Cycle Assessment—principles and frameworks, ISO 14040:2006(E) (2006), "considers the entire life cycle of a product from raw material extraction and acquisition, through energy and material production and manufacturing, to use and end-of-life treatment and final disposal."

An LCA study has the following four phases:

The **goal and scope definition**—aims and objectives of the study

Life Cycle Inventory (LCI) analysis—compilation and quantification of inputs and outputs for a product through its life cycle

Life Cycle Impact Assessment (LCIA)—understanding and evaluating the magnitude and significance of the potential environmental impacts for a product system throughout the life cycle of the product

Life Cycle Interpretation—the findings of the LCI or LCIA or both are evaluated in relation to the defined goal and scope in order to reach conclusions and recommendations

Although the concept of the LCA is simple, the analysis is quite complex in reality, primarily due to the difficulty in establishing the correct system boundaries, obtaining accurate data, and interpreting the results correctly (Porritt 2005).

8.4   ENVIRONMENTAL IMPACT CLASSIFICATION FACTORS

Most of the literature on LCA studies rarely considers the full eight environmental impact classification factors (EICF) listed below as outlined in ISO/TR 14047/2003 (2003). The EICF reviewed in this chapter are the following:

1. Acidification potential (AP)
2. Aquatic toxicity potential (AqTP)
3. Human toxicity potential (HTP)
4. Eutrophication potential (EP)
5. Global warming potential (GWP)
6. Nonrenewable/abiotic resource depletion potential (NRADP)
7. Ozone depletion potential (ODP)
8. Photochemical oxidants creation potential (POCP)

These are outlined in ISO/TR14047:2003 (2003) and also closely echoed by Azapagic et al. (2003, 2004). The British Standard BS8905:2011(2011) adds land use to the list to acknowledge the growing debate around food, fuel, feedstock, and fiber (Kern 2002). The European parliament has recently debated whether LCA should be voluntary or mandatory and what methodology should be used for collecting data (Banks 2012).

A key issue in LCA is a well-defined goal and scope with clear allocation (apportionment) of the inputs and outputs. This is especially true for agricultural systems where waste streams may have a role as raw materials for other industries (e.g., in natural fiber production, short fibers can go to papermaking or animal bedding and dust can be briquetted as fuel). Given that the yield of flax fiber is typically <20% of the harvested biomass, the calculated environmental burdens if allocated solely to the fiber will be greater than the values of the burdens if apportioned amongst all the components of the plant mass. The former scenario emphasizes issues and is useful in identifying more environmentally friendly production routes. The latter scenario with burdens disproportionately allocated to co-products and waste streams, can provide a much stronger "green marketing" position for the primary product. An intermediate position might use the relative economic values of each components, but this is less reliable over time in markets with volatile prices.

Ekvall and Finnveden (2001) presented a critical review of allocation in the context of ISO14041 (the standard has been superseded by ISO14044) with the view that the value of LCA depends on the extent to which it permits us to anticipate the environmental consequences of manipulation of technological systems. They recommend that "all of the environmental burdens of a multifunction process be allocated to the product investigated."

A good LCA requires consideration of at least the eight EICF mentioned above (not just greenhouse gas (GHG) considerations). There is no agreed methodology for aggregation of the multiple factors into a single score to allow clear distinction of the greenness of competing products (unless one serendipitously has the lowest impact for all categories).

8.4.1 ACIDIFICATION

Acidification is a consequence of acids (and other compounds that can be transformed into acids) being emitted to the atmosphere and subsequently deposited in surface soils and water. Increased acidity of these environments can result in negative consequences for coniferous trees (forest dieback) and the death of fish in addition to increased corrosion of man-made structures (buildings, vehicles, etc.).

Acidification considerably reduces the fertility of the soil, mainly by affecting its biology, by breaking up organic matter, and causing loss of plant nutrients (Sensi undated). Growing plants remove alkalinity from the soil and the soil acidification increases when the harvested products are removed. Soil acidification is closely linked to water acidification, which can affect aquatic life, groundwater, and the related drinking water supply. Turunen and van der Werf (2006, 2008) found that acidification was largely (62%–79%) due to the yarn production stage in the LCA carried out for hemp yarn production. They also suggested that the three impacts, energy use, climate change, and acidification, are strongly interrelated, as energy demand is largely met by fossil fuels, the combustion of which results in the emissions of CO_2 and SO_x, which is also the case for the melting, refining, forming, and finishing phases in glass fiber production.

8.4.2 AQUATIC TOXICITY/ECOTOXICITY

Eco-toxicity results from persistent chemicals reaching undesirable concentrations in each of the three elements of the environment (air, soil, and water) leading to damage to animals, eco-systems, and aquatic systems. The modeling of toxicity in LCA is complicated by the complex chemicals involved and their potential interactions. The assessment of potential toxic effects from pesticides in the aquatic environment on nontarget organisms, such as aquatic biota and soil micro-organisms, is becoming increasingly important (Amoros et al. 2007). Herbicides constitute >50% of pesticide production in Denmark, France, and the United Kingdom (Petit et al. 1995).

Glyphosate, *N*-(phosphonomethyl) glycine [$H_2O_3P-CH_2-NH-CH_2-COOH$], is a systemic, broad-spectrum, nonselective herbicide used to kill broad-leaved grasses and sedge species. It is one of the most used xenobiotics in modern agriculture (Peixoto 2005) and is used as a desiccant for flax. Petit et al. (1995) stated that it is completely degraded by microorganisms in soil, with a half-life of 60 days, primarily to nontoxic aminomethylphosphoric acid ([$H_2O_3P-CH_2-NH_2$]), which is further degraded in soil. Glyphosate decomposition occurs under both aerobic and anaerobic conditions. It is also inactivated by adsorption to clay in benthic or suspended sediment.

Eriksson et al. (2007) proposed a scientifically justifiable list of "selected storm water priority pollutants (SSPP)" to be used, for example, in evaluating the chemical risks posed by different water handling strategies. Glyphosate was included with the selected herbicides as it is extensively used in urban areas and along motorways in the United Kingdom, France, Denmark, and Sweden. They considered 15 routes for the removal of glyphosate from storm water and predicted that infiltration basins were best management practice, followed by subsurface flow constricted wetlands and porous pavements. Glyphosate is most readily removed by adsorption and microbial degradation. Settlement tanks were judged to be the least efficient technique for the removal of glyphosate.

The commercial herbicide Roundup® is a formulation of 36% glyphosate (normally considered the active ingredient) as the isopropylamine salt with a surfactant, polyoxyethyleneamine (POEA) (Amoros et al. 2007). Martinez and Brown (1991) found that just one-third of the amount of Roundup® relative to glyphosate alone was required to kill rats suggesting that ingredients labeled as "inert" should be considered otherwise. Peixoto (2005) clearly demonstrated the ability of Roundup® to impair mitochondrial bioenergetic reactions, especially at higher concentrations, while glyphosate alone did not affect mitochondrial respiration and membrane energization. They attribute the effect to nonspecific membrane permeabilization probably as a result of other ingredients (suspecting the POEA surfactant) or a synergy between components of the formulation.

Sandermann (2006) reported that a 1997 review (Heap 1997) listed 183 herbicide-resistant biotypes in 42 countries, including the first observation of glyphosate resistance (plants that survive and set seed). By 2004, 11 weed biotypes had been recorded as glyphosate resistant. The risk assessment for herbicide-resistant transgenic plants, notably Roundup-Ready® crops (e.g., alfalfa, canola, cotton, maize, and soybean), may well underestimate the long-term effects of dependence on glyphosate as a herbicide.

Petit et al. (1995) also considered Diquat [dipyrido (1,2-a:2′,1-c) pyrazinediium-6,7-dihyro-dibromomonohydrate] and Paraquat [4,4′-bipyridium-1,1′-dimethyldichloride]. Both these herbicides undergo photodegradation with half-lives of 2–11 and 1.5 days, respectively. These herbicides also undergo biodegradation (Diquat half-life = 15–32 days; Paraquat not given) and adsorption to clay.

In an environmental comparison of China Reed fiber as a substitute for glass fiber in plastic transport pallets, Corbière-Nicollier et al. (2001) found that cultivation of the reed had a dominant role in the factors for terrestrial ecotoxicity, human toxicity (when crop rotation has edible foods following the China Reed), and eutrophication due to (1) heavy metal emissions to soil and (2) phosphate emissions (from manure and fertilizer) to water.

8.4.3 HUMAN TOXICITY

Human and eco-toxicity results from the persistent chemicals reaching undesirable concentrations in each of the three segments of the environment, air, soil, and water, leading to damage in humans, animals, and ecosystems. The modeling of toxicity in LCA is complicated by the complex chemicals involved and their potential interactions.

Abrahams (2002) reviewed the effect of soils on the health of humans and concluded that contaminants are a known global problem that needs further research with respect to their behavior and the pathways to humans.

Using diesel in transportation and running machinery for both natural fiber and glass fiber production has potential contaminants of concern including carbon monoxide (CO), nitrogen dioxide (NO_2), and sulfur dioxide (SO_2). The direct impacts of these gases and pollutants include toxicity, global warming, acidification, and ozone layer depletion. The toxicological factors are calculated using scientific estimates for the acceptable daily intake or tolerable daily intake of the toxic substances. The human toxicological factors are still at an early stage of development so that HTP can only be taken as an indication and not as an absolute measure of the toxicity potential. Carbon monoxide has a human toxicology classification factor of 0.012. For agricultural chemicals, the human toxicology classification factors are 0.0017– 0.020 for ammonia (from fertilizer), 1.4 for arsenic as solid (herbicide), 0.00078 for nitrates (fertilizer), 0.00004 for phosphates (fertilizer), and 0.14 for pesticides (Azapagic et al. 2003, 2004).

8.4.4 EUTROPHICATION/NITRIFICATION

Eutrophication is defined as the potential for nutrients to cause overfertilization of water and soil, which in turn can result in increased growth of biomass.

Nitrogen (constituent of amino acids, nucleotides, and chlorophyll), carbon (major cellular constituent), and phosphate (constituent of ATP [adenosine triphosphate—one adenosine attached to three phosphate groups], ADP [adenosine diphosphate], and phospholipids) are essential nutrients for algal growth (Lebret et al. 2009).

Algae refer to a diverse group of eukaryotic (containing a nucleus enclosed within a well-defined nuclear membrane) microorganisms that share similar characteristics. They range from unicellular to multicellular plants that occur in freshwater, marine water, and damp environments, and range in size from minute phytoplankton to giant marine kelp. Algae possess chlorophyll, the green pigment essential for photosynthesis, and often contain additional pigments that mask the green color (e.g., fucoxanthin brown and phycoerythin red) (Horne and Goldman 1994; Wetzel 1983).

Typically, algae are autotrophic (derive cell carbon from inorganic carbon dioxide), photosynthetic (derive energy for cell synthesis from light), and contain chlorophyll. They are also chemotrophic in terms of night time respiration, for example, metabolism of molecular oxygen (O_2). Algae utilize photosynthesis (solar energy) to convert simple inorganic nutrients into more complex organic molecules.

Photosynthetic processes result in surplus oxygen and nonequilibrium conditions by producing reduced forms of organic matter, such as biomass containing high-energy bonds made with hydrogen and carbon, nitrogen, sulfur, and phosphorus compounds.

The organic matter produced serves as an energy source for nonphotosynthetic or heterotrophic organisms (animals, including most bacteria, which subsist on organic matter). Heterotrophic organisms tend to restore equilibrium by catalytically decomposing these unstable organic products of photosynthesis, thereby obtaining a source of energy for their metabolic needs. The organisms use this energy both to synthesize new cells and to maintain old cells already formed (Stumm and Morgan 1996). From the point of overall reactions, these heterotrophic organisms only act as reduction–oxidation catalysts—they only mediate the reaction (or more specifically the electron transfer). Oxidation may produce several intermediate reduction–oxidation states prior to reaching a fully oxidized state (e.g., inorganic state) (Stumm and Morgan 1996).

Phosphorus is an essential element in biological systems as it is a constituent of nucleic acid in phospholipids of cell membranes and ATP and ADP, which are involved in energy exchange in biological systems. ATP is the energy carrier in all cells. The energy produced by respiration is kept in these molecules and is stored as phosphate bonds (Todar undated).

Orthophosphate (PO_4^{3-}) is the only form of P that autotrophs can assimilate. The results are excessive production of autotrophs, especially algae and cyanobacteria. This high productivity leads to high bacterial populations and high respiration rates, leading to hypoxia or anoxia in poorly mixed bottom waters and at night in surface waters during calm, warm conditions. Low dissolved oxygen levels cause the loss of aquatic animals and release of many materials normally bound to bottom sediments including various forms of P. This release of P reinforces the eutrophication (Correll 1998).

Reddy et al. (1999) and Stoate (2007) reported that phosphorus is the nutrient that limits plant growth in most fresh (river/pond/lake) waters. Nitrogen contamination of fresh water does not necessarily result in a significant eutrophication hazard (Schindler 1978) although it may be the limiting nutrient in coastal waters (Reddy et al. 1999; Stoate 2007). van Dolah (2000) stated that increasing industrial and agricultural activities result in enhanced discharges of nitrate and phosphate in coastal waters, which have been correlated with strong algal growth. Phosphorus binds tightly to soil particles and hence is a particular problem when sediment enters watercourses as it leads to excessive algal growth. In turn, the algal blooms reduce light penetration into the water and hence inhibit the growth of macrophytes and their invertebrate predators. Furthermore, as the excessive plant biomass decomposes, it consumes dissolved oxygen in the water initially affecting top predators and hence disrupting the ecosystem.

According to the LCA carried out by Turunen and van der Werf (2006, 2008), the emissions from the soil (N and P) contributed about 90% of the eutrophication with the remaining 10% coming from diesel combustion in field operations. The use of fertilizers in agriculture is perceived as a major cause of eutrophication. However, Stoate (2007) suggested that phosphorus and other nutrients from village sewage works and from septic tanks at isolated dwellings also contribute to depletion of downstream invertebrate communities. He reported that the Game Conservancy Trust Allerton project revealed that phosphorus concentrations from septic tanks were more than 10 times those from arable field drains.

The nitrate anion, NO_3^-, has high solubility in water and is not significantly adsorbed onto most soils (Abrahams 2002). When both soil nitrate levels and water movement are high, leaching and run-off may be significant contributors to the nitrogen load in watercourses. While fertilizers are a source of nitrates, Abrahams (2002) suggested that in the climatic conditions of NW Europe, mineralization of soil organic matter (SOM) and crop/animal residues is a more significant source of leached nitrates because mineralization is not well synchronized with nitrogen uptake of the crop.

Crews and Peoples (2004) reviewed the use of legume-derived nitrogen against synthetic fertilizer-derived nitrogen and concluded that "the ecological integrity of legume-based agroecosystems is [only] marginally greater than that of fertilizer-based systems." This advantage could be eroded where best management practices are considered. However, nitrogen biologically fixed by legumes is derived from solar energy, whereas synthetic fertilizer is a heavy user of (fossil fuel) commercial energy and hence is likely to become less attractive in the future (especially in the context of GWP). Leguminous cover crops allowed to grow throughout the fallow season can substantially reduce nitrogen leaching as they scavenge nitrogen available in the soil. Furthermore, nitrogen synchrony (available during crop nitrogen uptake) can be increased by planting cover crops in the off-season and ploughing-in crops in spring rather than autumn.

Bradshaw et al. (2004) reported that a major eutrophication of Dallund Sø (a lake in Denmark) occurred as a result of the changing agricultural system and retting of flax and hemp during the Mediæval period (AD 1050–1536). Turunen and van der Werf (2006, 2008) found that the water-retting process of hemp has contributed 13% of the total eutrophication, which is higher than other retting processes such as bio-retting and stand/dew retting. Fiber processing operations also contributed to eutrophication through the emissions from electricity generation.

In an environmental comparison of China Reed fiber as a substitute for glass fiber in plastic transport pallets, Corbière-Nicollier et al. (2001) found that China Reed fiber was the better option for all factors except eutrophication. The Centrum voor Milieuwetenschappen Leiden (CML) and eco-indicator measures confirm the CST95 results (China Reed is considered better on all factors relative to glass fiber) except for eutrophication. The China Reed has a better score from the former two methods as they consider NO_x emissions to contribute to eutrophication, whereas CST95 considers that only phosphates contribute to eutrophication in Europe where lakes are normally phosphorus limited.

Cuttle et al. (2006) reviewed the methods available for the control of diffuse water pollution from agriculture. At the retting stage, eutrophication can be significantly reduced by stand/dew retting rather than immersion/water retting.

8.4.5 GLOBAL WARMING/CLIMATE CHANGE

Global warming is caused by the ability of the Earth's atmosphere to reflect some of the heat radiated from the Earth's surface back to the ground. This reflected radiation is increased by GHGs in the atmosphere. Increased emission of GHGs (CO_2, N_2O,

CH_4, and volatile organic compounds [VOCs]) will change the heat balance of the Earth and result in future climate change.

An increase of weather extremes has been a fundamental prediction of climate science for decades. Basic physics suggests that as the Earth warms, precipitation extremes will become more intense, winter and summer, simply because warmer air can carry more water vapor. Weather statistics confirm that this has begun to happen (Gillis 2011).

The worldwide climate change signs include the following (National Geographic 2007):

- The rising average temperature on Earth, which has climbed up by 0.8°C since 1880 with much of this is in recent decades.
- The increased rate of warming: The twentieth century's last two decades were the hottest in 400 years and possibly the warmest for several millennia, according to a number of climate studies. The United Nations' Intergovernmental Panel on Climate Change (IPCC) reports that 11 of the 12 years before 2007 were among the dozen warmest since 1850 (Metz et al. 2007).
- The temperatures in the Arctic: Average temperatures in Alaska, western Canada, and eastern Russia rose at twice the global average, according to the multinational Arctic Climate Impact Assessment report compiled between 2000 and 2004 (Hassol 2004).
- Rapid disappearance of the Arctic ice: The region may have its first completely ice-free summer by 2040 or earlier. Polar bears and indigenous cultures are already suffering from the sea-ice loss.
- Rapidly melting glaciers and mountain snows: For example, Montana's Glacier National Park had 150 glaciers in 1901 and now has only 27. In the Northern Hemisphere, thaws are now a week earlier in spring and freezes begin a week later. Sea levels are also expected to rise by 90–880 mm over the next century, mainly from melting glaciers and expanding seawater. Warmer ocean water may result in more intense and frequent tropical storms and hurricanes.
- Coral reefs are highly sensitive to small changes in water temperature. In 1998, they suffered the worst recorded bleaching or die-off in response to stress, with some areas seeing bleach rates of 70%. Experts expect these sorts of events to increase in frequency and intensity in the next 50 years as sea temperatures rise.
- Effects on biodiversity: The wildlife and species that cannot survive in warmer environments may become extinct. Human health is also at risk, as global climate change may result in the spreading of certain diseases such as malaria, the flooding of major cities, a greater risk of heat stroke for individuals, and poor air quality.

Industrialization, deforestation, and pollution have greatly increased atmospheric concentrations of water vapor, carbon dioxide, methane, and nitrous oxide, all GHGs that cause global warming/climate change. Solomon et al. (2007) reported that global warming could lead to large-scale food and water shortages and have consequent catastrophic effects on wildlife.

The IPCC report (Solomon et al. 2007) concluded that *"The global increases in carbon dioxide concentration are due primarily to fossil fuel use and land-use change, while those of methane and nitrous oxide are primarily due to agriculture."* The report goes on to note that these findings come with a *"very high confidence rate* [words emphasized in italics in the report summary] *that the globally averaged net effect of human activities since 1750 has been one of warming."*

Primary sources of GHG in the context of production of fibers include the following:

1. For glass fibers
   a. Production energy used in glass melting and spinning
   b. Production energy used in fiber forming and curing
   c. Emissions from glass melting, VOCs, raw material particles, and small amounts of CO, NO_x, SO_x, and fluorides
2. For natural fibers
   a. Production energy used to power agricultural equipment
   b. Production energy used to produce and apply fertilizers and pesticides
   c. Releases of CO_2 from decomposition and oxidation of soil organic carbon (SOC) following soil disturbance
   d. CO_2 and CH_4 (methane) from retting
   e. Production energy used in fiber processing in addition to emissions resulting from transport for both types of fiber

The Stern Review (2007) lists fertilizer manufacture as the fourth most energy intensive industry with energy consuming 13.31% of total costs (after electricity production and distribution: 26.70%; gas distribution: 42.90%; and refined petroleum: 72.83%). The U.K. production sectors were also ranked in terms of carbon intensity by detailed case study considering direct and indirect carbon costs applied to various fossil fuel inputs (oil, gas, and coal) and traced through production process to final goods prices by using an illustrative carbon price of £70 per tonne of carbon. The manufacture of fertilizer is ranked as the fifth in 123 U.K. production sectors at 4.61% immediately behind cement, lime (also used in agriculture), and plaster at 9% and its use produces both methane and nitrous oxide emissions.

Smil (2001) suggested that while there have been significant reductions in the energy required for industrial fixation of ammonia, the energy costs remain very high (27GJ/t NH_3 in the most efficient plants operating in the late 1990s versus >80 GJ/t NH_3 before 1955). Globally, around 1.3% of all energy produced is used for fertilizers. da Silva and Kulay (2005) performed an environmental comparison between two phosphate fertilizers, fused magnesium phosphate (FMP) and triple superphosphate (TSP), used in Brazil with six of the ISO/TR 14047:2003 (2003) EICF (i.e., excluding NRADP and POCP). In both cases, energy was the main negative environmental impact arising from the extensive electricity dependence for FMP and from transport distances for the TSP. Eutrophication was also an issue due to phosphate losses during the manufacture of FMP or to leaching of the phosphogypsum by-product in the case of TSP.

Abrahams (2002) stated that there "is an appreciable flux of CO_2 from the oxidation of SOM, while soils are also important sources of the GHGs CH_4 and N_2O." The management of the land and any future climatic warming may increase these emissions.

Lal (2004) reviewed the available information on energy use in farm operations and converted the data into kilograms of carbon equivalent (kg CE). The kg CE value is directly related to the rate of enrichment of atmospheric CO_2. The energy use in irrigation is the highest among the farm operations. Lift heights obviously depend on the water depth and the energy differs with the pumping pressures. The values quoted, for water from deep wells used for surface irrigation, include the energy in terms of electricity, diesel, gasoline, natural gas, LPG, and installation costs.

West and Marland (2002) reported that average data for the United States suggest that conversion from conventional tillage to no till will result in sequestration of 337 ± 108 kg C/ha y in agricultural soils to a depth of 300 mm. Furthermore, following such a change, the total change in flux of CO_2 to the atmosphere on nonirrigated crops is expected to be about 368 kg C/ha y.

8.4.6 DEPLETION OF RESOURCES

ISO/TR 14047:2003(E) (2003) includes both abiotic (nonbiological) and biotic resources within this category. Van Oers et al. (2002) considered abiotic resources to include both nonrenewable and renewable resources and defined an abiotic depletion potential (depletion of availability) for nonrenewable resources as "not replenished or broken down by geologic forces within a period of 500 years." They divide abiotic resources into the following three categories:

1. Deposits (resources that are not regenerated within human lifetimes, for example, minerals, sediments, clays, and fossil fuels)
2. Funds (resources that can be regenerated within human lifetimes, for example, groundwater and some soils)
3. Flows (resources that are constantly regenerated, e.g., solar energy, wind and river water)

They suggest that it "is debatable whether all the three types of abiotic resource can or should be aggregated into one measure for abiotic depletion"

They further divide deposits into Groups:

- Primary materials for industrial processing, including both (1) the atomic elements and (2) compounds (called "configurations" in their paper, which have specific physical–chemical composition, e.g., silicon oxide, feldspar, and gypsum)
- Primary materials for building applications (e.g., stone and construction sand) and energy carriers (e.g., oil and natural gas)

Group I materials may have many heterogeneous functions (e.g., transparency or electrical conductivity), including some not anticipated by the current generation that may arise from future technological developments. Groups II and III can be regarded as "substitutes": if one form is scarce, then another material can be used (e.g., brick as a replacement for stone or alloys used as structural materials can be replaced by composite materials containing polymer and resins).

Van Oers et al. (2002) suggested that reserves may be found in nature, in the economy (e.g., within consumer goods), and in landfills. Materials in the economy (e.g., copper in electrical wires) can be an appropriate source of materials, minimizing the depletion of primary stocks, subject to economic conditions. They redefine abiotic resource depletion as "the decrease of availability of *functions* of resources, both in the environment *and the economy*." Materials in landfill may be a richer source of particular elements than those in nature but may be less accessible as a resource due to technological and economic conditions or due to the potential for greater environmental impacts than when extracting the elements directly from primary resources. Loss in quality of materials should be considered as a loss of function.

Sand used in glass fiber production is naturally occurring, and although abundant, it is a nonrenewable resource that should be conserved by reduced use (Waste & Resources Action Programme 2006). Both quartz and feldspar are used in the manufacture of glass, and feldspar is used in the manufacture of fertilizers. The ultimate reserves of these minerals within the top 1 km of the Earth's crust are 5×10^{20} and 1×10^{20} kg, respectively (van Oers et al. 2002). The relative contribution (based on the normalization data for the global extraction of elements and compounds) to the depletion of abiotic resources is effectively zero.

Depletion of biotic resources is a complex issue closely related to ecotoxicity. At its most dramatic, this issue manifests itself as loss of biodiversity and in the limit as the extinction of plant and animal species. However, as with abiotic resources, loss in quality of life forms should be considered as one part of the depletion of the total resource; reduced fertility of a species may ultimately lead to extinction when other external factors combine to exert undue pressure. The direct loss of habitat is associated with some dredging operations in sand extraction in production of glass fiber. Dredging activities are known to affect the reproductive success of some species, such as herring, which are known to spawn on gravel sediments in a very few, highly specific locations (Waste & Resources Action Programme 2006). They are thus vulnerable to the impacts of aggregate extraction. Clearing of extraction areas and also the operations as a whole for the on-land processes significantly reduce habitats and biodiversity for the duration of the operations.

Soil consists of a mixture of small particles of various minerals together with organic materials and micro-organisms. Soil also contains water and air in variable amounts. The proportions of each component vary with geographical location and over time. The fertility of the soil can also be influenced by the presence of larger animals (from earthworms to burrowing mammals). Soil depletion hence crosses the boundary between abiotic and biotic resources. Intensive farming using inorganic fertilizers (mainly of nitrogen, phosphorus, and potassium, often referred to as NPK) with some lime (Ca) and iron (Fe) may not replace key trace elements in the soil. Herbicides and insecticides may inadvertently affect beneficial organisms (e.g., plants with nitrogen-fixing nodules on the roots or microorganisms).

Zwerman and De Haan (1973) reviewed the significance of the soil in environmental quality improvement outlining the plant and animal ecology and the impact of agriculture and industry upon them. They discussed remedial measures necessary to maintain and/or improve this environment. They stated that data exist to show that overfertilization actually decreases crop yield under a wide range of conditions.

O'Sullivan and Simota (1995) reviewed the problems of combining soil compaction models with crop production and environmental impact models. Håkansson and Lipiec (2000) reviewed the use of relative bulk density values in the study of soil structure and compaction. The "degree of compactness" was found to be more useful than bulk density or porosity parameters when studying biological effects of soil compaction. This parameter facilitates modeling of soil and crop responses to machinery traffic.

Reeves (1997) reviewed the lessons learnt from long-term continuous cropping systems and found that maintenance and improvement of soil quality was critical in the context of sustaining agricultural productivity and environmental quality for future generations. Even with crop rotation and manure additions, continuous cropping results in a decline in SOC—a key soil quality indicator. Loveland and Webb (2003) undertook a review of the critical level of SOM (with an equivalence of 2% SOC $\approx$ 3.4% SOM) in agricultural soils in temperate regions but their analysis proved inconclusive.

Reeves (1997) suggested that long-term conservation tillage (practices that reduce losses of soil and water when compared with conventional unridged or clean tillage) can sustain or actually increase SOC in intensive cropping systems within climatic limits. Agronomic productivity and economic sustainability have a more critical requirement for sound rotation practices in conservation tillage systems relative to conventional tillage systems. Sturz et al. (1997) observed that conservation tillage tends to concentrate plant debris and consequently microbial biomass in the top 50–150 mm of soil and could promote the survival of pathogens in humid climates. They reviewed plant diseases, pathogen interactions, and microbial antagonisms in these circumstances and found that while conventional wisdom still favors the expectation of increased plant disease development, the literature is contradictory in respect of promotion or suppression of the problem. Rasmussen (1999) reviewed the impact of ploughless soil tillage (PST) on yield (wheat, rape, and potatoes) and soil quality in the four Scandinavian countries. PST resulted in a decrease in the volume of macropores (drainable pores) and an increase in the volume of medium (water holding) pores with no significant effect on small (nonavailable water) pores. PST resulted in reduced infiltration of air and water, more plant residues on or near the soil surface (hence higher water content in the upper soil layer, lower evapotranspiration, lower soil temperature, and more stable soil aggregates), increased activity and biomass of some earthworms, and reduced erosion. There was also a long-term reduction in soil pH.

Karlen et al. (2001) defined a framework for soil quality that requires identification of critical soil functions, selection of meaningful indicators for those functions, development of appropriate scoring functions to interpret the indicators for various soil resources, and combination of that information into values that can be tracked over time to determine whether the soil resources are being sustained, degraded, or aggraded. Their review provides background for land managers, resource conservationists, ecologists, soil scientists, and others seeking tools to help ensure that land-use decisions and practices are sustainable.

8.4.7 OZONE DEPLETION

Ozone is formed and depleted naturally in the Earth's stratosphere (between 15 and 40 km above the Earth's surface). As the ozone in the stratosphere is reduced, more of the ultraviolet rays in sunlight can reach the Earth's surface affecting the health of humans, animals, and plants (Solomon 1999). Large ozone depletions in the stratospheric zone were first reported by Farman et al. (1985). The depletion of the global ozone layer emerged as one of the major global scientific and environmental issues of the twentieth century.

Halocarbon compounds are persistent synthetic halogen containing organic molecules that can reach the stratosphere leading to more rapid depletion of the ozone. Chlorofluorocarbons (CFCs) are a family of nonreactive, nonflammable gases and volatile liquids, which were widely used in refrigerators, air-conditioners, and spray cans. The Montreal Protocol banning CFCs was signed by leading industrial nations in 1987, based on negotiations started between European–Scandinavian countries and the United States over CFCs in aerosol sprays in 1983. The protocol has gone through a series of revisions (each one named after the city where the revision committee met) as new information from science and industry has become available. Scientists believe that the dramatic reduction in the production of CFCs will eventually reduce the ozone hole and a complete recovery is possible with time (Connor 2000; Sparling 2001).

Ravishankara et al. (2009) reported that anthropogenic N_2O is now the single most important ozone-depleting emission and is expected to remain the largest throughout the twenty-first century. N_2O is unregulated by the Montreal Protocol. Limiting future N_2O emissions would enhance the recovery of the ozone layer from its depleted state and would also reduce the anthropogenic forcing of the climate system.

The United Nations Environment Program following the Scientific Assessment of Ozone Depletion 2010 report (van Mael et al. 2010) highlights the effects of climate change on the ozone layer, as well as the impact of ozone changes on the Earth's climate. It also claims that the Montreal Protocol is a success and it has protected the stratospheric ozone layer from much higher levels of depletion by phasing out production and consumption of ozone-depleting substances. Changes in climate are expected to have an increasing influence on stratospheric ozone in the coming decades, and these changes derive principally from the emissions of long-lived GHGs, mainly carbon dioxide, associated with human activities.

The key findings on the ozone layer from the scientific assessment were

- Over the past decade, global, Arctic, and Antarctic ozone is no longer decreasing but is not yet increasing.
- As a result of the phase-out of ozone-depleting substances under the Montreal Protocol, the ozone layer outside the Polar Regions is projected to recover to its pre-1980 levels some time before the middle of this century.

- In contrast, the springtime ozone hole over the Antarctic is expected to recover much later.
- The impact of the Antarctic ozone hole on surface climate is becoming evident, leading to important changes in surface temperature and wind patterns.
- It is reaffirmed that at mid-latitudes, surface UV radiation has been about constant over the last decade.
- In Antarctica, large UV levels continue to be seen when the springtime ozone hole is large.

8.4.8 PHOTOCHEMICAL OXIDANTS

Photochemical ozone formation results from the degradation of VOCs in the presence of sunlight and the oxides of nitrogen (NO_x). Excess ozone can lead to damaged plant leaf surfaces, discoloration, reduced photosynthetic function, and ultimately death of the leaf and finally the whole plant. In animals, it can lead to severe respiratory problems and eye irritation.

8.5 LIFE CYCLE INVENTORY AND LIFE CYCLE IMPACT ASSESSMENT

LCI analysis is a process of quantifying energy and raw material requirements, atmospheric emissions, emissions into air, co-products, waste, and other releases for the entire life cycle of the product.

Diener and Siehler (1999) reported values for the energy used to produce either glass fiber yarn or a glass fiber mat as an accumulation of the energy in raw materials, mixture, transport, melting, spinning and mat production (Table 8.1). These data were subsequently used by Joshi et al. (2004) to compare natural fiber composites and glass fiber composites from an environmental view point. The educational software CES 2010 EDUPACK (Granta Materials Inspiration 2010a) reports embodied energy in the primary production of E-glass (0.4–12 μm monofilament) and the CO_2 footprint (Table 8.1). The CES report does not clearly define the reinforcement format of the glass fiber. Estimates of the embodied energy in the primary production of flax in the unwoven state are also included in Table 8.1 (Granta Materials Inspiration 2010b).

Dissanayake (2011) compiled an LCI for flax fibers, assuming a yield of fiber in a form ready for use as the reinforcement in composite at 1 tonne/ha. The analysis was separated into three scenarios based on different tillage and retting methods as follows:

Scenario 1: No-till and water retting (i.e., lowest energy)
Scenario 2: Conservation till and stand/dew retting (i.e., intermediate energy)
Scenario 3: Conventional till and bio-retting (i.e., highest energy)

Data are compiled for both flax fiber sliver (pre-spun) and yarn (post-spun) production as shown in Table 8.2. As recommended by Ekvall and Finnveden, all burdens were assigned to the fiber as the primary product. The published values (Diener and

Siehler 1999) for glass fiber reinforcements are 31.7 GJ/tonne for continuous fiber (which is equivalent to a yarn) and 54.7 GJ/tonne for mat (which is equivalent to the sliver). Based on the energy analysis, flax sliver from the low-energy route has an embodied energy directly comparable to the above value for glass fiber mat. Continuous glass fiber reinforcement has a significantly lower energy requirement than spun flax yarn.

TABLE 8.1
CO_2 Footprint and Energy Usage/Embodied Energy in Production of Fiber Products (Field to Factory Gate)

Glass Fiber Footprint (kg CO_2/kg)	Glass Fiber (GJ/tonne)	Glass Fiber Mat (GJ/ tonne)	References
	31.7	54.7	Diener and Siehler (1999)
2.65	45		Le Duigou et al. (2011)
4.26–4.71	67.7–74.9		CES EduPack 2010 (Granta 2010a)
Flax Fiber Footprint (kg CO_2 eq./kg)	**Flax/Jute Fiber (GJ/tonne)**	**Flax Fiber Mat (GJ/tonne)**	
	−19.5		Heat of combustion of natural fibers in Labouze et al. (2007)
−1.45	4.4		Scutched flax in Le Duigou et al. (2011)
−2.4	3.75–8.02		Jute fiber cultivation (excluding field labor, retting, and decortication) in van Dam and Bos (2004)
−1.4	11.7		Hackled flax in Le Duigou et al. (2011) allocated by mass
4.27–4.72	67.9–75.1		CES EduPack 2010 (Granta 2010b)
	80.4 (yarn)	54.2 (sliver)	Dissanayake (2011); full allocation to product

TABLE 8.2
LCI Data for Three Scenarios in Flax Fiber Sliver and Yarn Production

	Scenario 1		Scenario 2		Scenario 3	
	Sliver	Yarn	Sliver	Yarn	Sliver	Yarn
Agricultural operations	5	5.1	12.6	13.3	6.5	6.8
Fertilizer and pesticides	37.5	39.2	81.4	84.6	33.3	34.5
Fiber processing	11.7	36.1	19.1	43.7	79.5	106.6
Total (GJ/tonne of sliver/yarn)	54.2	80.4	113.1	141.6	119.3	147.9

Source: Dissanayake, N.P.J. (2011) *Life Cycle Assessment of flax fibres for the reinforcement of polymer matrix composites*, PhD thesis, University of Plymouth: Plymouth. pp. 1–264. http://pearl.plymouth.ac.uk/handle/10026.1/969.

Dissanayake (2011) extracted LCIA data for glass fiber production from EcoInvent v2.0 for 6 EICF (GWP, AP, EP, HTP, ODP, and POCP) and the values were normalized and plotted against the LCIA results of the three scenarios of flax fiber sliver production. The data are presented as a radar plot in Figure 8.1. These values in EcoInvent are mainly derived from CML 2001 and for glass fiber production in a European situation.

Scenario 1 used the least energy in the production of flax sliver/yarn while scenario 3 reported the highest energy use, as a consequence of the definitions of the respective scenarios. The GWP, ODP, and POCP are at their lowest in scenario 1 and the lowest AP, EP, HTP, and AqTP were reported from scenario 3. The spinning process inevitably raises the embodied energy for yarn. As a consequence, the substitution of glass fibers by natural fibers is truly dependent on the chosen reinforcement form and associated process. Flax sliver as reinforcement is comparable in energy terms with the glass fiber mat when no-till and water retting is used in the production.

Le Duigou et al. (2011) conducted an environmental impact analysis (Table 8.3) on French flax fibers using a different set of underlying assumptions to those of Dissanayake et al. for U.K. fibers but concluded that "without the allocation procedure the results from the two studies would be similar." The key differences were

- U.K. plants desiccated at mid-point flowering but French plants allowed to set seed
- U.K. yield only 6000 kg/ha but French yield 7500 kg/ha at harvest
- U.K. study excluded photosynthesis and CO sequestration
- Higher level of nuclear power in the French energy mix
- U.K. study allocated all burdens to fiber, whereas French study allocated on the basis of mass of product and co-products

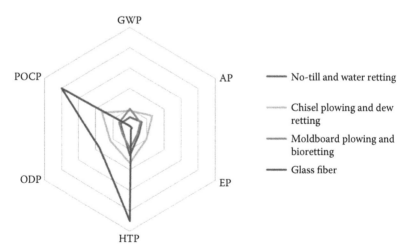

FIGURE 8.1 Representation of Life Cycle Impact Assessment for glass fibers (EcoInvent) versus flax sliver production. AP, acidification potential; HTP, human toxicity potential; EP, eutrophication potential; GWP, global warming potential; ODP, ozone depletion potential; POCP, photochemical oxidants creation potential.

TABLE 8.3
Environmental Impacts for the Production of Fibers

Impact Category	Glass Fibers	Scutched Flax with Photosynthesis (by Mass)	Hackled Flax with Photosynthesis (by Mass)	Hackled Flax without Photosynthesis (by Mass)	Hackled Flax (by Economic Value)	Units
Abiotic depletion	19	1.3	1.7	1.7	6.5	$\times 10^{-3}$ kg Sb eq/kg
Acidification	16	1.8	2.2	2.1	8.5	$\times 10^{-3}$ kg SO_2 eq/kg
Eutrophication	1.2	1.4	1.4	1.4	6.2	$\times 10^{-3}$ kg PO_4 eq/kg
Global warming (GWP100)	2.65	−1.45	−1.4	0.3	−6.4	kg CO_2 eq/kg
Ozone layer depletion (ODP)	200	21	24	24	98	$\times 10^{-9}$ kg CFC-11 eq/kg
Human toxicity	9.1	0.15	0.215	0.21	0.77	kg 1,4-DB eq/kg
Fresh water aquatic ecotoxicity	170	54	59	59	240	$\times 10^{-3}$ kg 1,4-DB eq/kg
Photochemical oxidation	600	58	73	73	270	$\times 10^{-6}$ kg C_2H_4 eq/kg
Terrestrial ecotoxicity	42	2.6	8.7	8.7	22	$\times 10^{-3}$ kg 1,4-DB eq/kg
Nonrenewable energy consumption	45	4.4	11.7	—	—	GJ/tonne
Land use	7	970	850	850	3800	$\times 10^{-3}$ m²/year/kg

Source: Le Duigou, A., Davies, P. and Baley, C. *J. Biobased Mater. Bioenergy*, 5, 153–165, 2011.

8.6 DISCUSSION

There are many environmental impacts that cannot be quantified but still should be considered in a Life Cycle Analysis. The environmental impacts that cannot be measured by established methods include implications for biodiversity caused by agriculture and the appearance of the landscape. These potential environmental impacts can only be discussed qualitatively in environmental assessments. The impacts associated with air and water can have significant effects on the regions adjacent to the source causing many more interrelated impacts that makes the quantification very dependent on the defined scope for the analysis.

Pesticides used in the cultivation of flax may contaminate water and thus impact on biodiversity and humans. Some of these impacts can be measured in terms of eutrophication, acidification, human toxicity, and aquatic toxicity, but the extension of these impacts to loss of aquatic life or contamination of drinking waters is very complex to quantify.

There are many other environmental factors that are not covered in the above eight environmental classification factors such as

1. Noise and vibration arising from farm machinery and glass fiber production process
2. Soil erosion and soil compaction from agricultural methods or sand dredging
3. Water and land use

Soil erosion is mainly caused by the action of water and wind in arable lowland and by frost and animals in the uplands. The main impacts of soil erosion include reduced soil water storage capacity and loss of nutrients affecting soil fertility. Soil erosion is significantly increased where the crops are drilled up and down a slope although this is not normal practice as it results in reduced yield (Skinner et al. 1997). No-tillage has been successfully adopted in the Southeastern United States to control soil erosion and to enhance soil water conservation, especially on the sloping land (Lindwall et al. 1994). Soil compaction can result from the use of heavy machinery in agricultural operations such as plowing, sowing, and harvesting and affects soil biodiversity and soil structure. Soil compaction can also lead to problems such as water logging (European Environment Agency 2006).

Intensification of agriculture (monoculture cropping, soil compaction or erosion, inorganic fertilizer and pest control, agrochemicals leaching to ground and surface waters, and overabstraction of water) can directly result in loss of crop and natural biodiversity (Mozumder and Berrens 2007), for example, loss of grass land, field boundaries, and tree lines.

The photosynthetic CO_2 fixation in fiber plants has a positive effect on the CO_2 balance of the environment. There are several negative contributors in the production phase, such as fertilizers, pesticides, and fossil fuel use in agricultural activities. The stored CO_2 remains locked within the fibers throughout their use phase. The end-of-life disposal method and the durability of the natural fiber product are critical components in assessing the sequestered CO_2 within the plant or fiber as there is a possibility of returning this CO_2 into the atmosphere (Murphy and Norton 2008).

In general, bast crops absorb 1.7–1.9 tonnes of CO_2 per tonne of bast crop cellulose produced (Nimbin Wave 2010). CO_2 sequestration in soil can be increased by avoiding emissions into the atmosphere. Lal (1997) stated that the carbon sequestration in conservation tillage practices is higher than that in the conventional tillage practices.

West and Marland (2002) stated that changing from conventional tillage to no-till enhances the C sequestration in soil and decreases the CO_2 emissions. The no-till/minimum tillage route is the best way to maintain the existing CO_2 storage in soil together with other practices such as efficient use of pesticides, irrigation, and heavy agricultural machinery. Based on the energy analysis, continuous glass fiber reinforcement appears to be superior to spun flax yarn and the gains from substitution of glass fibers by natural fibers are dependent on the chosen reinforcement form with respect to the environmental benefits. According to the Dissanayake LCIA, the production of glass fibers has higher values for ODP and POCP and lower values for AP, EP, and HTP than flax fiber production in any of the three scenarios. There is no agreed methodology for cross comparison of the eight EICF, so the levels of the ODP and POCP might be excessive or negligible relative to the other six EICF.

8.7 CONCLUSIONS

The major environmental impacts associated with flax fiber and glass fiber production cannot be treated in isolation as they are frequently interrelated. Some of the impacts are unseen for several years and the increasing environmental pressures on agriculture means that new environmental techniques need to be adopted.

Environmentally orientated agriculture is the best way forward to reduce the potential environmental impacts, which includes preserving and improving water quality and resources, controlling soil erosion/compaction, preserving physical/chemical/biological soil quality and air quality, preserving biodiversity and reducing energy consumption by using renewable energy sources, etc. Environmental credentials for flax fiber production can be improved by adopting no-till for preparing the ground, using organic agrochemicals, biological control of pests with traditional water retting, and using sliver as reinforcement rather than yarn.

Alternative bast fibers such as hemp and nettle not only require less agrochemical input per tonne of green stem but may also have higher yield per hectare and higher percentage of long fiber. Therefore, other bast fibers could be "greener" than flax fibers and environmental benefits can be further improved by adopting sustainable agriculture. Considering the sliver as reinforcement in polymer matrix composites will make the natural fibers superior to man-made glass fibers.

ACKNOWLEDGMENTS

NPJD is grateful to University of Plymouth, School of Marine Science and Engineering for their contribution to research degree fees. The authors are grateful to Martin Ansell, Stephen Grove, Wayne Hall, Ross Pomeroy, and Miggy Singh for their respective valuable inputs to this study.

GLOSSARY

(from the Concise Oxford Dictionary [1977] except where stated otherwise)

Aggrade: fill and raise the level of (the bed of a stream) by deposition of sediment (Farlex 2012)

Antagonism: active interference with the action of another

Benthic: from the depths of the sea

Ciliate: with short hair like vibrating structures on the surface

Composting: decay of organic materials within a layered structure of (usually) plant waste

Cultivars: plant varieties produced by plant breeding (Bartle 2007)

Desiccation: application of a hygroscopic (moisture-absorbing) substance as a drying agent

Enzyme: a protein catalyst of a specific biochemical reaction (e.g., cellulase or pectinase digest cellulose or pectin, respectively)

Feces: waste matter discharged from bowels

Filament: a fiber of indefinite length (Farnfield and Alvey 1975)

Flagellate: protozoa with one or more long lash-like appendages

Fumigant: fumes used to disinfect or purify

Fungi: unicellular, multicellular, or multinucleate nonphotosynthetic organisms that feed on organic matter, including mushrooms, toadstools, yeasts, and molds

Handle: the subjective assessment of a textile material obtained from the sense of touch (Farnfield and Alvey 1975)

Macrophyte: an individual alga large enough to be seen easily with the unaided eye

Mineralization: partial or complete conversion to a mineral

Mitochondria: specialized structures within cells, where genetic material is contained within a distinct organelle, containing enzymes for respiration and energy production (Bartle 2007)

Organelle: organized or specialized structure within a cell

Pesticide: substance for destroying pests: used here as a generic term to include fungicides, herbicides, and insecticides

Protozoa: microscopic unicellular organisms

Regurgitant: swallowed food returned to the mouth

Scutching: decortication of flax

Shive: thin piece or fragment; specifically, a piece of the central woody part (pith) of the flax stem removed by the operation of decortication (Bartle 2007)

Spinning/Spun: the process, or processes, used in the production of yarns or filaments (Farnfield and Alvey 1975)

Symbionts: organisms living in symbiosis, that is to their mutual advantage

Tillage: the preparation of land for cropping (Bartle 2007)

Transgenic: plant (or animal) with genetic material introduced from another species

Twist: the spiral deposition of the components of a yarn, which is usually the result of relative rotation of the extremities of the yarns (Farnfield and Alvey 1975)

Windrows (composting): system of composting involving the aeration of horizontally extended piles generally 1.5–3 m in height with length limited by the size of the composting site. Aeration can be achieved by mechanical turning and/or the delivery of air from the base of the windrow (New South Wales Government Environment and Heritage 2012)

Windrows (flax): part of the harvesting process during which rows of mown (cut) and drying plant material lie in the field awaiting collection by baling and subsequent transportation (Bartle 2007)

Xenobiotic: chemicals, especially pesticides, for elimination of specific biological species

Yarn: a product of substantial length and relatively small cross section of fibers and or filaments with or without twist (Farnfield and Alvey 1975).

REFERENCES

Abrahams, P.W. 2002. Soils: their implications to human health. *The Science of the Total Environment* 291: 1–32.

Amoros, I., Alonso, J.L., Romaguera, S. and Carrasco, J.M. 2007. Assessment of toxicity of a glyphosate-based formulation using bacterial systems in lake water. *Chemosphere* 67: 2221–2228.

ASTM International. 2003. *Standard Test Method for Determining Aerobic Biodegradation of Plastic Materials under Controlled Composting Conditions.* West Conshohocken, PA: ASTM D5338-98.

ASTM International. 2004. *Standard Specification for Compostable Plastics.* West Conshohocken, PA: ASTM D6400-04.

Azapagic, A., Emsley, A. and Hamerton, I. 2003. *Polymers, the Environment and Sustainable Development.* Chichester: John Wiley & Sons.

Azapagic, A., Perdan, S. and Clift, R. 2004. *Sustainable Development in Practice—Case Studies for Engineers and Scientists.* Chichester: John Wiley & Sons.

Banks, M. 2012. The cycle of life. *Parliament Magazine* 23–26.

Bartle, I. 2007. Private communication with the LINK Non-Food Crops Programme Co-ordinator.

Bradshaw, E.W., Rasmussen, P. and Odgaard, B.V. 2004. Mid- to late-Holocene land-use change and lake development at Dallund Sø, Denmark: synthesis of multiproxy data, linking land and lake. *The Holocene* 15: 1152–1162.

British Standard. 2000. *Packaging: Requirements for Packaging Recoverable through Composting and Biodegradation - Test Scheme and Evaluation Criteria for the Final Acceptance of Packaging.* BS EN 13432:2000. ISBN 0-580-36765-7.

British Standard. 2011. *BS8905:2011 Framework for the Assessment of the Sustainable Use of Materials—Guidance.*

Carson, R. 1962. *Silent Spring.* London: Penguin Classics. ISBN 978-0-14118494-4.

Concise English Dictionary. 1977. Oxford University Press: Oxford.

Connor, S. 2000. *CFC ban "should close up hole in ozone layer" Independent, London.* http://www.independent.co.uk/news/science/cfc-ban-should-close-up-hole-in-ozone-layer-626298.html, accessed on 23 January 2013.

Corbière-Nicollier, T., Gfeller Laban, B., Lundquist L., Leterrier, Y., Månson, J.-A.E. and Jolliet, O. 2001. Life cycle assessment of biofibres replacing glass fibres as reinforcement in plastics. *Resources, Conservation and Recycling* 33: 267–287.

Correll, D.L. 1998. The role of phosphorus in the eutrophication of receiving waters: a review. *Journal of Environmental Quality* 27: 261–266.

Crews, T.E. and Peoples, M.B. 2004. Legume versus fertilizer sources of nitrogen: ecological trade-offs and human needs. *Agriculture, Ecosystems & Environment* 102: 279–297.

Cuttle, S.P., Macleod, C.J.A., Chadwick, D.R., Scholefield, D., Haygarth, P.M., Newell-Price, P., Harris, D., Shepherd, M.A., Chambers, B.J. and Humphrey, R. 2006. *An Inventory of Methods to Control Diffuse Water Pollution from Agriculture (DWPA)*, IGER/ADAS Report prepared as part of Defra Project ES0203.

da Silva, G.A. and Kulay, L.A. 2005. Environmental performance comparison of wet and thermal routes for phosphate fertilizer production using LCA - a Brazilian experience. *Journal of Cleaner Production* 13: 1321–1325.

Diener, J. and Siehler, U. 1999. Okologischer vergleich von NMT-und GMT-Bauteilen. *Die Angewandte Makromolekulare Chemie* 272: 1–4.

Dissanayake, N.P.J. (2011) *Life Cycle Assessment of flax fibres for the reinforcement of polymer matrix composites*, PhD thesis, University of Plymouth: Plymouth. pp. 1–264. http://pearl.plymouth.ac.uk/handle/10026.1/969.

Ekvall, T. and Finnveden, G. 2001. Allocation in ISO14041—a critical review. *Journal of Cleaner Production* 9: 197–208.

Eriksson, E., Baun, A., Scholes, L., Ledin, A., Ahlman, S., Revitt, M., Noutsopoulos, C. and Mikkelsen, P.S. 2007. Selected stormwater priority pollutants—a European perspective. *Science of the Total Environment* 383: 41–51.

European Environment Agency. 2006. *How Much Bioenergy Can Europe Produce Without Harming the Environment?* Report No 7/2006, Copenhagen DK: EEA. pp. 1–67. ISBN 92-9167-849-X.

Farlex. 2012. http://www.thefreedictionary.com/aggrade, accessed on 23 January 2013.

Farman, J.C., Gardiner, B.G. and Shanklin, J.D. 1985. Large losses of total ozone in Antarctica reveal seasonal ClO_x/NO_x interaction. *Nature* 315: 207–210.

Farnfield, C.A. and Alvey, P.J. 1975. *Textile Terms and Definitions*, 7th ed. Manchester: The Textile Institute.

George, J., Sreekala, M.S. and Thomas, S. 2001. A review on interface modification and characterization of natural fiber reinforced plastic composites. *Polymer Engineering and Science* 41: 1471–1485.

Gillis, J. 2011. Heavy rains linked to humans. *The New York Times*. http://www.nytimes.com/2011/02/17/science/earth/17extreme.html, accessed on 23 January 2013.

Goldstein, N. 2005. Source separated organics as feedstock for digesters. *BioCycle*. 46(8): 42.

Govaerts, B., Verhulst, N., Castellanos-Navarrete, A., Sayre, K.D., Dixon, J. and Dendooven, L. 2009. Conservation agriculture and soil carbon sequestration: between myth and farmer reality. *Critical Reviews in Plant Science*. 28: 97–122.

Granta Materials Inspiration. 2010a. *Glass, E Grade (0.4-12 micron monofilament, f), Information from Educational Software*—CES 2010 EduPack.

Granta Materials Inspiration. 2010b. *Flax, Information from Educational Software*—CES 2010 EduPack.

Greenham, L. and Walsh, P. 2004. Carbon dioxide detectors for health and safety applications. *Petro Industry News*, 34–35.

Håkansson, I. and Lipiec, J. 2000. A review of the usefulness of relative bulk density values in studies of soil structure and compaction. *Soil and Tillage Research* 53: 71–85.

Harwood, J., McCormick, P., Waldron, D. and Bonadei, R. 2008. Evaluation of flax accessions for high value textile end uses. *Industrial Crops and Products* 27: 22–28.

Hassol, S.J. 2004. *Impacts of a Warming Arctic—Arctic Climate Impact Assessment*. New York: ACIA. pp. 1–160. http://amap.no/workdocs/index.cfm?dirsub=%2FACIA%2Foverview, accessed on 23 January 2013.

Heap, I.M. 1997. The occurrence of herbicide-resistant weeds worldwide. *Pesticide Science* 51(3): 235–243.

Henfaes Research Centre. 2004. *Flax and Hemp Project: Guidelines for Growing Flax*. Bangor, Gwynedd: University of Wales, 1–2.

Horne, A.J. and Goldman, C.R. 1994. *Limnology*, 2nd ed. New York: McGraw-Hill.

ISO (International Organisation for Standards). 1999. *ISO 14852: Determination of the ultimate aerobic biodegradability of plastic materials in an aqueous medium—Method by analysis of evolved carbon dioxide.* Geneva, Switzerland: ISO.

ISO (International Organisation for Standards). 2003. *Technical Report: Environmental Management—Life Cycle Assessment—Examples of application of ISO 14042*, Geneva, Switzerland: ISO/TR 14047.

ISO (International Organisation for Standards). 2005. *Determination of the Ultimate Aerobic Biodegradability and Disintegration of Plastic Materials under Controlled Composting Conditions: Method by Analysis of Evolved Carbon Dioxide—Part 1: General Method.* Geneva, Switzerland: ISO 14855-1:2005. ISBN 0-580-32875-9.

ISO (International Organisation for Standards). 2006. *Environmental Management—Life Cycle Assessment—Principles and frameworks.* Geneva, Switzerland: ISO 14040:2006.

Jana, S., Chakrabarty, N.R. and Sarkar, S.C. 2001. Removal of carbon dioxide from biogas for methane generation. *Journal of Energy in Southern Africa* 12: 412–414.

Joshi, S.V., Drzal, L.T., Mohanty, A.K. and Arora, S. 2004. Are natural fiber composites environmentally superior to glass fiber reinforced composites? *Composites Part A: Applied Science and Manufacturing* 35: 371–376.

Kabir, M.M., Wang, H., Lau, K.T. and Cardona, F. 2012. Chemical treatments on plant-based natural fibre reinforced polymer composites: an overview. *Composites Part B: Engineering* 43: 2883–2892.

Kalia, S., Kaith, B.S. and Kaur, I. 2009. Pretreatments of natural fibers and their application as reinforcing material in polymer composites—a review. *Polymer Engineering and Science* 49: 1253–1272.

Karlen, D.L., Andrews, S.S. and Doran, J.W. 2001. Soil quality: current concepts and applications. *Advances in Agronomy* 74: 1–40.

Kern, M. 2002. Food, feed, fibre, fuel and industrial products of the future: challenges and opportunities—understanding the strategic potential of plant genetic engineering. *Journal of Agronomy and Crop Science* 188: 291–305.

Labouze, E., Le Guern, Y. and Petiot, C. 2007. *Analyse de cycle de vie comparée d'une chemise en lin et d'une chemise en coton—Bio Intelligence Service SAS Final report after critical review.* http://issuu.com/linenandhemp/docs/rapport_acv, accessed on 23 January 2013.

Lal, R. 1997. Residue management, conservation tillage and soil restoration for mitigating greenhouse effect by CO_2-enrichment. *Soil and Tillage Research* 43: 81–107.

Lal, R. 2004. Carbon emission from farm operations. *Environment International* 30: 981–990.

Le Duigou, A., Davies, P. and Baley, C. 2011. Environmental impact analysis of the production of flax fibres to be used as composite material reinforcement. *Journal of Biobased Materials and Bioenergy* 5: 153–165.

Lebret, K., Thabard, M. and Hellio, C. 2009. Chapter 4: Algae as marine fouling organisms: adhesion damage and prevention. In: *Advances in Marine Antifouling Coatings and Technologies.* Eds. Hellio, C. and Yebra, D.M. Cambridge, MA: Woodhead Publishing Limited, 80–112.

Lewington, A. 2003. *Plants for People.* London: Eden Project Books/Transworld.

Li, X., Tabil, H.G. and Panigrahi, S. 2007. Chemical treatments of natural fibre for use in natural fibre-reinforced composites: a review. *Journal of Polymers and the Environment* 15: 25–33.

Lindwall, C.W., Larney, F.J., Johnston, A.M. and Moyer, J.R. 1994. Crop management in conservation tillage systems. In: *Managing Agricultural Residues.* Ed. Unger, P.W. Boca Raton, FL: CRC Press, 185–209.

Loveland, P. and Webb, J. 2003. Is there a critical level of organic matter in the agricultural soils of temperate regions: a review. *Soil and Tillage Research* 70: 1–18.

Malthus, T. 1798. *An Essay on the Principle of Population, as It Affects the Future Improvement of Society with Remarks on the Speculations of Mr. Godwin, M. Condorcet, and Other Writers.* London: Johnson.

Martinez, T.T. and Brown, K. 1991. Oral and pulmonary toxicology of the surfactant used in Roundup herbicide. *Proceedings of the Western Pharmacology Society* 34: 43–46.

Metz, B., Davidson, O., Bosch, P., Meyer, L. and Dave, R. 2007. *Climate Change 2007—Mitigation of Climate Change.* Cambridge, MA: Cambridge University Press.

Moll, S., Skovgaard, M., Schepelmann, P. and Kaźmierczyk, P. 2005. *Sustainable Use and Management of Natural Resources,* EEA Report No 9/2005, ISSN 1725-9177.

Mozumder, P. and Berrens, R.P. 2007. Inorganic fertilizer use and biodiversity risk: an empirical investigation. *Journal of Ecological Economics* 62: 538–543.

Murphy, R. and Bartle, I. 2004. *Biodegradable Polymers and Sustainability: Insights from Life Cycle Assessment.* London: National Non-Food Crops Centre.

Murphy, R.J. and Norton, A. 2008. *Life Cycle Assessment of Natural Fibre Insulation Materials* (Funded by DEFRA). London: National Non Food Crop Centre (NNFCC), 1–79.

National Geographic. 2007. *Global Warming Fast Facts.* http://news.nationalgeographic.com/news/2004/12/1206_041206_global_warming.html, accessed on 23 January 2013.

Net Composites. (undated). *Glass Fibre/Fiber.* http://www.netcomposites.com/guide/glass-fibrefiber/32, accessed on 23 January 2013.

New South Wales Government Environment and Heritage. 2012. *Environmental Issues: Glossary.* http://www.environment.nsw.gov.au/waste/envguidlns/compostingglossary.htm, accessed on 23 January 2013.

Nimbin Wave. 2010. *Cannabis Carbon Credits.* http://nimbinwave.com/64, accessed on 23 January 2013.

O'Sullivan, M.F. and Simota, C. 1995. Modelling the environmental impacts of soil compaction: a review. *Soil and Tillage Research* 35: 69–84.

Peixoto, F. 2005. Comparative effects of the roundup and glyphosate on mitochondrial oxidative phosphorylation. *Chemosphere* 61: 1115–1122.

Petit, V., Cabridenc, R., Swannell, R.P.J. and Sokhi, R S. 1995. Review of strategies for modelling the environmental fate of pesticides discharged into riverine systems. *Environment International* 21: 167–176.

Pickering, S.J. 2006. Recycling technologies for thermoset composite materials—current status. *Composites Part A: Applied Science and Manufacturing* 37: 1206–1215.

Porritt, J. 2005. *Capitalism as if the World Matters.* London, UK and Sterling, USA: Earthscan.

Rasmussen, K.J. 1999. Impact of ploughless soil tillage on yield and soil quality: a Scandinavian review. *Soil and Tillage Research* 53: 3–14.

Ravishankara, A.R., Daniel, J.S. and Portmann, R.W. 2009. Nitrous Oxide (N_2O): the dominant ozone-depleting substance emitted in the 21st century. *Science* 326: 123–125.

Reddy, K.R., Kadlec, R.H., Flaig, E.H. and Gale, P.M. 1999. Phosphorus retention in streams and wetlands: a review. *Critical Reviews in Environmental Science and Technology* 29: 83–146.

Reeves, D.W. 1997. The role of soil organic matter in maintaining soil quality in continuous cropping systems. *Soil and Tillage Research* 43: 131–167.

Sandermann, H. 2006. Plant biotechnology: ecological case studies on herbicide resistance. *Trends in Plant Science* 11: 324–328.

Schindler, D.W. 1978. Factors regulating phytoplankton production and standing crop in the world's freshwaters. *Limnology and Oceanography* 23: 478–486.

Sensi, A. (undated). *Agriculture and Acidification,* http://ec.europa.eu/agriculture/envir/report/en/acid_en/report.htm, accessed on 23 January 2013.

Sharma, H.S.S., Mercer, P.C. and Brown, A.E. 1989. Review of recent research on retting of flax in Northern Ireland. *International Biodeterioration* 25: 327–342.

Skinner, J.A., Lewis, K.A., Bardon, K.S., Tucker, P., Catt, J.A. and Chambers, B.J. 1997. An overview of the environmental impacts of agriculture in the UK. *Journal of Environmental Management* 50: 111–128.

Smil, V. 2001. *Enriching the World: Fritz Haber, Carl Bosch, and the Transformation of World Food Production*. Cambridge, MA: MIT Press. ISBN 026219449x.

Solomon, S. 1999. Stratospheric ozone depletion: a review of concepts and history. *Review of Geophysics* 37: 275–316.

Solomon, S., Qin, D., Manning, M., Chen, Z., Marquis, M., Averyt, K.B., Tignor, M. and Miller, H.L. 2007. *Summary for Policymakers, in Climate Change 2007: The Physical Science Basis, Group I to the Fourth Assessment Report of the Intergovernmental Panel on Climate Change*. Cambridge, MA, and New York: Cambridge University Press.

Sparling, B. 2001. *Ozone Depletion, History and politics/Ozone History.* http://www.nas.nasa.gov/About/Education/Ozone/history.html, accessed on 18 June 2010, now withdrawn.

Stern, N. 2007. *The Economics of Climate Change - The Stern Review 2007*. Cambridge, MA: Cambridge University Press.

Stoate, C. 2007. The Eye Brook – a multifunctional approach to catchment management. *British Wildlife* 18: 478–486.

Stumm, W. and Morgan, J.J. 1996. *Aquatic Chemistry–Chemical Equilibria and Rates in Natural Waters*. New York: John Wiley.

Sturz, A.V., Carter, M.R. and Johnston, H.W. 1997. A review of plant disease, pathogen interactions and microbial antagonism under conservation tillage in temperate humid agriculture. *Soil and Tillage Research* 41: 169–189.

Summerscales, J. 2013. *Virtual Book on Bast Fibres and Their Composites.* http://www.tech.plym.ac.uk/sme/mats324/bast_book.htm, accessed on 23 January 2013.

Todar, K. (undated). *The Microbial World.* http://textbookofbacteriology.net/themicrobial-world/environmental.html, accessed on 23 January 2013.

Tottle, C.R. 1984. *An Encyclopædia of Metallurgy and Materials*. Plymouth, MN: Macdonald and Evans.

Turner, J.A. 1987. *Linseed Law: A Handbook for Growers and Advisers*. Hadleigh: BASF (UK) Limited.

Turunen, L. and van der Werf, H.M.G. 2006. *Life Cycle Analysis of Hemp Textile Yarn: Comparison of Three Hemp Fibre Processing Scenarios and a Flax Scenario*. France: French National Institute for Agronomy Research.

Turunen, L. and van der Werf, H.M.G. 2008. The environmental impacts of the production of hemp and flax textile yarn. *Industrial Crops and Products* 27: 1–10.

van Dam, J.E.G. and Bos, H.L. 2004. Consultation on natural fibres: the environmental impact of hard fibres and jute in non-textile industrial applications, in ESC-Fibres Consultation no 04/42004, Rome.

van Dolah, H.M. 2000. Marine algal toxins: origins, health effects, and their increased occurrence. *Environmental Health Perspectives* 108S1: 133–141.

van Mael, C.R., Nullis, C. and Nuttal, N. 2010. *New Report Highlights Two-Way Link Between Ozone Layer and Climate Change.* http://www.unep.org/Documents.Multilingual/Default.asp?DocumentID=647&ArticleID=6751&l=en&t=long, accessed on 23 January 2013.

van Oers, L., de Koning, L., Guinée, J.B. and Huppes, G. 2002. *Abiotic Resource Depletion in LCA—Improving Characterisation Factors for Abiotic Resource Depletion*, in Dutch LCA Handbook. The Hague, Netherlands: Directoraat-Generaal Rijkswaterstaat.

von Carlowitz, H. 1713. *Sylvicultura oeconomica, oder haußwirthliche Nachricht und Naturmäßige Anweisung zur wilden Baum-Zucht [Economic Woodland Management: Information and Instructions for the Husbandry of Natural Wild Trees]*. Leipzig: Braun.

Waste & Resources Action Programme. 2006. *An Environmental Impact Assessment of the Use of Recycled Glass as a Tertiary Filter Medium.* WRAP: Banbury. ISBN 1-84405-269-9. http://www2.wrap.org.uk/downloads/Environmental_Impact_Report_-_Final_Report_July_2006.1830565b.3017.pdf, accessed on 23 January 2013.

West, T.O. and Marland, G. 2002. A synthesis of carbon sequestration, carbon emissions, and net carbon flux in agriculture: comparing tillage practices in the United States. *Agriculture, Ecosystems & Environment* 91: 217–232.

Wetzel, R.G. 1983. *Limnology.* 2nd ed. Philadelphia: Saunders College Publishing.

World Commission on Environment and Development. 1987. *Our Common Future* (The Brundtland Report). Oxford: Oxford Paperbacks. ISBN 978-0-19-282080-8.

Zafeiropoulos, N.E. (Ed). 2011. *Interface Engineering of Natural Fibre Composites for Maximum Performance.* Cambridge, MA: Woodhead Publishing.

Zwerman, P.J. and De Haan, F.A.M. 1973. Significance of the soil in environmental quality improvement—a review. *The Science of the Total Environment* 2: 121–155.

9 Effect of Halloysite Nanotubes on Water Absorption, Thermal, and Mechanical Properties of Cellulose Fiber–Reinforced Vinyl Ester Composites

Abdullah Alhuthali and It Meng Low

CONTENTS

9.1   INTRODUCTION

Halloysite nanotubes (HNTs) are derived from naturally deposited aluminosilicate $(Al_2Si_2O_5(OH)_{4n}H_2O)$ and are chemically similar to kaolin (Joussein et al. 2005; Ismail et al. 2008). Structurally, due to mismatch between tetrahedral and octahedral internal components, HNTs take on a cylindrical shape, forming tubes that are typically between 1 and 15 μm in length (Lecouvet et al. 2011). These tubes have dimensions of between 50 and 70 nm for the outer diameters and between 10 and 30 nm for the inner diameters. As the tubes are hollow, they allow HNTs to have very high surface area with a high aspect that promotes excellent interaction between the filler and the matrix (Handge et al. 2010; Liu et al. 2007). Tensile, fracture, and impact strength as well as other mechanical and thermal properties are believed to be dramatically improved when HNTs are added to epoxy, polystyrene, polypropylene, polyvinyl alcohol, and other polymers (Deng et al. 2008; Du et al. 2006; Rooj et al. 2010; Zhao and Liu 2008).

Natural fibers are eco-friendly, commercially viable fillers that have excellent modulus to weight ratios and are capable of forming polymer composites with excellent toughness properties (Jawaid and Abdul Khalil 2011; Liu and Hughes 2008; Sgriccia et al. 2008). Natural fibers are biodegradable unlike some plastics and have a production that is energy-efficient and often less expensive than production with synthetic counterparts (Venkateshwaran et al. 2011). Natural fibers are lighter than synthetic materials and having an excellent modulus to weight ratio are ideal for stiffness-critical designs needed in the construction, automotive, and even aerospace industries (Anuar and Zuraida 2011; Hossain et al. 2011; Kumar et al. 2010). Acoustic damping properties of natural fibers make them suitable for use in components in the internal areas of automobiles (Sgriccia et al. 2008). Compared to synthetic fibers, many natural fibers have elastic modulus and specific modulus comparable or better than synthetic fiber composites (Haq et al. 2008).

So far studies have established that the addition of natural fibers to polymer matrices as a microscale reinforcement material provides toughness and strength to these composites (Akil et al. 2011; Ku et al. 2011). Barrier, thermal, and mechanical properties of polymeric composites can also be improved by adding a low concentration of nanofiller particles as a nanoscale reinforcement material (Bruzaud and Bourmaud 2007; Hossain et al. 2011). The combination of multiscale reinforcement materials in polymer matrix can provide the reinforcement properties at two scales. However, incorporation of nanofiller particles to natural fiber–polymer composites is an area that has not been widely investigated.

The selection of HNTs as the nanofiller to be incorporated in natural fiber–polymer composites for the purposes of development and characterization is a novel proposition and one that it seems has yet to be attempted by the scientific community at this time. This study, therefore, aims to investigate the addition of HNTs to vinyl ester reinforced with recycled cellulose fiber (RCF) and characterize the properties of the resulting composite in terms of water absorption, mechanical and thermal properties, and flammability.

9.2 EXPERIMENT

9.2.1 MATERIALS

RCF in sheet form, grade 80GSM, 100 µm thickness, supplied by Todae Company, NSW, Australia; a general purpose vinyl ester resin, supplied by Fibreglass & Aesin Sales Pty Ltd, WA, Perth, Australia; and ultrafine HNTs with 50.4%; SiO_2, 35.5%; Al_2O_3, 0.25; Fe_2O_3, and 0.05; TiO_2 (wt%), supplied by NZCC, New Zealand, were used in this study.

9.2.2 SAMPLE PREPARATION

Control samples of pure vinyl ester were fabricated to provide the baseline data. The vinyl ester resin (VER) was mixed with 1.0 wt% catalyst (methyl ethyl ketone peroxide [MEKP]) to prepare samples. To ensure that no air bubbles are formed within the matrix, the mixture was slowly and thoroughly mixed, then poured into silicon molds, and left under low vacuum (20 kPa) for 2 hours and later at room temperature for 24 hours to cure. To fabricate eco-composites (VER/RCF) the RCF sheets were dried for 60 minutes at 150°C, then fully soaked in the vinyl ester system, and then laid up in a close silicon mold and pressed under 20-kg load and under vacuum of 60 kPa for 2 hours. The samples were later left to cure for 24 hours at room temperature giving a 40% weight percentage of fibers in these eco-composites. To fabricate eco-nanocomposites the nano-mixtures containing 1%, 3%, and 5% concentration of HNTs were first prepared. The HNTs were dried for 60 minutes at 150°C, followed by mixing with VER for 30 minutes using a high-speed electrical mixer. To remove air bubbles, the mixtures were left under vacuum (60 kPa). To avoid creating air bubbles inside the composites, a catalyst was added and mixed manually. RCF sheets were dried for 60 minutes at 150°C, then fully soaked in the mixtures, and pressed together under 20 kg and under vacuum of 60 kPa for 2 hours. Samples were left to cure at room temperature for 24 hours, giving a 40% weight percentage of fibers in these eco-nanocomposites, and the resultant eco-nanocomposites labeled VER/RCF/1% HNTs, VER/RCF/3% HNTs, and VER/RCF/5% HNTs.

9.2.3 PHASE COMPOSITION AND MICROSTRUCTURE

A combined small-angle and wide-angle X-ray scattering experiment was conducted to characterize the phase composition of prepared samples. Measurements were carried out at the SAXS/WAXS beamline of the Australian Synchrotron in Melbourne, Australia. A beam energy of 20 keV (wavelength of 0.62 Å) was used in the range of $2\theta = 0.29$–$30.00°$.

To study the morphologies of the HNTs and their dispersion inside the vinyl ester matrix, a transmission electron microscope (TEM) (JEOL JEM2011, Japan) was used. A NEON 40ESB scanning electron microscope (SEM) (ZEISS, UK) operating at accelerating voltage of 5 kV under secondary electron mode was used to examine the microstructure of HNTs and fracture surfaces of the samples. In order to avoid charging, all samples were coated with platinum.

9.2.4 WATER UPTAKE TEST

Samples with rectangular-shaped dimensions (20 mm × 20 mm × 6 mm) were used. Individual samples were soaked in tap water at room temperature. At prescribed intervals, individual samples were removed from water. After excess water was removed, they were weighed and immediately returned to the water. The amount of moisture uptake or moisture absorbed (M_A) by the samples over a period of 120 days was determined using the following equation (Dhakal et al. 2007):

$$M_A = \frac{M_T - M_D}{M_D} \times 100 \qquad (9.1)$$

where M_D is the dry mass and M_T is the mass of sample soaked for time t.

The diffusion coefficient behavior of samples has been calculated by using the following equation (Dhakal et al. 2007):

$$D = \frac{\pi}{16} \left(\frac{M_t/M_\infty}{\sqrt{t/h}} \right)^2 \qquad (9.2)$$

where M_∞ is maximum water uptake, M_t is water uptake at time t, h is sample thickness, and D is diffusion coefficient.

9.2.5 MECHANICAL TESTS

9.2.5.1 Flexural Strength

Rectangular bars of 60 mm × 10 mm × 6 mm were cut from the fully cured samples for three-point bend tests with a span of 40 mm to evaluate the flexural strength. A LLOYD Material Testing Machine (5–50 kN) with a displacement rate of 1.0 mm/min was used to perform the test. Five samples of each batch were used to evaluate flexural strength. The values were recorded and analyzed with the machine software (NEXYGEN*Plus*), and value averages were calculated.

9.2.5.2 Impact Strength

To determine impact strength σ_i, a Zwick Charpy impact tester with a 2.0 J pendulum hammer was used. Five 40-mm span bar samples were assessed. Impact strength was calculated using the following equation (Low et al. 2007):

$$\sigma_i = \frac{E}{A} \qquad (9.3)$$

where E is the impact energy to break a sample with a ligament of area A.

9.2.5.3 Fracture Toughness

For the fracture toughness (K_{IC}) measurement, the ratio of notch length to width of sample (a/w) used was 0.4, and a sharp razor blade was used to initiate a sharp crack. The flexural tests were performed with a LLOYD Material Testing Machine using a displacement rate of 1.0 mm/min; five samples of each composition were used for the measurements. The value of (K_{IC}) was computed using the following equation (Low et al. 2007):

$$K_{IC} = \frac{p_m S}{WD^{2/3}} f\left(\frac{a}{w}\right) \tag{9.4}$$

where p_m is the load at crack extension, S is the span of the sample, D is the specimen thickness, w is the specimen width, a is the crack length, and $f(a/w)$ is the polynomial geometrical correction factor given as (Low et al. 2007)

$$f(a/w) = \frac{3(a/w)^{1/2}\left[1.99 - (a/w)(1-a/w)(2.15 - 3.93\,a/w + 2.7\,a^2/w^2)\right]}{2(1+2a/w)(1-a/w)^{2/3}} \tag{9.5}$$

9.2.5.4 Impact Toughness

To determine impact toughness G_{IC}, a Zwick Charpy impact tester with a 2.0 J pendulum hammer was used. Five 40-mm span bar samples with varying notch lengths and razor cracks were used. The impact toughness value was calculated using the following equation (Low et al. 2007):

$$U = G_{IC}BD\varphi + U_0 \tag{9.6}$$

where U is the measured energy, U_0 is the kinetic energy, D is the specimen thickness, B is the specimen breadth, and φ is the calibration factor for the geometry used (Handge et al. 2010).

9.2.6 THERMAL AND FLAMMABILITY TEST

A thermogravimetric analyzer (TGA-DTA; Instrument: 2960 SDT V3.0F) was used to examine the thermal behavior of composites. At a rate of 20°C/min, composites were heated from room temperature to 800°C. Thermal decomposition temperatures of the composites were examined under 20 ml/min of nitrogen using platinum pans. Horizontal burning test determined flammability in terms of ignition time, burning-out time, and fire velocity. For each composite, three samples (100 mm × 10 mm × 10 mm) were prepared and hung on a retort stand. A stop watch was used to record times, and a constant flame source was applied.

9.3 RESULTS AND DISCUSSION

9.3.1 STRUCTURE AND MORPHOLOGY OF HNT AND VER/HNT COMPOSITES

9.3.1.1 SAXS/WAXS Analysis

The diffraction patterns for the various composite samples are shown in Figure 9.1. A diffraction peak at 2θ at approximately $5°$ is related to the (001) plane of HNTs (Figure 9.1a), which corresponds to a basal spacing of 0.721 nm. A highly disordered tubular morphology featuring interstratification of layers with various hydration states and small crystal size is indicated by this basal reflection (Du et al. 2010; Rooj et al. 2010). The broad and diffuse peak at $2\theta = 6–10°$ reflects the amorphous nature of vinyl ester matrix. The diffraction patterns for VER/50 wt% RCF/5 wt% HNT and VER/50 wt% RCF composites are shown in Figure 9.1b and c, respectively.

9.3.1.2 Microstructures

SEM and TEM images of HNTs reveal that the majority of HNTs exist in a tubular shape, and this is evidenced in Figure 9.2. The presence of short tubular HNTs,

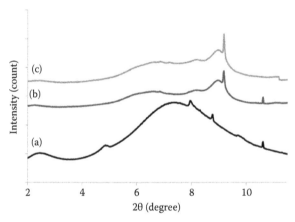

FIGURE 9.1 SAXS/WAXS plots for various samples: (a) VER/5 wt% HNTs, (b) VER/ RCF/5 wt% HNTs, and (c) VER/50 wt% RCF. HNT, halloysite nanotube; RCF, recycled cellulose fiber; VER, vinyl ester resin.

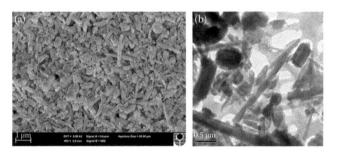

FIGURE 9.2 (a) SEM micrograph of halloysite nanotube (HNT) clusters and (b) TEM micrograph of HNT clusters. SEM, scanning electron microscope; TEM, transmission electron microscope.

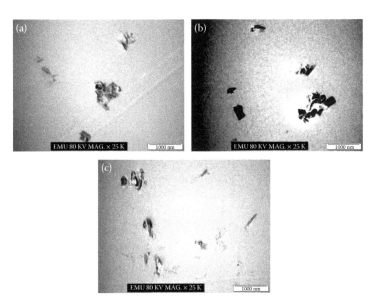

FIGURE 9.3 Dispersion of halloysite nanotube clusters within composites with nanoclay loadings: (a) 1 wt%, (b) 3 wt%, and (c) 5 wt%.

semi-rolled HNTs, and pseudo-spherical HNTs is evident. A mean particle size of 2 μm and a length of HNTs ranging from 300 nm to 3 μm was determined. The HNT aspect ratio varies between 1.5, 3, and 10. The inner diameters range from 10 to 50 nm, whereas the outer diameters of the HNTs range from 30 to 200 nm. The TEM micrographs (Figure 9.3) show that there is an acceptable degree of dispersion for the 1%, 3%, and 5% of HNTs within the composites.

9.3.2 WATER UPTAKE

Water absorption curves of the eco-composites and eco-nanocomposites are shown in Figure 9.4. When the samples were first exposed to water, the process of water absorption occurred rapidly then gradually the absorption rate slowed down until equilibrium, these behaviors follow Fickian diffusion behaviors (Dhakal et al. 2007; Ladhari et al. 2010). Clearly, increasing HNTs in addition to the system resulted in a reduction in the uptake of water. It is the large aspect ratio of HNTs that is believed to interfere with water molecule transfer paths causing the path of direct-fast diffusion into the polymer matrix to alter into a path that is torturous or maze-like, reducing the overall uptake of water (Chang et al. 2010; Xie et al. 2011). The impermeability of nanocomposites provided by HNTs prevents their complete saturation and causes maximum water uptake to be lower (Liu et al. 2005).

The maximum water uptake M_∞ and diffusion coefficient D values for all composites are shown in Table 9.1. The amount of water absorbed is decreased, as HNT loading is increased, which reveals the effect of HNTs on water absorption in composites; a desirable property for commercial applications. The content and diffusivity values of the HNTs do not produce any observable trend. Although, as

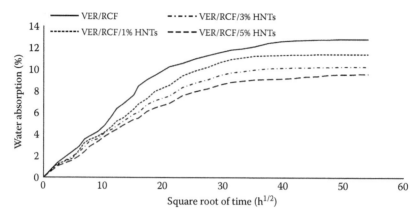

FIGURE 9.4 Water absorption behavior of VER, VER/RCF, VER/RCF/1% HNTs, VER/ RCF/3% HNTs, and VER/RCF/5% HNTs. HNT, halloysite nanotube; RCF, recycled cellulose fiber; VER, vinyl ester resin.

TABLE 9.1
Maximum Water Uptake (M_∞) and Diffusion Coefficients (D) of VER-Eco and Eco-Nanocomposites

Samples	M_∞ (%)	D (mm^2/s)
VER/RCF	12.83	2.81×10^{-6}
VER/RCF/1% HNTs	11.44	2.56×10^{-6}
VER/RCF/3% HNTs	10.28	2.16×10^{-6}
VER/RCF/5% HNTs	9.58	2.99×10^{-6}

HNT, halloysite nanotube; RCF, recycled cellulose fiber; VER, vinyl ester resin.

---

the loading of HNTs is increased, no statistically significant change in the diffusion behaviors of the composites based on the diffusion coefficient results is observed.

9.3.3 MECHANICAL PROPERTIES

9.3.3.1 Strength

Figures 9.5 and 9.6 show the results for flexural strength and impact strength, measures of strength properties. In comparison with pure VER, eco-composites and eco-nanocomposites had greater strength. The flexural strength of the eco-composite was over three times that of the pure sample (Figure 9.5). The impact strength of pure VER was 2.6 kJ/m^2, whereas eco-composites were found to be six times stronger at 15.9 kJ/m^2 (Figure 9.6). The enhanced flexural properties of the eco-composite are believed to be underpinned by the high strength and modulus of the cellulose fibers as well as the good matrix–fiber interfacial adhesion (Anuar and Zuraida 2011; Bakare et al. 2010; Espert et al. 2004). Other studies have documented the improvements

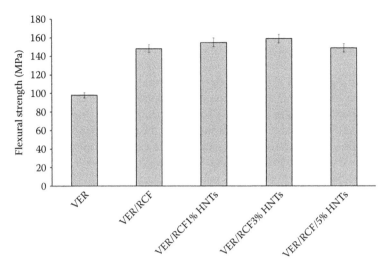

FIGURE 9.5 Flexural strength of vinyl ester eco-composites and eco-nanocomposites.

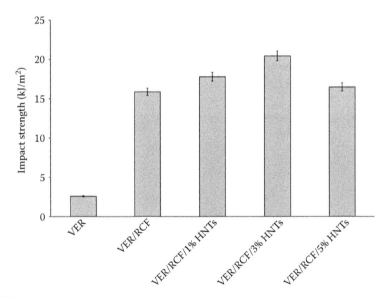

FIGURE 9.6 Impact strength of vinyl ester eco-composites and eco-nanocomposites.

in strength properties observed in polymer reinforced with natural fiber composite (Bax and Müssig 2008; Low et al. 2009; Wambua et al. 2003). A remarkable increase in flexural strength and impact strength was achieved for samples with 1 and 3 wt% HNTs.

A primary determinant of composite quality is the quality of adhesion between the fiber and matrix. The role of the matrix, in a fiber-reinforced composite, is to transfer load to stiff fibers through shear stresses at the interface. A good bond between the polymeric matrix and the fibers is essential for this process (Bakare et al. 2010; Chen et al. 2009). Using SEM, the fracture surfaces of eco-nanocomposites and eco-composites and the effect of the addition of HNTs on fiber–matrix adhesion can be studied.

SEM micrographs in Figure 9.7 show the fracture surfaces of all samples. The investigation of fiber–matrix adhesion quality is based on fiber pull-out from the matrix, disparity in length of fibers, fiber surfaces, and fiber–matrix gaps apparent in each of the composites. Figure 9.7a shows that the pull-out lengths, the extent that individual fibers are isolated, were greater in the eco-composites compared to the eco-nanocomposites as shown in Figure 9.7b through d. Moreover, Figure 9.7a shows that the number of fibers that were pulled out was greater in the eco-composites compared to the eco-nanocomposites as shown in Figure 9.7b through d. This phenomenon of greater pull-out lengths and greater number of fibers pulled-out in eco-composites is a consequence of poor adhesion between the fibers and the matrix (Silva et al. 2006; Stocchi et al. 2007). Another distinguishing feature of the eco-composites is the clean appearance of the fiber surfaces. This clean appearance is another indicator of poor adhesion between matrix and fibers present (de Rosa et al. 2010; Mylsamy and Rajendran 2011). This cleanness of the eco-composite fiber surfaces can be clearly contrasted with the rough appearance of fiber surfaces in eco-nanocomposites, indicating strong fiber–matrix adhesion in the latter. Finally, matrix–fiber gaps appear larger in the eco-composites compared to the matrix–fiber gaps seen in the eco-nanocomposites that appear smaller. These observations indicated that the addition of HNTs leads to stronger fiber–matrix adhesion, which results

FIGURE 9.7 SEM images showing the fracture surfaces of (a) eco-composite and eco-nanocomposite with various nanoclay loadings: (b) 1 wt%, (c) 3 wt%, and (d) 5 wt%.

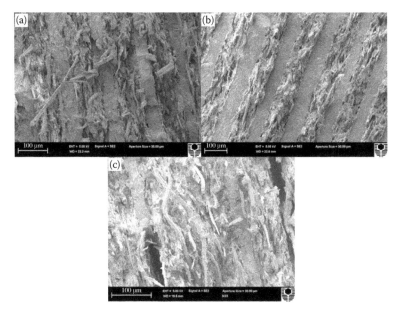

FIGURE 9.8 Fracture surfaces of eco-nanocomposites with various nanoclay loadings: (a) 1 wt%, (b) 3 wt%, and (c) 5 wt%.

in higher strengths (Franco-Marquès et al. 2011; Kumar et al. 2010; Suppakarn and Jarukumjorn 2009).

With regard to the addition of 5 wt% HNTs, it was found that this concentration leads to less improvement in strength properties compared to 1 and 3 wt% HNTs. Processing events are believed to underpin the failure of 5 wt% HNT addition to further enhance strength properties. When there is a high loading of HNTs, the viscosity increases during mixing of resin and HNTs rendering degassing insufficient before curing. It is vital that during processing a complete degassing process is ensured for the composite to minimize void formation. The formation of voids in composites, even on exposure to low strain, can cause specimen failure in these composites thereby decreasing the strength properties of the materials. Figure 9.8 provides evidence for this, highlighting the existence of voids in the 5 wt% HNTs eco-nanocomposites and the absence of such voids in 1 and 3 wt% eco-nanocomposites.

Poor fiber–matrix adhesion is another one of the undesirable consequences of the highly viscous mixture that results when 5 wt% HNT is added to polymer. High viscosity causes a reduction in wettability, and that leads to reduction in the interfacial adhesion between matrix and fibers, causing further decreasing of strength properties of the resulting composite (Ashori and Nourbakhsh 2009; Avella et al. 2009).

9.3.4 TOUGHNESS

Figures 9.9 and 9.10 show that, compared to pure samples, the eco-composites have much greater fracture toughness and impact toughness properties. High crack deflection, energy dissipation, and fracture resistance properties are believed to be provided

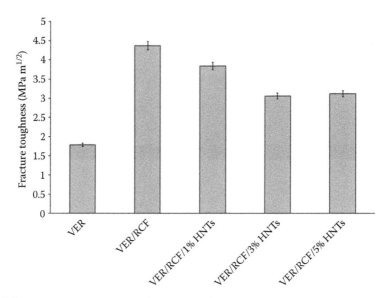

FIGURE 9.9   Fracture toughness of eco-composites and eco-nanocomposites.

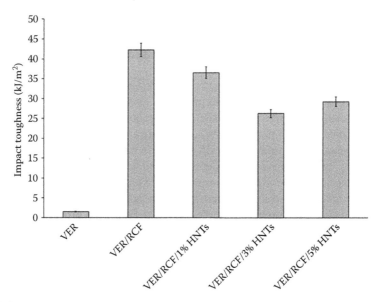

FIGURE 9.10   Impact toughness of eco-composites and eco-nanocomposites.

to these composites through the interaction of cellulose fibers with the matrix (Low et al. 2009; Stocchi et al. 2007; Wong et al. 2010). The favorable fracture toughness properties that natural fiber–polymer composites typically display are believed to be determined through crack-deflection, de-bonding between fiber and matrix, the pull-out effect, and a fiber-bridging mechanism. Addition of HNTs, however, leads to a reduction in the toughness properties but causes improvements in fiber–matrix adhesion, which leads to increased strength properties in eco-nanocomposites, as previously described.

These improvements of fiber–matrix adhesion make the eco-nanocomposite brittle and that prevents fiber pull-outs and fiber de-bonding. As these are the material's major energy absorption mechanisms, their prevention causes the composite to become brittle. Thus, without mechanisms to absorb energy, the toughness properties of the eco-nanocomposites are reduced (Venkateshwaran et al. 2011; Wong et al. 2010). The fibers in the eco-composite can be seen in Figure 9.7a to have slid out from the matrix in greater lengths and in a greater number. In contrast, with the addition of HNTs, the fibers pulled-out are shorter and in less number as shown in Figure 9.7b through d, indicating the strong interfacial adhesion of the eco-nanocomposites (Mylsamy and Rajendran 2011).

9.3.5 THERMAL STABILITY AND FLAMMABILITY

The thermogravimetric analysis (TGA) curves for vinyl ester, eco-composites, and eco-nanocomposites are shown in Figure 9.11. Here, the temperature range used was from room temperature to 800°C. Thermal degradation, for vinyl ester, occurred in a single stage at around 360°C. The samples reinforced with RCF sheets, compared with the pure samples, showed a marginally higher thermal stability that could be attributed to the higher and longer thermal resistance of the cellulose fibers (Curvelo et al. 2001; Ma et al. 2005). A slight weight loss occurred between 60°C and 100°C in all composites due to the release of moisture. The degradation profile of the composites started at approximately 260°C. At around 230°C–260°C, the degradation profile of the composites started according to TGA. Between 270°C and 480°C degradation of the eco-composites followed relating to constituent decomposition. Continued decomposition was evident from 380°C until the temperature reached nearly 500°C at which point a constant mass was achieved. For the eco-nanocomposites the thermal stability followed a similar trend albeit

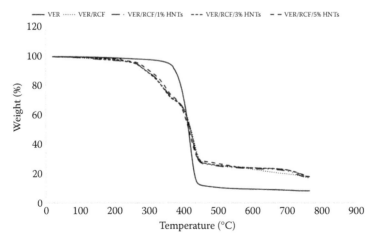

FIGURE 9.11 Thermogravimetric analysis curves of eco-composites and eco-nanocomposites.

requiring a marginally higher temperature whereby eco-nanocomposites of 1, 3, and 5 wt% HNTs required 374°C, 379.3°C, and 384.1°C, respectively.

In terms of weight loss, from 100°C to 200°C, eco-composites and eco-nanocomposites gave a percentage mass drop of approximately 2.62% according to TGA. At 300°C, 11.8% drop in mass was recorded. However, at the same temperature, the eco-nanocomposites gave approximately 11.4%, 10.7%, and 9.5% degradation for composites with 1% HNTs, 3% HNTs, and 5% HNTs, respectively. Above 700°C, the residual weight of eco-composites was 16.3% of the original. While for the pure sample, only 8.3% of the residual weight had remained. For the eco-nanocomposites with 1%, 3%, and 5% HNTs, the residual weight was 17.7%, 18.2%, and 18.6%, respectively. A clear relationship between the addition of HNTs and increases in thermal stability can be seen according to the TGA.

The improvement of the adhesion between RCF and VER as a result of the addition of HNTs can lead to the enhancement of thermal stability (de Rosa et al. 2010; Yu et al. 2010). Nano-fillers such as HNTs are believed to provide, first, a thermal barrier, which prevents heat transfer inside polymer matrix (Du et al. 2006; Paul and Robeson 2008), and second, a mass transport barrier, which during the process of degradation forms a char that hinders the escape of the volatile products (Leszczyńska et al. 2007). The hollow tubular structure of HNTs is also reported to be another factor that leads to enhanced thermal stability. The hollow tubular structure of HNTs is believed to enable the entrapment of degradation products inside the lumens, causing effective delay in mass transfer that leads to improved thermal stability (Ismail et al. 2008; Lecouvet et al. 2011). The presence of iron oxides, Fe_2O_3, in silicate fillers is also believed to be a possible flame retardant, enhancing thermal stability of composites by trapping radicals during the process of degradation (Kashiwagi et al. 2002; Zhu et al. 2001).

Flammability tests conducted at ambient conditions included burning out rate, ignition time, and fire spreading speed determinations. Pure VER samples were found to burn out and ignite faster than VER/HNTs composites as seen in Table 9.2. Calculations imply that fire spreads through pure VER at nearly twice the rate of 5 wt% VER/HNT composite, thus highlighting the favorable flammability resistance of the composites. The barrier effects of HNTs are believed to be the

TABLE 9.2
Flammability Properties of Vinyl Ester Resin and Its Composites

Sample	Burning-Out Rate (g/min)	Ignition Time (S)	Fire-Spreading Speed (mm/min)
VER	0.67	10.77	31.5
VER/RCF	0.54	17.21	19.7
VER/RCF/1% NC	0.49	18.65	23.2
VER/RCF/3% NC	0.46	22.76	19.1
VER/RCF/5% NC	0.41	25.42	17.5

HNT, halloysite nanotube; RCF, recycled cellulose fiber; VER, vinyl ester resin.

dominant reasons for the reduction in flammability of VER/HNT composites. First, the presence of HNTs within composites provides a mechanism of insulation that protects the composites from catching fire. Furthermore, char formation of HNTs acts as a heat and fire retardant (Becker et al. 2004; Paul et al. 2003; Vyazovkin et al. 2004).

9.4 CONCLUSIONS

Vinyl ester eco-composites and vinyl ester eco-nanocomposites have been fabricated and studied in terms of water uptake, thermal, and mechanical properties. HNTs were found to be effective in reducing the water uptake in eco-nanocomposites. Strengths were also found to be enhanced due to the reinforcing effect of both RCF and HNTs. In particular, addition of HNT improved fiber–matrix adhesion in eco-nanocomposites and gave greater strength properties. Owing to the toughness mechanism provided by cellulose fibers, the presence of cellulose fiber increased the toughness properties of all composites. Addition of HNT, however, led to a reduction in the toughness properties due to the improvement of fiber–matrix adhesion. Thermal stability and flammability of the eco-nanocomposites were preferable to those of the eco-composites or pure VER. Addition of HNT increased both the thermal stability and fire-resisting properties of the eco-nanocomposites.

ACKNOWLEDGMENTS

We are grateful to our colleague Dr. Nobuo Tezuka for providing HNTs for this study, and we thank Mr. Andreas Viereckl from Mechanical Engineering at Curtin University for assistance with mechanical tests. We also thank Elaine Miller and Dr. Nigel Kirby of the Australian Synchrotron for their assistance with SEM imaging and SAXS/WAXS data collection, respectively. Peter Chapman and Kristy Blyth from the Chemistry Department are thanked for their support in the preparation of the measurements for both the FTIR and TGA experiments.

REFERENCES

Akil, H.M., Omar, M.F., Mazuki, A.A.M., Safiee, S., Ishak, Z.A.M. and Abu Bakar, A. 2011. Kenaf fiber reinforced composites: A review. *Materials and Design* 32: 4107–4121.

Anuar, H. and Zuraida, A. 2011. Improvement in mechanical properties of reinforced thermoplastic elastomer composite with kenaf bast fibre. *Composites Part B: Engineering* 42: 462–465.

Ashori, A. and Nourbakhsh, A. 2009. Characteristics of wood–fiber plastic composites made of recycled materials. *Waste Management* 29: 1291–1295.

Avella, M., Buzarovska, A., Errico, M., Gentile, G. and Grozdanov, A. 2009. Eco-challenges of bio-based polymer composites. *Materials* 2: 911–925.

Bakare, I.O., Okieimen, F.E., Pavithran, C., Abdul Khalil, H.P.S. and Brahmakumar, M. 2010. Mechanical and thermal properties of sisal fiber-reinforced rubber seed oil-based polyurethane composites. *Materials and Design* 31: 4274–4280.

Bax, B. and Müssig, J. 2008. Impact and tensile properties of PLA/Cordenka and PLA/flax composites. *Composites Science and Technology* 68: 1601–1607.

Becker, O., Varley, R.J. and Simon, G.P. 2004. Thermal stability and water uptake of high performance epoxy layered silicate nanocomposites. *European Polymer Journal* 40: 187–195.

Bruzaud, S. and Bourmaud, A. 2007. Thermal degradation and (nano)mechanical behavior of layered silicate reinforced poly(3-hydroxybutyrate-co-3-hydroxyvalerate) nanocomposites. *Polymer Testing* 26: 652–659.

Chang, P.R., Jian, R., Yu, J. and Ma, X. 2010. Fabrication and characterisation of chitosan nanoparticles/plasticised-starch composites. *Food Chemistry* 120: 736–740.

Chen, H., Miao, M. and Ding, X. 2009. Influence of moisture absorption on the interfacial strength of bamboo/vinyl ester composites. *Composites Part A: Applied Science and Manufacturing* 40: 2013–2019.

Curvelo, A.A.S., de Carvalho, A.J.F. and Agnelli, J.A.M. 2001. Thermoplastic starch–cellulosic fibers composites: Preliminary results. *Carbohydrate Polymers* 45: 183–188.

Deng, S., Zhang, J., Ye, L. and Wu, J. 2008. Toughening epoxies with halloysite nanotubes. *Polymer* 49: 5119–5127.

De Rosa, I.M., Santulli, C. and Sarasini, F. 2010. Mechanical and thermal characterization of epoxy composites reinforced with random and quasi-unidirectional untreated Phormium tenax leaf fibers. *Materials and Design* 31: 2397–2405.

Dhakal, H.N., Zhang, Z.Y. and Richardson, M.O.W. 2007. Effect of water absorption on the mechanical properties of hemp fibre reinforced unsaturated polyester composites. *Composites Science and Technology* 67: 1674–1683.

Du, M., Guo, B. and Jia, D. 2006. Thermal stability and flame retardant effects of halloysite nanotubes on poly(propylene). *European Polymer Journal* 42: 1362–1369.

Du, M., Guo, B. and Jia, D. 2010. Newly emerging applications of halloysite nanotubes: A review. *Polymer International* 59: 574–582.

Espert, A., Vilaplana, F. and Karlsson, S. 2004. Comparison of water absorption in natural cellulosic fibres from wood and one-year crops in polypropylene composites and its influence on their mechanical properties. *Composites Part A: Applied Science and Manufacturing* 35: 1267–1276.

Franco-Marquès, E., Méndez, J.A., Pèlach, M.A., Vilaseca, F., Bayer, J. and Mutjé, P. 2011. Influence of coupling agents in the preparation of polypropylene composites reinforced with recycled fibers. *Chemical Engineering Journal* 166: 1170–1178.

Handge, U.A., Hedicke-Höchstötter, K. and Altstädt, V. 2010. Composites of polyamide 6 and silicate nanotubes of the mineral halloysite: Influence of molecular weight on thermal, mechanical and rheological properties. *Polymer* 51: 2690–2699.

Haq, M., Burgueño, R., Mohanty, A.K. and Misra, M. 2008. Hybrid bio-based composites from blends of unsaturated polyester and soybean oil reinforced with nanoclay and natural fibers. *Composites Science and Technology* 68: 3344–3351.

Hossain, M.K., Dewan, M.W., Hosur, M. and Jeelani, S. 2011. Mechanical performances of surface modified jute fiber reinforced biopol nanophased green composites. *Composites Part B: Engineering* 42(6): 1701–1707.

Ismail, H., Pasbakhsh, P., Fauzi, M.N.A. and Abu Bakar, A. 2008. Morphological, thermal and tensile properties of halloysite nanotubes filled ethylene propylene diene monomer (EPDM) nanocomposites. *Polymer Testing* 27: 841–850.

Jawaid, M. and Abdul Khalil, H.P.S. 2011. Cellulosic/synthetic fibre reinforced polymer hybrid composites: A review. *Carbohydrate Polymers* 86: 1–18.

Joussein, E., Petit, S., Churchman, J., Theng, B., Righi, D. and Delvaux, B. 2005. Halloysite clay minerals: A review. *Clay Minerals* 40: 383–426.

Kashiwagi, T., Grulke, E., Hilding, J., Harris, R., Awad, W. and Douglas, J. 2002. Thermal degradation and flammability properties of poly(propylene)/carbon nanotube composites. *Macromolecular Rapid Communications* 23: 761–765.

Ku, H., Wang, H., Pattarachaiyakoop, N. and Trada, M. 2011. A review on the tensile properties of natural fiber reinforced polymer composites. *Composites Part B: Engineering* 42: 856–873.

Kumar, R., Yakabu, M.K. and Anandjiwala, R.D. 2010. Effect of montmorillonite clay on flax fabric reinforced poly lactic acid composites with amphiphilic additives. *Composites Part A: Applied Science and Manufacturing* 41: 1620–1627.

Ladhari, A., Ben Daly, H., Belhadjsalah, H., Cole, K.C. and Denault, J. 2010. Investigation of water absorption in clay-reinforced polypropylene nanocomposites. *Polymer Degradation and Stability* 95: 429–439.

Lecouvet, B., Gutierrez, J.G., Sclavons, M. and Bailly, C. 2011. Structure–property relationships in polyamide 12/halloysite nanotube nanocomposites. *Polymer Degradation and Stability* 96: 226–235.

Leszczyńska, A., Njuguna, J., Pielichowski, K. and Banerjee, J.R. 2007. Polymer/montmorillonite nanocomposites with improved thermal properties: Part I. Factors influencing thermal stability and mechanisms of thermal stability improvement. *Thermochimica Acta* 453: 75–96.

Liu, M., Guo, B., Du, M. and Jia, D. 2007. Properties of halloysite nanotube–epoxy resin hybrids and the interfacial reactions in the systems. *Nanotechnology* 18: 455703, 455709.

Liu, Q. and Hughes, M. 2008. The fracture behaviour and toughness of woven flax fibre reinforced epoxy composites. *Composites Part A: Applied Science and Manufacturing* 39: 1644–1652.

Liu, W., Hoa, S.V. and Pugh, M. 2005. Fracture toughness and water uptake of high-performance epoxy/nanoclay nanocomposites. *Composites Science and Technology* 65: 2364–2373.

Low, I.M., McGrath, M., Lawrence, D., Schmidt, P., Lane, J. and Latella, B.A. 2007. Mechanical and fracture properties of cellulose-fibre-reinforced epoxy laminates. *Composites Part A: Applied Science and Manufacturing* 38: 963–974.

Low, I.M., Somers, J., Kho, H.S., Davies, I.J. and Latella, B.A. 2009. Fabrication and properties of recycled cellulose fibre-reinforced epoxy composites. *Composite Interfaces* 16: 659–669.

Ma, X., Yu, J. and Kennedy, J.F. 2005. Studies on the properties of natural fibers-reinforced thermoplastic starch composites. *Carbohydrate Polymers* 62: 19–24.

Mylsamy, K. and Rajendran, I. 2011. The mechanical properties, deformation and thermomechanical properties of alkali treated and untreated Agave continuous fibre reinforced epoxy composites. *Materials and Design* 32: 3076–3084.

Paul, D.R. and Robeson, L.M. 2008. Polymer nanotechnology: Nanocomposites. *Polymer* 49: 3187–3204.

Paul, M.A., Alexandre, M., Degée, P., Henrist, C., Rulmont, A. and Dubois, P. 2003. New nanocomposite materials based on plasticized poly(l-lactide) and organo-modified montmorillonites: Thermal and morphological study. *Polymer* 44: 443–450.

Rooj, S., Das, A., Thakur, V., Mahaling, R.N., Bhowmick, A.K. and Heinrich, G. 2010. Preparation and properties of natural nanocomposites based on natural rubber and naturally occurring halloysite nanotubes. *Materials and Design* 31: 2151–2156.

Sgriccia, N., Hawley, M.C. and Misra, M. 2008. Characterization of natural fiber surfaces and natural fiber composites. *Composites Part A: Applied Science and Manufacturing* 39: 1632–1637.

Silva, R.V., Spinelli, D., Bose Filho, W.W., Claro Neto, S., Chierice, G.O. and Tarpani, J.R. 2006. Fracture toughness of natural fibers/castor oil polyurethane composites. *Composites Science and Technology* 66: 1328–1335.

Stocchi, A., Bernal, C., Vázquez, A., Biagotti, J. and Kenny, J. A. 2007. Silicone treatment compared to traditional natural fiber treatments: Effect on the mechanical and viscoelastic properties of jute-vinylester laminates. *Journal of Composite Materials* 41: 2005–2024.

Suppakarn, N. and Jarukumjorn, K. 2009. Mechanical properties and flammability of sisal/PP composites: Effect of flame retardant type and content. *Composites Part B: Engineering* 40: 613–618.

Venkateshwaran, N., ElayaPerumal, A., Alavudeen, A. and Thiruchitrambalam, M. 2011. Mechanical and water absorption behaviour of banana/sisal reinforced hybrid composites. *Materials and Design* 32: 4017–4021.

Vyazovkin, S., Dranca, I., Fan, X. and Advincula, R. 2004. Kinetics of the thermal and thermo-oxidative degradation of a polystyrene–clay nanocomposite. *Macromolecular Rapid Communications* 25: 498–503.

Wambua, P., Ivens, J. and Verpoest, I. 2003. Natural fibres: Can they replace glass in fibre reinforced plastics? *Composites Science and Technology* 63: 1259–1264.

Wong, K.J., Zahi, S., Low, K.O. and Lim, C.C. 2010. Fracture characterisation of short bamboo fibre reinforced polyester composites. *Materials and Design* 31: 4147–4154.

Xie, Y., Chang, P.R., Wang, S., Yu, J. and Ma, X. 2011. Preparation and properties of halloysite nanotubes/plasticized Dioscorea opposita Thunb. starch composites. *Carbohydrate Polymers* 83: 186–191.

Yu, T., Ren, J., Li, S., Yuan, H. and Li, Y. 2010. Effect of fiber surface-treatments on the properties of poly(lactic acid)/ramie composites. *Composites Part A: Applied Science and Manufacturing* 41: 499–505.

Zhao, M. and Liu, P. 2008. Halloysite nanotubes/polystyrene (HNTs/PS) nanocomposites via in situ bulk polymerization. *Journal of Thermal Analysis and Calorimetry* 94: 103–107.

Zhu, J., Uhl, F.M., Morgan, A.B. and Wilkie, C.A. 2001. Studies on the mechanism by which the formation of nanocomposites enhances thermal stability. *Chemistry of Materials* 13: 4649–4654.

10 Eco-Friendly Fiber-Reinforced Natural Rubber Green Composites
A Perspective on the Future

Raghavan Prasanth, Ravi Shankar, Anna Dilfi,
Vijay K. Thakur, and Jou-Hyeon Ahn

CONTENTS

10.1 INTRODUCTION

Plant-based natural fibers are often used as a reinforcing material for environmentally friendly green composites. A composite material can be defined as a macroscopic combination of two or more distinct materials of continuous and discontinuous medium, having a recognizable interface between them. The discontinuous medium that is stiffer and stronger than the continuous phase is called the reinforcement and the so-called continuous phase is referred to as the matrix. A composite material can provide superior and unique mechanical and physical properties, because it combines the most desirable properties of its constituents while suppressing the least desirable properties. The properties of a composite are dependent on the properties of the constituent materials, and their distribution and interaction. At present, composite materials play a key role in aerospace industry, automobile industry, and other engineering applications as they exhibit outstanding strength to weight and modulus to weight ratio. Based on the matrix material, which forms the continuous phase, the composites are broadly classified into metal matrix, ceramic matrix, and polymer matrix composites. Among these, polymer matrix composites are much easier to fabricate than the rest, due to relatively low processing temperature. Several researchers investigated the structure, properties, and applications of various composites (Amash and Zugenmaier 1998; Gonzalez and Yong 1998; Kishore 1998; Oya and Hamada 1998; Raghavan and Wool 1999; Srivastava and Hogg 1998; Thomas and Winstone 1998; Tjong and Meng 1999; Zheng et al. 1999). In recent decades, natural fibers as an alternative reinforcement in polymer composites have attracted wide attention due to their advantages over conventional glass and carbon fibers (Saheb and Jog 1999). Natural fibers are environmentally friendly not only for their savings in process energy, which is an unavoidable problem for synthetic fibers, but also for their renewable and biodegradable characteristics to protect the environment.

Flax, hemp, jute, sisal, kenaf, coir, kapok, banana, and many other fibers are used for reinforcing composite (Cyras et al. 2004). Both natural fibers and their polymer composites are neutral with respect to CO_2 emissions that cause the Earth's greenhouse effect, a major factor for global warming (Gore 2006), and are referred to as green composites.

10.2 CLASSIFICATION OF GREEN COMPOSITES

10.2.1 PARTICULATE REINFORCEMENT

Particulate natural fillers, such as saw dust and coconut shell powder, are employed to improve high temperature performance, increase wear resistance, and reduce friction and/or shrinkage. In many cases, particulate fillers are used to reduce cost; under these conditions, the additive is called a cost-effective filler (simply filler), whereas when a considerable change in the properties of the composite occurs, the additive is known as a reinforcement filler (simply reinforcement). The particles also share the load with the matrix, but to a lesser extent than fibers. Therefore, in most cases, particulate reinforcement improves stiffness, not strength. Hard particles in a brittle matrix cause localized stress concentrations in the matrix, which reduce the overall impact strength and make the composite economical (Richardson 1987).

10.2.2 FIBROUS REINFORCEMENT

Fibrous reinforcement represents a physical rather than a chemical means of changing a material to suit various engineering applications (Warner 1995). The component materials of fiber-reinforced composites are fibers and matrix and the composite exhibit anisotropy in properties. The measured strength of most synthetic materials is much less than predicted theoretically due to the flaws in the form of cracks perpendicular to the applied load that are present in bulk materials. Natural fibers have much higher longitudinal strengths, because the larger flaws are not generally present in such small cross-sectional areas. In the composite, generally fibers are the major load-carrying members while the surrounding matrix keeps them in the desired location and orientation. Furthermore, the matrix acts as a load transfer medium and protects the fibers from environmental damages due to elevated temperature and humidity. The nature of the interface between them is important as far as the properties of the composites are concerned. The fibers dispersed in the composite matrix can be continuous or discontinuous. In continuous fiber reinforcement, the transfer of the load from matrix to the fibers will be easy and very effective, whereas in discontinuous (or short fiber) fiber reinforcement, the fibers must have sufficient length (critical fiber length) to transfer the load effectively. The high strength and moduli of these composites can be tailored to the high load directions. They exhibit high internal damping and better dimensional stability over a wide range of temperature due to their lower coefficient of thermal expansion and insulating nature. This leads to better vibration energy absorption within the material and results in reduced transmission of noise and vibration to neighboring structures.

10.2.3 Hybrid Composites

Green composites incorporated with two or more different types of fillers, especially fibers in a single matrix, are commonly known as hybrid composites. Hybridization is commonly used to improve the properties and to lower the cost of conventional composites. There are different types of hybrid composites according to the way in which the component materials are incorporated. Hybrid composites are generally classified as (1) sandwich type, (2) interply, (3) intraply, and (4) intimately mixed (Mallick 1988). In sandwich hybrids, one material is sandwiched between the layers of another, whereas in interply, alternate layers of two or more materials are stacked in a regular manner. Rows of two or more constituents are arranged in a random manner in intraply hybrids, whereas in the intimately mixed type, these constituents are mixed as much as possible so that no concentration of either type is present in the composite material. A laminate is fabricated by stacking a number of laminas in the thickness direction. Generally, three layers are arranged alternatively for better bonding between reinforcement and the polymer matrix, for example, plywood and paper. These laminates can have unidirectional or bidirectional orientation of the fiber reinforcement according to the end use of the composite. A hybrid laminate can also be fabricated by the use of different constituent materials or of the same material with a different reinforcing pattern. In many laminate composites, natural fibers are used due to their good combination of physical, mechanical, and thermal behavior.

10.3 ADVANTAGES OF NATURAL FIBERS AS REINFORCEMENT IN COMPOSITES

Natural fibers, as a substitute for glass components, have gained immense interest in the last decade, especially in the housing sector. The moderate mechanical properties of natural fibers prevent them from being used in high-performance applications where carbon fiber-reinforced composites would be utilized, but for many reasons, they can compete with glass fiber. The low specific weight, which results in a higher specific strength and stiffness than glass, is a benefit. The use of renewable natural fibers contributes to sustainable development. Nowadays, natural fiber–reinforced polymer composites come prior to synthetic fiber–reinforced composites in properties such as biodegradability, combustibility, weight, nontoxicity, decreased environmental pollution, low cost, and ease of recyclability. These advantages place the natural fiber composites among the high-performance composites having economical and environmental advantages. The versatile high-performance applications of natural fiber composites, which can replace glass and carbon fibers, were listed in an article by Hill (1997). The vegetable fiber has a density of about half that of glass fiber. During the processing of natural fiber composites, there is no abrasion of the processing machines and the fibers can withstand processing temperatures up to 250°C (Nielsen 1974; Satyanarayana and Pillai 1990; The Wealth of India 1959). They are completely combustible without the production of either toxic gases or solid residues.

Wright and Mathias (1993) succeeded in preparing lightweight materials from balsawood and polymer. Investigations have been carried out by Hednberg and Gateholm (1995) in recycling the plastic and cellulose waste into composite. Systematic investigations on wood flour–reinforced polystyrene composites have been carried out by Maldas and Kokta (1991). The effects of hybridization of saw dust with glass and mica and of the surface treatment of the reinforcing filler on the mechanical properties were studied (Maldas and Kokta 1990). Natural fibers, such as sisal, coir, oil palm, bamboo, have been proved to be a better reinforcement in rubber matrix (Ismail et al. 1997a, 2002b; Prasanthkumar and Thomas 1995a; Seki 2009). Incorporation of natural fibers resulted in better long-term mechanical performance of elastomers. The poor reinforcing effect of these cellulosic fibers in elastomers was overcome by giving specific modifications. The use of natural fibers in the automobile industry has grown rapidly over the last 5 years. The range of products in the automobile industry based on natural fibers is based on polymers, such as plastics and elastomers, and fibers, such as flax, hemp, and sisal. Recently, value-added composite materials were developed from neisan jute fabric and polypropylene having enhanced mechanical properties and reduced hydrophilicity (Harikumar et al. 1999). The following sections briefly discuss the type and properties of natural fibers used for preparing natural rubber (NR) composites and their properties.

10.4 FIBER-REINFORCED NATURAL RUBBER COMPOSITE: MATRIX AND FIBER

In recent years, fiber-reinforced elastomers have received a lot of attention from the composite industry due to their easy processing and low cost coupled with high strength. These composite materials bridge the gap between conventional elastomers and fibers by combining the stiffness and load-bearing capacity of fibers with the elasticity and impact strength of rubber for engineering applications. Due to the high modulus, high strength and low creep, short glass fiber–reinforced rubber composites have been extensively studied. The studies showed that performance of fiber-reinforced rubber composite greatly depends on the amount of fiber loaded in the matrix and the uniformity in dispersion, fiber orientation, adhesion between fiber and rubber, and the length to diameter ratio (aspect ratio) of the fiber. The major applications of these composites are in tires, roofing, hoses, sheeting, conveyer belts, industrial rubber products, and complex-shaped articles. Apart from glass fiber–reinforced rubber composites, considerable studies are reported for natural fiber–reinforced rubber composites (Bhagavan et al. 1987; Coran et al. 1974; Geethamma et al. 1998; Varghese et al. 1991, 1994). The major components of these composites are natural fibers and rubbers. Rubbers include synthetic rubbers and NR which act as the matrix in the composite. Rubber is a unique, versatile, and adaptable material that has been successfully used as matrix for composite preparation. Rubber is defined as a material that is capable of recovering from larger elastic deformations quickly and forcibly. They can be modified to a state in which it is essentially insoluble but can swell in solvents such as benzene, toluene, methyl ethyl ketone, and so on (Vajrasthira et al. 2003). Its elastic strain and impact strength are much higher than metals and plastics. Hence, it can function at high strains. It is stretched rapidly even under small load to about 1000% elongation. On releasing the applied forces, rubber retracts rapidly almost fully.

10.4.1 NATURAL RUBBER

Since fibers cannot transfer loads from one to the other, they have limited use in engineering applications. When they are embedded in a matrix material to form a composite, the matrix serves to bind the fibers, transfer loads to the fibers, and protect them against environmental attack and damage due to handling. The matrix has a strong influence on several mechanical properties of the composite such as modulus and strength, shear properties, and properties in compression. Physical and chemical characteristics of the matrix such as melting or curing temperature, viscosity, and reactivity with fibers influence the choice of fabrication process. The matrix material for a composite system is selected, keeping in view all of the above factors. The commonly used matrix materials are polymers, metals, and ceramics. Among the polymers, NR composite gained attention due to its high performance and environmental friendliness. NR is a high-molecular-weight polymer of isoprene in which essentially all the isoprenes have the *cis* 1–4 configuration. In its very natural state, rubber exists as a colloidal suspension in water called latex of rubber-producing plants—*Hevea brasiliensis*. Among various rubbers (synthetic and natural rubber), NR is very important because it possess the general features of other rubbers in addition to the following highly peculiar characteristics (Geethamma et al. 2005). The properties of NR are presented in Table 10.1. Because it is of biological origin, it is renewable, inexpensive, and creates no health hazards. It possesses high tensile strength due to strain-induced crystallization, superior building tack, which is essential in many products such as tires, hoses and belts, and so on, and good crack propagation resistance. The field latex is concentrated by centrifugation, creaming, or electrodecantation. Generally, the latex is coagulated by formic or acetic acid. Technical grading of rubber is done according to the composition of rubber, source material, initial plasticity, and so on. The main criterion is the dirt content, that is the residue left after the rubber sample was dissolved in an appropriate solvent, washed through a 45-μm sieve, and dried. In addition to the different grades, certain modified forms of NR such as deproteinized NR that is having high reproducibility, oil extended NR, also called freeze-resistant rubber, and cyclic NR, which has a high adhesive property, are also available. The fibers used for NR composites are discussed in the following section.

TABLE 10.1
Properties of Natural Rubber

Properties	
Dirt content (% by mass)	0.03
Volatile mass (% by mass)	0.50
Nitrogen (% by mass)	0.30
Ash (% by mass)	0.40
Initial plasticity number (P_0)	38
Plasticity retention index	78

A considerable amount of research work has been reported on plant fiber–reinforced elastomer composites. The properties of cellulosic fiber–elastomer composites have been studied and it was found that the aspect ratio of the fiber plays has a major role on composite properties. The effects of particulate fillers on these composites have also been reported. It was found that fiber–matrix adhesion in this system could be promoted by the addition of definite proportions of silica/resorcinol/hexamethylenetetramine. The addition of either carbon black alone or both silica and carbon black to a rubber compound containing resorcinol and hexamethylenetetramine was associated with the achievement of a good adhesion between fiber and rubber matrix and it was found that silica–carbon black system exhibits better adhesion. It was also reported that processing properties such as green strength and mill shrinkage were improved by the addition of fiber. Also, the fiber addition also improved the tear strength by obstructing the development of the tear path.

10.4.2 Fibers

"Fiber" is defined as any single unit of matter characterized by flexibility, fineness, and high aspect ratio (Nielsen 1974). It is a slender filament that is longer than 100 μm or has an aspect ratio greater than 10. Fibers have a fine hair-like structure and they are of animal, vegetable, mineral, or synthetic origin (Cook 1968). Fibers are broadly classified as natural and manmade or synthetic.

10.4.2.1 Natural Fibers

These are one of the major renewable resource materials throughout the world. There are about 2000 species of useful fiber plants in various parts of the world and these are used for many applications. In the past few decades, environmental issues concerning global scale pollution and climate changes renewed the interest in natural materials including cellulose-rich fibers extracted from cultivated plants, also known as lignocellulosic fibers, and their polymer composites (Bledzki and Gassan 1999; Crocker 2008; Mohanty et al. 2000, 2002, 2006; Monteiro et al. 2009a, 2011; Netravali and Chabba 2003; Saheb and Jog 1999; Satyanarayana et al. 2007). Natural fibers have different origins such as wood, pulp, cotton, bark, nut shells, bagasse, corncobs, bamboo, cereal straw, and vegetable (e.g., flax, jute, hemp, sisal, and ramie) (Bledzki et al. 2008; Li et al. 2000; Sabu and Pothan 2009; Xiao et al. 2003). These fibers are mainly made of cellulose, hemicelluloses, lignin, and pectins, with a small quantity of extractives. Compared with conventional inorganic fillers such as glass fiber and carbon fibers, natural fibers provide many advantages: (1) abundance and therefore low cost, (2) biodegradability, (3) flexibility during processing and less resulting machine wear, (4) minimal health hazards, (5) low density, (6) desirable fiber aspect ratio, and (7) relatively high tensile and flexural modulus (Abdelmouleh et al. 2007; Bledzki and Gassan 1999; Malkapuram et al. 2008; Tserki et al. 2005; Wambua et al. 2003; Xie et al. 2010). The renewable and biodegradable characteristics of natural fibers facilitate their ultimate disposal by composting or incineration, options not possible with most industrial fibers. The fibers also contain sequestered atmospheric carbon dioxide in their structure and are invariably of lower embodied energy compared with industrially produced glass fibers.

10.4.3 Chemical Structure of Natural Fibers

The chemical composition and cell structure of natural fibers are quite complex. Each fiber is essentially a composite, in which rigid cellulose microfibrils are immersed in a soft lignin and hemicellulose matrix. The chemical composition of natural fibers varies depending on the type of the fiber. Primarily, fibers contain cellulose, hemicellulose, and lignin. The properties of each constituent contribute to the overall properties of the fiber. Cellulose resists alkalies and most of the organic acids but can be destroyed by strong mineral acids. Generally, a single fiber has a diameter of around 10–20 µm. From the living cell, cellulose is produced as microfibrils of 5 nm diameters; each is composed of 30–100 cellulose molecules in extended chain conformation and provides mechanical strength to the fiber. A good orientation of microfibrils along with high cellulose content is essential for obtaining a fiber with good mechanical properties. Hemicellulose is responsible for the biodegradation, moisture absorption, and thermal degradation of the fiber, and it shows the least resistance whereas lignin is thermally stable but is responsible for ultraviolet (UV) degradation.

The term cellulose was first used by Payen in 1938. Since then, it has been generally accepted that cellulose is a linear condensation polymer consisting of D-anhydro glucopyranose units joined together by β-1-4 glycosidic bonds. Each unit is rotated through 180° with respect to its neighbors so that the structure repeats itself for every two units. The pair of units is called cellobiose, and since cellulose is made up of cellobiose units, cellulose is technically a polymer of cellobiose rather than α-D-glucose. The chemical character of a cellulose molecule is determined by the sensitivity of β-glucosidic linkages, between the glucose repeating units to hydrolytic attack and by the presence of three hydroxyl groups, one primary and two secondary in each of the base units. These reactive hydroxyl groups are able to undergo esterification and etherification reactions. The main cause of the relative stiffness and rigidity of the cellulose molecule is the intramolecular hydrogen bonding, which is reflected in its high viscosity in solution, its high tendency to crystallize, and the ability to form fibrillar strands. The β-glycosidic linkages further favor the chain stiffness. The molecular structure of cellulose is responsible for its supramolecular structure and this in turn determines many of the physical and chemical properties of the fiber. The mechanical properties of natural fibers also depend on the cellulose type, because each type of cellulose has its own cell geometry and the geometrical constitution determines the mechanical properties.

The microstructure of natural fibers is comprised of different hierarchical structures and so is extremely complicated. Each fiber cell is constituted of four concentric layers, the primary wall, outer secondary wall, middle secondary wall, and inner secondary wall. The primary wall is porous and these pores act as diffusion paths of water through the wall. It is initially cellulosic but becomes lignified on growth and consists of pectin and other noncarbohydrates. The secondary wall is developed on the inner surface of the primary wall, which is comprised of a number of cylindrical and anisotropic cellulose microfibrils. These are surrounded and joined by a loose and complicated macromolecular network of lignin–hemicelluloses matrix. The microfibrils present in the inner secondary wall are spirally arranged about the fiber axis at an angle called the microfibrillar angle, which varies from fiber to fiber. The lumen in the center of the fiber contributes to the water uptake properties of the fiber (Cook 1968).

10.4.4 PROPERTIES OF NATURAL FIBERS

Natural fibers enjoy the potential for utilization in composites due to their adequate tensile strength and good specific modulus, thus ensuring a value-added application avenue. Plant-based composites have been widely used in construction; the ancient Egyptians used them to reinforce clay walls. To eliminate problems resulting from the incorporation of synthetic fibers such as high abrasiveness, health hazards, and disposal problems, incorporation of natural fibers is proposed. They are abundant, renewable, cheap, and have low density. Material scientists focus their attention on natural composites reinforced with fibers, such as jute, sisal, coir, pineapple, and banana, primarily to cut down the cost of raw materials. The properties of natural fibers are tabulated in Table 10.2 and their mechanical properties are compared

TABLE 10.2
Properties of Natural Fibers

Fiber	Type	Fiber Length (cm)	Strength/Flexibility
Alpaca	Animal	~8	Elastic and strong
Camel	Animal	2.5–30	Strong, tensile strength 1.78 g/d, elongation 39%–40%
Horse hair	Animal	7.5–90	Stiff and elastic, cannot be spun
Mohair	Animal	25–50	Resilient, twice as strong as wool, elongation 30%
Wool	Animal	3.8–37.5	Low tensile strength, good elasticity, elongation 25%–35%
Hemp	Bast	100–200	Durable and strong but weaker than flax
Jute	Bast	150–300	Weaker than hemp or flax, elongation 1.7% (dry)
Linen	Bast	0.6–6.5	Stronger than cotton, elongation 1.8% (dry), 2.2% (wet)
Ramie	Bast	15–20	Stronger than flax or hemp
Abaca	Leaf	100–500	Hard and strong
Pineapple	Leaf	10–20	Strong and flexible
Sisal	Leaf	60–120	Weaker and less flexible than hemp
Silk	Moths	25000–75000	Excellent tensile strength, good elasticity, elongation 20%–25% (dry), 30% (wet)
Coir	Seed hair	12–20	Stiff and elastic (like horse hair)
Cotton	Seed hair	1.6–6.0	High tensile strength, poor elasticity, elongation 5%–10%
Kapok	Seed hair	2–3.2	Resilient, lightweight

TABLE 10.3
Comparison of Mechanical Properties of Natural Fibers
with Glass Fiber

Fiber	Specific Gravity	Tensile Strength (MPa)	Modulus (GPa)	Specific Modulus
Jute	1.3	393	55	38
Sisal	1.3	510	28	22
Flax	1.5	344	27	50
Sunhemp	1.07	389	35	32
Pineapple	1.56	170	62	40
Glass fiber-E	2.5	3400	72	28

Source: Saheb, D.N. and Jog, J.P. *Adv. Polym. Technol.,* 18, 351–363, 1999.

in Table 10.3 (Saheb 1999). Natural fibers are known for their good absorbency, mechanical strength, abrasion resistance, dimensional stability, wicking ability, resistance to acids and alkalis, electrical insulation, and complete combustion without any environmental hazards. Most of these fibers have poor resilience and elastic recovery, damaged by strong mineral acids and dry heat.

10.5 MAJOR DISADVANTAGES OF NATURAL FIBERS

10.5.1 MOISTURE ABSORPTION OF FIBERS

The lignocellulosic natural fibers are hydrophilic and absorb moisture. The swelling behavior of natural fibers is generally affected by their morphology as well as physical and chemical structures. Biofibers change their dimensions with varying moisture content, because the cell wall of polymers contains hydroxyl and other oxygen containing groups, which attract moisture through hydrogen bonding. The hemicelluloses are mainly responsible for moisture absorption, and water penetration through natural fibers can be explained by capillary action (Feughelman and Nordan 1962). The waxy materials present on the surface help to retain the water molecules on the fiber. The porous nature of the natural fiber accounts for the large initial uptake at the capillary region. The hydroxyl group (–OH) in the cellulose, hemicellulose, and lignin build a large amount of hydrogen bonds between the macromolecules in the plant fiber cell wall. Subjecting the plant fibers to humidity causes the bonds to break. The hydroxyl group then forms a new hydrogen bond with water molecules, which induces swelling (Joly et al. 1996). Generally, moisture content in natural fibers varies between 5% and 10%. This can lead to dimensional variations in composites and also affect the mechanical properties of composites. Therefore, the removal of moisture from fibers is very essential before the preparation of the composites. During processing of composites, the moisture content can lead to poor processability and may result in porous products. The moisture absorption of natural fibers can be reduced by proper surface modifications with chemicals or grafting of vinyl monomers. Table 10.4 summarizes the various chemical treatments and

TABLE 10.4
Chemical Treatments Used for Modification of Natural Fibers

Fiber	Chemical Treatments	Coupling Agents/ Compatibilizers
Wood flour	Succinic acid, EHMA, styrene, urea–formaldehyde, *m*-phenylene bismaleimide, acetic anhydride, maleic anhydride, itaconic anhydride, polyisocyanate, linoleic acid, abietic acid, oxalic acid, rosin	Maleated PP, acrylic acid grafted PP, Epolene C-18, Silane A-172, A-174 and A-100, PMPPIC, zirconates, titanates
Jute	Phenol–formaldehyde, malemine–formaldehyde, cardanol–formaldehyde	–
Sisal	NaOH, isocyanate, sodium alginate, *N*-substituted methacrylamide	–
Pineapple	*p*-Phenylene diamine	–
Banana	Sodium alginate	–
Coir	Sodium alginate, sodium carbonate	

Source: Saheb, D.N. and Jog, J.P. *Adv. Polym. Technol.,* 18, 351–363, 1999.

coupling agents used for the modification of natural fibers to improve fiber properties such as fiber finish, compatibility with matrix, and chemical bonding between fiber and matrix (Saheb and Jog 1999). Chemical treatments such as dewaxing (defatting), delignification, bleaching, acetylation, and chemical grafting are used for modifying the surface properties of the fibers and enhancing their performance.

10.5.2 Thermal Stability of Natural Fibers

Natural fibers are complex mixtures of organic materials, and as a result, thermal treatment leads to a variety of physical and chemical changes. The limited thermal stability of natural fiber is one of its drawbacks. The thermal stability of natural fibers can be studied by thermogravimetric analysis (TGA). As mentioned above, natural fiber is composed of mainly cellulose, hemicellulose, and lignin. Each of the three major components has its own characteristic properties with respect to thermal degradation, which are based on the polymer composites. However, the microstructure and the three-dimensional nature of natural fiber are variables, which also play important roles in terms of their effects on combustion behavior. Thus, the individual chemical components of the fiber behave differently if they are isolated or intimately combined within each single cell of the fiber structure (Schniewind 1986).

Lignin, specifically the low-molecular-weight protolignin, degrades first at a slower rate than the other constituents. This process has been described by Shukry and Girgis (1992) and Beall (1986), who also presented an analysis of the products of degradation. The thermal degradation of cellulose-based fibers is greatly influenced by their structure and chemical composition. The natural fiber starts degrading at about 240°C. The thermal degradation of lignocellulosic materials has been reviewed by Tinh et al. (1981a,b) in detail for modified and unmodified materials. Gassan and Bledzki (2001) studied the thermal degradation pattern of jute and flax and found that at temperatures below 170°C, fiber properties were affected only slightly while at temperatures above 170°C, significant drop in tenacity and higher degree of degradation was observed. Because of chain scission, a slight increase in the degree of crystallinity was observed. The thermal degradation patterns of other cellulosic fibers, such as oil palm, sisal, banana, coir, hemp, and jute, was also reported (Ansell et al. 2001; Baiardo et al. 2002; Geethamma et al. 1995; George et al. 1998c; John and Anandjiwala 2008; Pothen et al. 1999). It was reported that chemical modification improved the thermal stability of their composites. Chemically modified fibers showed a satisfactory thermal stability at processing temperatures for potential composites. Thermal degradation of natural fibers is a two-stage process. The low temperature degradation (80–180°C) process is associated with degradation of lignin, whereas the high temperature degradation (280–380°C) process is due to cellulose. The degradation of natural fibers is a crucial aspect in the development of natural fiber composites and thus has a bearing on the curing temperature in the case of elastomers and thermosets and extrusion temperature in thermoplastic composites (Alvarez and Vazquez 2004; Ge et al. 2004). Also, thermal degradation of the fibers results in production of volatiles at processing temperatures above 200°C. This can lead to porous polymer products with lower densities and inferior mechanical properties.

10.5.3 BIODEGRADATION AND PHOTODEGRADATION OF NATURAL FIBERS

The lignocellulosic natural fibers are degraded biologically by very specific enzymes capable of hydrolyzing the cellulose, especially hemicellulose present in the cell wall, into digestible units (Hatakeyama et al. 2005). Lignocellulosic NRs exposed outdoors undergo photochemical degradation caused by UV light. Resistance to biodegradation and UV radiation can be improved by bonding chemicals to the cell wall or by adding polymer to the cell matrix. Biodegradation of cellulose causes weakening of the natural fiber. Photodegradation primarily takes place in the lignin component that is responsible for color changes (Rowell 1985). The surface becomes richer in cellulose content as the lignin degrades. In comparison with lignin, cellulose is less susceptible to UV degradation.

10.6 FIBER–MATRIX INTERFACE AND INTERFACIAL MODIFICATIONS

The incorporation of hydrophilic natural fibers in polymers leads to heterogeneous systems whose properties are inferior due to lack of adhesion between the fibers and the matrix. Thus, the treatment of fibers for improved adhesion is a critical step in the development of such composites. The treatment of the fibers may be bleaching, delignification, grafting of monomers, acetylation, and so on. In addition to the surface treatment of fibers, use of a compatibilizer or a coupling agent for effective stress transfer across the interface can also be explored (Abdulmouleh et al. 2004; Harper and Wolcott 2004; Lu et al. 2005; Thielemans and Wool 2004). The term interface has been defined as the two-dimensional boundary regions between two phases in contact. The interphase in a composite is the matrix surrounding a fiber, and there is a gradient in properties observed between matrix and interphase (Liao et al. 1997). The composition, structure, and properties of the interface may be variable across the region and may also differ in composition, structure, or properties from either of the two contacting phases, fiber and matrix (Schwartz 1984).

This interfacial region exhibits a complex interplay of physical and chemical factors that exert a considerable influence in controlling the properties of reinforced composites. The interfacial interaction depends on the fiber aspect ratio, strength of interactions, fiber orientation and aggregation, and so on (Alvarez et al. 2005; Brahmakumar et al. 2005; Sreekala et al. 1996). Extensive research has been done to evaluate the interfacial shear strength (ISS) of man-made fibers (Wong et al. 2004; Hristov et al. 2004; Cheng et al. 1993; Drazal and Madukar 1993) and natural fibers (Felix and Gatenholm 1991, 1993; Khalil et al. 2001; Mani and Satyanarayan 1990) by using methods such as fiber pull-out tests, critical fiber length, and microbond tests. In fiber-reinforced composites, stresses acting on the matrix are transmitted to the fiber across the interface. For efficient stress transfer, the fibers have to be strongly bonded to the matrix. Composite materials with weak interface have relatively low strength and stiffness but high resistance to fracture, whereas materials with strong interface have high strength and stiffness but are very brittle. The effects are related to the ease of debonding and pull out of fibers from the matrix during crack propagation. The fiber–matrix interface adhesion can be explained by five main mechanisms.

10.6.1 Adsorption and Wetting

Adsorption and wetting is due to the physical attraction between the surfaces, which is better understood by considering the wetting of solid surfaces by liquids. Between two solids, the surface roughness prevents the wetting except at isolated points. When the fiber surface is contaminated, the effective surface energy decreases. This hinders a strong physical bond between fiber and matrix interface.

10.6.2 Interdiffusion

Polymer molecules can be diffused into the molecular network of the other surface such as fiber. The bond strength will depend on the amount of molecular conformation, constituents involved, and the ease of molecular motion.

10.6.3 Electrostatic Attraction

This type of linkage is possible when there is a charge difference at the interface. The anionic and cationic species present at the fiber and matrix phases will have an important role in the bonding of the fiber–matrix composites via electrostatic attraction. Introduction of coupling agents at the interface can enhance bonding through the attraction of cationic functional groups by anionic surface and vice versa.

10.6.4   CHEMICAL BONDING AND CHEMICAL TREATMENT

Natural fibers are hydrophilic and thus exhibit poor resistance to moisture, are incompatible with the hydrophobic polymer matrix, and have a tendency to form aggregates. To eliminate the problems related to high water absorption, treatment of fibers with hydrophobic, aliphatic, and cyclic structures containing reactive functional groups has been attempted (Anderson and Tillman 1989; Bledzki and Gassan 1996; Mannan-Kh and Latifa 1980; Mohanty and Singh 1987; Sahoo et al. 1986; Samal et al. 1989). Hence, these reactive functional groups imparted on the fiber surface are capable of bonding with compatible reactive groups in the matrix polymer, for example the carboxyl group of the polyester resin. Thus, modification of natural fibers makes the fibers hydrophobic to improve interfacial adhesion between the fiber and the matrix polymer. The type of bonding greatly influences the strength of the composite. Interfacial chemical bonding can increase the adhesive bond strength by preventing molecular slippage at a sharp interface during fracture and increasing the fracture energy by increasing the interfacial attraction. Chemical bonding can be achieved by modifying the surface of the fibers.

10.6.5   MECHANICAL ADHESION

The adhesion between the fiber and matrix is an important factor for the composite. Mechanical interlocking at the fiber–matrix interface is possible. The degree of roughness of the fiber surface is very significant in determining the mechanical and chemical bonding at the interface. This is due to the larger surface area available on a rough fiber. Surface roughness can increase the adhesive bond strength by promoting wetting or providing mechanical anchoring sites. The adhesion at the interface can be improved by modification of fiber surface. As discussed before, this can be achieved by changing the hydrophilic surface of the fiber to hydrophobic in nature. A compatibilizer or coupling agent is also used for improving the adhesion between the fiber and matrix (Abdulmouleh et al. 2004; Harper and Wolcott 2004; Lu et al. 2005; Thielemans and Wool 2004). The compatibilizer can be a polymer with functional groups grafted into the chain of the polymer. The coupling agents are tetrafunctional organometallic compounds based on silicon, titanium, or zirconium and are commonly known as silane, titanate, or zirconate coupling agents (Saheb and Jog 1999; Wang et al. 2007). Pedro et al. (1997) found that preimpregnation of cellulose fibers in low-density polyethylene (LDPE)/xylene solution and the use of a coupling agent result in a small increment in mechanical properties of LDPE reinforced with green cellulosic fiber composites that is attributed to an improvement in the interface between fibers and matrix. The fiber treatment also improved the shear properties of the composite and fiber dispersion in the matrix (Pedro et al. 1997). Ishak et al. (1998) used silane coupling agents and compatibilizers to improve the mechanical properties of oil palm fiber filled high-density polyethylene (PE) composites. In all cases, it seems that the mechanical properties of the composites have improved significantly. The use of silane coupling agent enhanced the tensile properties and tear strength of bamboo fiber filled rubber composites (Ray et al. 2001). The silane coupling agent is believed to improve the surface functionality of bamboo fibers and subsequently enable the bamboo fibers to bond chemically to the rubber matrix. The wetting of cellulosic fibers in rubber matrix is also improved by the use of a coupling agent.

Usually, natural fibers are treated with sodium hydroxide (NaOH) to remove lignin, pectin, and waxy substances. Alkalization gives rough surface topography to the fiber. NaOH is used in this method to remove the hydrogen bonding in the network structure of the fiber's cellulose, thereby increasing the fiber's surface roughness (Prasad et al. 1983). The increase in the percentage crystallinty index of alkali-treated fiber occurs because of the removal of cementing materials, which leads to better packing of the cellulose chain and an increase in molecular orientation. The elastic modulus of the fiber is expected to increase with increasing degree of molecular orientation (Gassan and Bledzki 1999b). Superior mechanical properties of alkali-treated jute-based biodegradable polyester composites were attributed to the fact that alkali treatment improves the fiber surface characteristics by removing the impurities of the fiber surface, thereby producing a rough surface morphology (Bisanda and Ansell 1991).

The effect of alkalization and fiber alignment on the mechanical and thermal properties of kenaf and hemp fiber–reinforced polyester composites was studied by Aziz and Ansell (2004). Samal et al. (1995) studied the effect of alkali treatment and cyanoethylation on coir fibers and found that the modified coir fiber showed significant hydrophobicity, improved tensile strength, and moderate resistance to chemical reagents. Hill and Khalil (2000a) studied the benefit of fiber treatment by chemical modification (acetylation) of the fibers and by the use of silane coupling agent on the mechanical properties of oil palm and coir fiber–reinforced polyester composites. They found that acetylation of coir and oil palm fibers results in the increase in the ISS between the fiber and the matrix and increase in the mechanical properties of the composites. George et al. (1998b) analyzed the improved interfacial interaction in chemically modified pineapple leaf fiber–reinforced PE composites. They used various reagents such as NaOH, silanes, and peroxides to improve the interfacial bonding. The influence of fiber surface modification on the mechanical performance of oil palm fiber–reinforced phenol–formaldehyde composites were studied by Sreekala et al. (2000). The effects of various chemical modifications of jute fibers as a means of improving its suitability as reinforcement in biopol-based composites were performed by Mohanty et al. (2000). Sisal fibers were chemically treated with a two-step treatment with sodium sulfate solution followed by acetic anhydride to promote adhesion to a polyester resin matrix (Calado et al. 2003). It was found that the chemical treatment improved the fiber–matrix interaction as revealed by the brittle behavior of the composites reinforced with treated fibers. Although the treatment improved the fiber behavior in relation to moisture, the water absorption capacity of the composites. This should be due to the failure to remove all the unreacted hydrophilic species left by treatment or to the formation of acetyl cellulose microtubes in the treated fiber.

Rozman et al. (2000) used lignin as compatibilizers in coconut fiber polypropylene composites. Since lignin contains both polar hydroxyl groups and nonpolar hydrocarbon and benzene ring, it can play a role in enhancing the compatibility between both components. The composite with lignin as a compatibilizer possessed higher flexural properties compared with the control composites. Lignin also reduces water absorption and swelling thickness of the composites. Thomas and coworkers (Prasanthkumar and Thomas 1995c; Geethamma et al. 1996; Sreekala and Thomas 2003; Joseph et al. 2002; George et al. 1998a; Pothen et al. 1997) have carried out systematic studies on the chemical modification of various natural fibers such as sisal, coir, oil palm, banana, and pineapples and its reinforcing effect on various rubbers, thermoplastics, and thermosets. In all cases, it was observed that the composite properties have greatly improved by using treated fibers due to better fiber–matrix interaction. The studies so far reported prove that utilization of natural fibers in polymeric matrices offers economical, environmental, and qualitative advantages. Owing to the uncertainties prevailing in the supply and price of petroleum-based products, it is highly important to use the naturally occurring alternatives. Proper utilization of indigenously available raw materials will open up new markets for these natural resources. Hence, studies on composites containing natural fibers are important.

10.7 NATURAL RUBBER–SISAL FIBER COMPOSITE

10.7.1 SISAL FIBER

Sisal fiber is a promising reinforcement and one of the most widely used natural fibers for use in composites on account of its low cost, low density, high specific strength and modulus, lack of health risk, easy availability in some countries, renewability, and very easy cultivation. It has short renewal times and grows wild in the hedges of fields and railway tracks (Mukherjee and Satyanarayana 1984). Nearly 4.5 million tons of sisal fiber is produced every year throughout the world. Tanzania and Brazil are the two main sisal-producing countries (Chand et al. 1988). Sisal fiber is a hard fiber extracted from the leaves of the sisal plant (*Agave sisalana*). Although native to tropical and subtropical North and South America, the sisal plant is now widely grown in the tropical countries of Africa, the West Indies, and the Far East (Behera et al. 2012). A sketch and photograph of the sisal plant and cross section of a sisal leaf are shown in Figure 10.1, and sisal fibers are extracted from the leaves. A sisal plant produces about 200–250 leaves and each leaf contains 1000–1200 fiber bundles, which are composed of 4% fiber, 0.75% cuticle, 8% dry matter, and 87.25% water (Mukherjee and Satyanarayana 1984). Normally, a leaf weighing about 600 g will yield about 3% by weight of fiber with each leaf containing about 1000 fibers. The sisal leaf contains three types of fibers (Behera et al. 2012): mechanical, ribbon, and xylem. The mechanical fibers are mostly extracted from the periphery of the leaf. They have a roughly thickened-horseshoe shape and seldom divide during the extraction processes. They are the most commercially useful of the sisal fibers. Ribbon fibers occur in association with the conducting tissues in the median line of the leaf. Figure 10.1a shows a cross section of a sisal leaf and indicates from where mechanical and ribbon fibers are obtained (Behera et al. 2012). The related conducting tissue structure of the ribbon fiber gives them considerable mechanical strength. They are the longest fibers

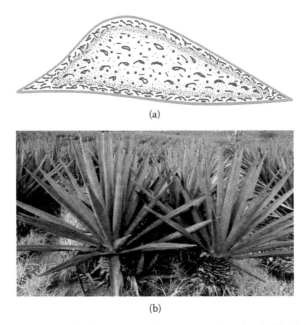

(a)

(b)

FIGURE 10.1 (a) Sketch of a sisal plant and the cross section of a sisal leaf. (b) Photograph of a sisal plant. (From www.brazilianfibres.com. With permission from Brazilian Fibres copyright 2012.)

and compared with mechanical fibers they can be easily split longitudinally during processing. Xylem fibers have an irregular shape and occur opposite to the ribbon fibers through the connection of vascular bundles. They are composed of thin-walled cells and are therefore easily broken up and lost during the extraction process. The processing methods for extracting sisal fibers have been described by Chand et al. (1988) and Mukherjee and Satyanarayana (1984). The methods include (1) retting followed by scraping and (2) mechanical means using decorticators. It is shown that the mechanical process yields about 2% ± 0.4% fiber (15 kg/8 h) with good quality having a lustrous color, while the retting process yields a large quantity of poor-quality fibers. After extraction, the fibers are washed thoroughly in plenty of clean water to remove the surplus wastes such as chlorophyll, leaf juices, and adhesive solids.

The chemical compositions of sisal fibers have been reported by several groups of researchers (Wilson 1971; Rowell et al. 1992; Joseph et al. 1996; Chand and Hashmi 1993; Jacob et al. 2004b) and are tabulated in Table 10.5 (Behera et al. 2012). For example, Wilson (1971) indicated that sisal fiber contains 78% cellulose, 8% lignin, 10% hemicelluloses, 2% waxes, and about 1% ash by weight; but Rowell et al. (1992) found that sisal contains 43%–56% cellulose, 7%–9% lignin, 21%–24% pentosan, and 0.6%–1.1% ash. More recently, Joseph et al. (1996) reported that sisal contains 85% ± 8.8% cellulose. These large variations in chemical compositions of sisal fiber are a result of different sources, ages, measurement methods, and so on. Indeed, Chand and Hashmi (1993) showed that the cellulose and lignin contents of sisal vary from 49.6%–60.95% and 3.75%–4.40%, respectively, depending on the age of the

TABLE 10.5

Proximate Analysis of Sisal Fibers

Properties		Sisal-01	Sisal-02
		Sisal Type	
Moisture (%)		10.11	9.89
Length (cm)		80–115	80–110
Width (mm)		0.3	0.25
Density	Pycnometer method	1.45	1.36
	Width method	1.65	1.54
Tensile strength, $B \times L$ (km)		6.1	5.86
Solubility in alkali (%)			
1% NaOH		22.7	22.76
3% NaOH		23.8	23.92
5% NaOH		30.1	29.72
7% NaOH		32.55	34.34
Solubility in organic solvents (%)			
Ethanol	Cold	0.65	0.66
	Hot	0.80	0.78
Benzene	Cold	1.15	1.22
	Hot	3.79	3.82
Ethanol–benzene (1:1)	Cold	0.85	0.85
	Hot	1.75	1.85
Water	Cold	0.25	0.33
	Hot	0.54	0.59
Acetone extractives (%)		4.79	5.23
Holocellulose (%)		87.05	86.34
α-Cellulose (%)		61.76	67.19
Lignin% (acid insoluble)		11.7	10.22
Ash (%)		1.13	1.27
Silica (%)		0.33	0.47
Pentosan (%)		22.34	19.49
Pectin (%)		1.2	0.9

Source: Behera, S., Sahu, S., Patel, S. and Mishra, B.K. *J. Indian Pulp Paper Tech. Assoc.,* 24, 37–43, 2012.

plant. The length of sisal fiber is between 1.0 and 1.5 m and the diameter is about 100–300 μm (Bisanda and Ansell 1991). The properties of sisal fibers in comparison with oil palm fiber are presented in Table 10.6. The fiber is actually a bundle of hollow subfibers. Their cell walls are reinforced with spirally oriented cellulose in a hemicellulose and lignin matrix. So, the cell wall is a composite structure of lignocellulosic material reinforced by helical microfibrillar bands of cellulose. The composition of the external surface of the cell wall is a layer of lignaceous material

TABLE 10.6
Properties of Sisal and Oil Palm Fiber

Chemical Constituents (%)	Sisal	Oil Palm
Cellulose	78	65
Hemicelluloses	10	–
Lignin	8	19
Wax	2	–
Ash	1	2
Physical Properties		
Diameter (μm)	120–140	150–500
Density (g/cm³)	1.45	0.7–1.55
Tensile strength (MPa)	530–630	248
Elongation at break (%)	3–7	14
Young's modulus (GPa)	17–22	6700
Microfibrillar angle (°)	20–25	46

and waxy substances that bond the cell to its adjacent neighbors. Hence, this surface will not form a strong bond with a polymer matrix. Also, cellulose is a hydrophilic glucan polymer consisting of a linear chain of 1,4-β-bonded anhydroglucose units (Li et al. 1987) and this large amount of hydroxyl groups will give hydrophilic properties to sisal fibers. This will lead to a very poor interface between sisal fiber and the hydrophobic polymer matrix and moisture absorption resistance. Although sisal fiber is one of the most widely used natural fibers, a large quantity of this economic and renewable resource is still underutilized. At present, sisal fiber is mainly used as ropes for the marine and agriculture industry (Mukherjee and Satyanarayana 1984). Other applications of sisal fibers include twines, cords, upholstery, padding and mat making, fishing nets, and fancy articles, such as purses, wall hangings, table mats, and so on (Chand et al. 1984). A new potential application is for manufacture of corrugated roofing panels that are strong and cheap with good fire resistance (Bisanda 1993).

10.7.2 NATURAL RUBBER–SISAL FIBER COMPOSITE AND PROPERTIES

Rubber is the second most widely used matrix for sisal fiber composites behind PE (Kumar and Thomas 1995a,b; Prasanth et al. 1995; Varghese et al. 1994a,b,c, 1993). Short fiber–reinforced elastomer composites are well known because of the easy processing and low cost coupled with high mechanical strength. They are also amenable to standard rubber processing techniques, such as injection molding, extrusion, calendering, and various types of molding techniques. Most of the properties of fiber-reinforced rubber composites strongly depend on microstructural parameters such as fiber diameter, fiber length, volume fraction of fiber, orientation of fiber, and the bonding between the fiber and the rubber. The fiber–rubber interfacial bonding can be improved by modifying the fiber surface by chemical treatment or by adding an external bonding agent. The length of the fiber should be above a critical length for effective stress transfer from the matrix to the fiber to occur, which in turn depends on the bonding between the fiber and the matrix. Similarly, the reinforcing effect of the fibers at low volume fraction is pronounced only when good mechanical bonding exists between the fiber and the matrix. In the composite, rubber matrices include NR and synthetic rubber such as styrene–butadiene rubber. The main research areas concern the effect of fiber length, orientation, loading, type of bonding agent, and fiber–matrix interaction on the properties of the composites, including mechanical properties, rheological behavior, thermal aging, γ-radiation, and ozone resistance.

Varghese et al. (1993, 1994b) studied the effect of acetylation and bonding agent on the aging properties of short sisal fiber–reinforced NR composites that include thermal aging, γ-radiation, and ozone resistance and rheological behavior. The resistance of NR to aging and ozone is poor due to the presence of the reactive double bond in the main chain. Sisal fiber possesses excellent ageing resistance. Moreover, in the presence of high-energy radiation, the reactive sites formed either on the rubber or fiber in the composite may cause increased bonding between the fiber and rubber. The studies found that high fiber loadings in to the composites showed better resistance to aging, especially in the presence of bonding agent. Acetylation of fiber and fiber orientation were also found to reduce the extent of degradation. The dosage of γ-radiation was found to influence in the extent of the retention/degradation process. The retention in tensile strength remains almost constant beyond 6 vol% loading of fiber for the systems with and without bonding agent at low irradiation. On prolonged irradiation (15 Mrad), degradation of polymer chains is the main reaction taking place at low fiber loading, whereas at higher fiber loading, retention of tensile properties is higher. In the case of thermal aging, the retention in tensile properties increases continuously with fiber loading for both systems. The increase in retention is more in composites incorporated with bonding agent. The ozone resistance of the samples is better at higher fiber loading, especially in the presence of bonding agent. In all the cases, the performance of composites incorporating acetylated fiber was better than composites incorporating untreated fiber. In summary, high fiber volume fraction shows better resistance to aging, especially with fiber surface treatment. Fiber orientation is also found to reduce the extent of degradation under these aging conditions. Increasing the dosage of γ-radiation was found to increase the extent of the ageing process.

Further, Varghese et al. (1991, 1992) studied the cure characteristics and mechanical properties of short sisal fiber–NR composite. The study includes the effects of acetylation, aspect ratio, volume loading of the fiber, and a dry bonding system on the above properties. The studies found that between acetylated and untreated short sisal fiber–reinforced NR composites; acetylation improves the adhesion between the rubber and fiber, which leads to higher tensile strength, modulus, and tear strength. The acetylation caused the fiber to fibrillate into the ultimate and the treated surface is highly rough. This facilitates mechanical bonding with rubber matrix. Also, the hydrogen atom on the acetyl carbon atom becomes more reactive due to the presence of the carbonyl group. This may form a chemical link with active sites on rubber, thereby improving adhesion.

It is reported that on increasing the fiber loading, the optimum cure time remains almost constant. This trend is observed for the mixes containing both acetylated and untreated fiber. In the presence of bonding agent, the optimum cure time is lower for mixes containing acetylated fiber when compared with that of compounds containing untreated fiber at the same loading. They have investigated the effect of different bonding agents on the physical and mechanical properties of sisal fiber–reinforced NR composites. The treatments employed included alkali immersion at high temperature and the use of bonding agent based on phenol-formaldehyde or resorcinol formaldehyde, hexamethylenetetramine and precipitated silica (HRH system) at different concentrations. Bonding by the HRH system involves a condensation reaction between resorcinol and the methylene donor (hexa), which takes place during vulcanization, and silica is reported to accelerate this reaction. The bonding resin acts as an intermediate in binding rubber and the fiber. Interestingly, the surface characterization by scanning electron microscope (SEM) images revealed that rubber particles are adhered to the fiber due to good bonding between the treated fiber and rubber. The presence of silica affects the cure time of the composite. Silica retards cure probably by absorbing the curatives. Incorporation of resorcinol and hexamethylenetetramine as the dry bonding agent does not affect cure time. However, in the case of both acetylated and untreated fiber-filled composites, the optimum cure time is slightly lower when the fiber loading is 40 phr (parts per hundred rubber).

The effect of critical fiber length was also studied with acetylated sisal fiber with different fiber lengths (6, 10, and 15 mm). Shear forces during mixing orient most of the fibers along the grain direction and cause breakage of fibers. The extent of fiber breakage was evaluated by dissolution of the compound in benzene, followed by the extraction of fiber and examination of fiber length by a polarizing microscope. The breakage analysis data showed that breakage increases with fiber length. However, no measurable change in average fiber diameter (0.103 mm) occurred during mixing. From the data, it is seen that the compound have mixed with 10 mm long fibers the majority of fibers had an aspect ratio of 20–60 (2–6 mm length) after breakage. Even though an aspect ratio of 100–200 is generally required for effective stress transfer in short fiber–elastomer composites. They reported that composite prepared with 10 mm fiber length showed higher tensile strength and modulus and had 22.4% of the fibers in the range of 4–6 mm length after breakdown. The observation indicates that the original fiber length of 10 mm is essential and sufficient for getting the reinforcement effect in NR—sisal short fiber composite. They concluded that alkalitreated fiber imparts better physical properties to the rubber mixes than untreated fiber and sisal fiber acts as a reinforcing agent only when added above a volume loading of 10 phr. The bonding between sisal fiber and rubber matrix is generally very poor but can be enhanced by resorcinol–formaldehyde pretreated silica. The aging resistance of the rubber–coir composites is excellent for a fiber loading of 30 phr with bonding agents, and mechanical anisotropy is observed at fiber loading in excess of 10 phr.

The effect of adhesion on the equilibrium swelling of short sisal fiber–reinforced NR composites in a series of normal alkanes such as pentane, hexane, heptane, and octane have been studied by Varghese et al. (1995). Their results showed that increased fiber content and the bonding agent reduced the swelling considerably.

It was also found that with improved adhesion between short fiber and rubber, the factor $(V_1 - V_f)/V_1$, decreases, where V_1 and V_f are the volume fraction of rubber in dry and swollen samples, respectively. Prasant and Thomas (1995a,b) also reported increased mechanical strength with fiber loading and bonding agent. As the concentration of fiber increases, the tear strength and tensile strength of sisal fiber styrene–butadiene rubber composite is increased in both longitudinal and transverse directions. The enhancement in tear and tensile strength was almost three to four times higher than those of the unfilled vulcanizates under similar conditions.

The increasing use of short fiber composites in static and dynamic applications led to the importance of stress relaxation measurements, as the behavior of the rubber–fiber interface can be easily detected by stress relaxation studies. Vulcanized rubbers when subjected to constant deformation undergo a marked relaxation of stress both at low and high temperature. The stress under a constant deformation decays by an amount substantially proportional to the logarithm of period in the deformed state. The stress relaxation behavior of short jute fiber-reinforced-nitrile rubber composites with respect to the effects of strain level, bonding agent, fiber content, fiber concentration, temperature, and prestrain has been studied in detail by Bhagavan et al. (1987). They reported that short fiber increases the rate of stress relaxation over the corresponding unfilled vulcanizate. Composites containing bonding agent exhibit slower relaxation than those without a bonding system, and the effect of fiber orientation on the relaxation behavior appears to be marginal. The prestrain decreases the stress relaxation rate considerably, particularly for composites without a bonding agent. Varghese et al. (1994c) studied the stress relaxation behavior of chemically treated (acetylated) short sisal fiber–reinforced NR composites with special reference to the effects of strain level, fiber loading, bonding agent, fiber orientation, and temperature. The sisal fibers were chopped to a length of 10 mm and were immersed in 18% aqueous sodium hydroxide solution followed by washing with water. The fiber is then soaked in glacial acetic acid and then in acetic anhydride containing two drops of concentrated sulfuric acid followed by washing and drying. Composite is prepared by mixing the fiber with NR and vulcanizers along with bonding agents. Experimental results show that, for the best balance of properties, the fiber length is about 6 mm. This is the same as the sisal–PE composites. Orientation effects are as expected. Addition of short sisal fibers to rubber offers good reinforcement, which can be further strengthened by a suitable coupling agent such as a resorcinol–hexa bonding system. They reported the existence of a single relaxation pattern in the unfilled stock and a two-stage relaxation mechanism for the acetylated fiber-filled NR composites. In the two-stage relaxation, the initial relaxation occurred quickly (<200 seconds), and the second-stage relaxation took much longer to complete the process. The initial mechanism is a result of the fiber–rubber attachments and the latter is due to the physical and chemical relaxation processes of the NR molecules. The relaxation process is also influenced by the bonding agent, which indicated that process involved in fiber–rubber interface. For the composites in the absence of bonding agent, the rate of relaxation increased with strain level, but in the presence of bonding agent, the relaxation rate is almost independent of strain level, because of the strong fiber–matrix interface. Gum vulcanite shows only one relaxation process, the rate which is almost independent of the strain level. It was also observed that, in the case of composites with no bonding agent, the modulus is lower and elongation is higher. The initial rate of the stress-relaxation process diminishes after aging (at 70 and 100°C for 4 days).

Mechanical properties and cure characteristics of NR composites reinforced with sisal/oil palm hybrid fibers was studied by Jacob et al. (2004a). Composite was prepared by reinforcing NR with untreated sisal and oil palm fibers chopped to different fiber lengths. The influence of fiber length and concentration (fiber loading) on the processability characteristics and mechanical properties of the hybrid composites was studied. The properties of short fiber–reinforced elastomer composites depend on the degree to which an applied load is transmitted to the fibers. The extent of load transmittance is a function of the fiber length and the magnitude of the fiber–matrix interaction. At a critical fiber length (l_c), the load transmittance from the matrix to the fiber is at a maximum. If l_c was greater than the length of the fiber, the stressed fiber debonds from the matrix and the composite fails at a low load. It is found that optimum lengths for the sisal and oil palm fibers were 10 and 6 mm, respectively, and were effective for reinforcement in the NR matrix. Increasing the fiber length resulted in a decrease in the properties. This was due to fiber entanglements prevalent at longer fiber lengths. The mechanical properties of the composites in the longitudinal direction were superior to those in the transverse direction. The longitudinal orientation of fibers resulted in a maximum tensile strength, and as the angle of orientation (transverse direction) of the fibers increased, the tensile strength decreased. When the fibers were longitudinally oriented, the fibers were aligned in the direction of force, and the fibers transferred stress uniformly. When transversely oriented, the fibers were aligned perpendicular to the direction of load, and they could take part in stress transfer. Generally, the tensile strength initially drops up to a certain amount of fiber and then increases. It was observed that tensile strength decreased initially up to 10 phr and then starts to increase to optimum with 30 phr fiber loading. Further increasing the concentration of fibers resulted in a reduction in the tensile strength properties and tear strength but an increase in the modulus of the composites. At low levels of fiber loading (in this case <10 phr), the orientation of the fibers is poor, and the fibers are not capable of transferring load to one another, and stress gets accumulated at certain points in the composite, which leads to a low tensile strength. At high levels of fiber loading, the increased population of fibers leads to agglomeration, and stress transfer gets blocked. At intermediate levels of loading (in this case 30 phr), the population of the fibers is just right for maximum orientation and the fibers actively participated in stress transfer. Fiber breakage analysis was performed and it was found that the extent of breaking of the sisal and oil palm fibers was low. The vulcanization parameters such as cure time and scorch time were found to be independent of fiber loading. The breakage of fibers due to high shear forces occurs during mixing. This indicates that control of the fiber length and aspect ratio of the fibers in a rubber matrix is difficult. The severity of fiber breakage depends mainly on the type of fiber, the initial aspect ratio, and the magnitude of stress and strain experienced by the fibers during processing.

Dynamic mechanical analysis (DMA) to determine the dynamic properties as a function of temperature for NR reinforced with different fiber loadings of sisal and oil palm fibers and the composites treated with different concentrations of NaOH solution was reported by Jacob et al. (2006). DMA is a very useful technique for investigating the mechanical behavior of a material on the basis of its viscoelasticity. Generally, DMA can provide more information about a given polymer matrix than other mechanical tests, since it can cover a wide range of temperatures and frequencies especially sensitive to all kinds of transitions and relaxation processes of matrix resin and also to the morphology of the composites. The three important parameters that can be obtained during a dynamic mechanical test are (1) storage modulus, which is a measure of the maximum energy, stored in the material during one cycle of oscillation. The test also gives an idea of the stiffness behavior and load-bearing capability of composite material, (2) loss modulus, which is proportional to the amount of energy that has been dissipated as heat by the sample, and (3) mechanical damping term tan δ, which is the ratio of the loss modulus to storage modulus and is related to the degree of molecular mobility in the polymer material. The other useful quantities that can be characterized by DMA are storage and loss compliance, dynamic and complex viscosity, creep compliance, and the stress relaxation modulus. The study showed that storage modulus and loss modulus, increases with increasing sisal and oil palm fiber content at all temperatures compared with the gum stock. The increment is prominent in the glassy state below glass transition temperature, T_g, although there is not much effect in the rubbery plateau region. It was noted that in the glassy region, the modulus values gradually increase, whereas in the rubbery region, there is not much change. In the glassy region, the components are in a frozen state and are highly immobile. In such a state, there exists a close and tight packing resulting in high modulus. As temperature increases, the components become more mobile and lose their close packing arrangement. As a result, in the rubbery region, there is no significant change in modulus. The dynamic mechanical properties of a composite material depend on the fiber content, the presence of additives such as filler, the compatibilizer, the fiber orientation, and the mode of testing. The loss modulus value increases with fiber loading reaching a maximum for a composite containing 50 phr loading. The maximum heat dissipation occurs at the temperature where E'' is at a maximum. The gum compound showed that the loss modulus peak was at −54°C, which was attributed to the mobility of the chains. An interesting observation is the slight broadening of the loss modulus peaks with increase in fiber loading. This could be due to the increase in energy absorption caused by the addition of fibers. Another reason is that the rubber matrix surrounding the fibers is in a different physical state to the rest of the matrix, thus hindering the molecular motion. The value of tan δ_{max} decreases with fiber loading. Incorporation of fibers acted as barriers to the mobility of rubber chains, leading to lower flexibility, lower degrees of molecular motion, and hence lower damping characteristics. Another reason for the decrease of tan δ_{max} is that there is less matrix material by volume to dissipate the vibration energy and also due to fiber agglomeration at high levels of fiber loading.

The effect of chemical treatment and bonding agent on the dynamic properties of composite was also studied. For that the composite is treated with different concentration of NaOH varying from 0.5% to 4%. Both storage modulus and loss modulus values increase with chemical treatment, while tan δ value decreased. Maximum E' is exhibited by the composite prepared with fibers treated with 4% NaOH, while the composite containing fibers treated with 2% NaOH exhibits the maximum loss modulus. In the case of untreated composites, the hydrophilic nature of sisal and oil palm induces poor wettability, and the presence of moisture at the fiber–rubber interface promotes the formation of voids at the interface and produces composites of lower stiffness and strength. For the alkali-treated composites, the moisture absorbing capacity of fibers is reduced leading to improved wetting, which produces a strong interfacial bond with the rubber matrix giving rise to a much stiffer composite with high modulus. Another reason for high storage modulus of the treated composites is that since treated fibers have more surface area, there is the formation of more cross links within the rubber matrix–fiber network leading to high storage modulus values. Among the treated composites, the height of the tan δ peak is somewhat the same for all the composites. This shows that all the treated composites possess the same order of damping capabilities. The reason why treated composites possess low tan δ is because of the strong and rigid fiber–matrix interface due to improved adhesion, which reduces molecular mobility in the interfacial zone. Another factor is that presence of bonding agent results in more cross links being formed, and alkali treatment of fibers leads to the strengthening of these cross links. As a result, molecular motion along the rubber macromolecular chain is severely hindered leading to low damping characteristics. Also, alkali treatment shifted T_g to higher temperatures. It is obvious that chemically treated composites have decreased tan δ value as compared to untreated composites.

10.8   NATURAL RUBBER–COIR FIBER COMPOSITE

Coir is an inexpensive fiber among the various natural fibers available in world. Coarse coir fiber is obtained from the fibrous mesocarp of coconuts, which is available in large quantities as residue from coconut production areas. Furthermore, it possesses the advantages of a lignocellulosic fiber. Coconut coir is an interesting product as it has the lowest thermal conductivity and bulk density as compared with other natural fibers. It is not as brittle as glass fibers, amenable to chemical modification, nontoxic, and poses no waste disposal problems. Because of its hard-wearing quality, durability, and other advantages, it is used for making a wide variety of floor furnishing materials, yarn, rope, and so on, which consume only a small percentage of the total coir production (Satyanarayna et al. 1981). Coir fiber has certain advantages over other natural fibers. It possesses high weather resistance due to a higher amount of lignin and absorbs water to a lesser extent due to its lower cellulose content. The fiber can be stretched beyond its elastic limit without rupture due to the helical arrangement of microfibrils at 458°. Hence, research has been undertaken to identify new fields of application for coir such as reinforcement in fiber-reinforced polymer composites (Rout et al. 2001a,b). The addition of coconut coir reduced the thermal conductivity of the composite specimens and yielded a lightweight product. Unfortunately, the performance of coir as a reinforcement in polymer composites is unsatisfactory and not comparable even with other natural fibers. This inferior performance of coir is due to various factors such as its low cellulose content, high lignin and hemicellulose content, high microfibrillar angle, and large and variable diameter. The various physical properties of coir fibers are compared with those of other natural fibers in Table 10.7 (Satyanarayana et al. 1982). Development of

TABLE 10.7

Comparison of Cost and Properties of Coir Fiber with Other Natural Fibers

Properties	Coir	Banana	Sisal	Pineapple	Jute
Diameter (μm)	100–460	80–250	50–200	20–80	–
Density (g/cm³)	1.15	1.35	1.45	1.44	1.45
Microfibrillar angle (°)	30–49	11	10–22	14–18	8.1
Cellulose/lignin content (%)	43/45	65/5	67/12	81/12	63
Elastic modulus (GPa)	4–6	8–20	9–16	34–83	
Tenacity (MPa)	131–175	529–759	568–640	413–1627	533
Elongation (%)	15–40	1.0–3.5	3–7	0.8–1.6	
Cost (relative to coir)	1	3	1.5	1.5	2

Source: Satyanarayana, K.G., Pillai, C.K.S., Sukumaran, K., Pillai, S.G.K., Vijayn, K. and Rohatgi, P.K. *J. Material Sci.,* 17, 2453–2462, 1982.

composite materials for buildings using coconut coir with low thermal conductivity is an interesting alternative that would solve environment and energy concerns.

Investigations carried out by Hill and Khalil (2000b) and Monteiro et al. (2005, 2006a,b) showed that coir fibers are not an effective reinforcement for polymer matrix composites. The water adsorbed into the lignocellulosic surface of the hydrophilic coir fiber apparently prevents an efficient adhesion to the hydrophobic polymer matrix, which also happens in other natural fiber composites (Vasquez et al. 1999). As a consequence, the incorporation of coir fiber tends to decrease the mechanical strength of polymer composites for any volume fraction of fiber (Monteiro et al. 2005). In principle, there are ways to reverse this decreasing mechanical properties condition. Therefore, Monteiro et al. (2006a) proposed a strong alkali treatment of coir fiber, which improves the adhesion of coir fiber to the polymer matrix and thus increases the composite's strength by approximately 50% for a volume fraction of 30% of coir fiber–reinforced polyester composite. Another possibility of effective reinforcement to a polymer matrix could be obtained through the selection of thinner coir fiber. A considerable amount of research has been carried out on coir fiber–reinforced polymer composites (Biswas et al. 2011; Gassan and Bledzki 1999a; Hill and Khalil 2000b; Monteiro et al. 2005, 2006a,b; Prasad et al. 1983; Rout et al. 2003; Satyanarayana et al. 1986; Vasquez et al. 1999). Composite properties of coir fiber–reinforced polyester (Prasad et al. 1983; Monteiro et al. 2005, 2006a; Rout et al. 2003; Satyanarayana et al. 1986; Santafé Júnior et al. 2010; Monteiro et al. 2009b) and epoxy (Biswas et al. 2011) have been studied and tested as helmets, roofing, and postboxes (Satyanarayana et al. 1986). These composites, with coir loading ranging from 9 to 15 wt%, have a flexural strength of about 38 MPa. Mechanical properties (tested in tension, flexure, and notched Izod impact) of coir–polyester composites with untreated and treated coir fibers, and with fiber loading of 17 wt%, were reported (Rout et al. 2003). Recently, the effect of fiber length on the mechanical behavior of coir fiber–reinforced epoxy composites was reported (Biswas et al. 2011). The hardness of the composite decreases with increase in fiber length up to 20 mm. However, further increase in fiber length increases the microhardness value. The resistance to impact loading and flexural strength of coconut coir fiber–reinforced epoxy composites improves with increase in fiber length. It is interesting to note that flexural strength and tensile moduli increases with increase in fiber length. Tensile properties showed a gradual increase with increase in fiber length reaching a maximum at about 30 mm (13.05 MPa). The study has confirmed that coconut coir fiber–reinforced epoxy composites have better tensile strength (13.05 MPa), tensile modulus (2.064 GPa), and flexural strength (35.42 MPa). However, less attention has been paid to the incorporation of coir fibers to form elastomeric composites.

An effect of loading, fiber length, orientation, and alkali treatment on short coir fiber–reinforced NR composites and their dynamic mechanical behavior was studied by Geethamma et al. (2005). The vulcanization parameters, processability characteristics, and stress–strain properties of these composites were analyzed. Monsanto R-100 rheometer studies showed maximum torque value (measure of cure characteristics), T_{max}, decreases gradually with increase in fiber length. The highest viscosity (T_{max}) of the mix containing coir fibers of lower length is attributed to the increased number of fiber ends, which enhance the overall friction of the compound. The mechanical properties of the composites in the longitudinal direction were superior to those in the transverse direction; the optimum length for coir fiber in NR system was found to be 10 mm in order to achieve good reinforcement in NR composites. Tensile moduli in the longitudinal direction increased rapidly and linearly as the length of the fibers were increased up to 10 mm length. The modulus at 20% elongation was almost constant after 10 mm length. But the modulus at 10% elongation reached a maximum for the composite containing coir fibers of length 10 mm and then decreased with increasing fiber length. However, the tensile modulus in the direction perpendicular to the orientation of fibers decreased considerably and more or less linearly as the length of the coir fibers increased. As compared with that of the gum compound (compound without fiber), the tensile strength decreased drastically up to the composite containing coir fiber of length 6 mm and then increased gradually reaching a maximum for the composite containing 10 mm long fibers. Further increase in fiber length reduces the tensile strength due to fiber entanglement. This suggested that the critical fiber length is 10 mm for effective tensile stress transfer between fiber and matrix. The same trend can be seen in the case of tear strength properties where the values show maxima at 10 mm fiber length in transverse and longitudinal directions. Due to the brittle nature of the fiber, addition of coir fiber decreases the elongation at break values of NR composites. The higher mechanical properties in the longitudinal direction show the anisotropic behavior of coir fiber–reinforced NR composites. When the fibers are oriented in the longitudinal direction, the mechanical property values are much higher than those in the transverse direction. The breakage of fiber in the composite containing the original fiber length of 10 mm was determined and it was observed that majority of fibers retain a length in the range of 6–10 mm after mixing in a two roll mill. Due to the intrinsically flexible nature of cellulosic fibers, they can bend and curl during milling and thus avoid severe breakage. Hence, the breakage of coir fibers is comparatively low when it is used as a reinforcing fiber in NR composites.

   The composite properties were enhanced by the chemical treatment of coir fiber with 5% sodium hydroxide (Geethamma et al. 1998, 1995) and sodium carbonate solution (Geethamma et al. 1998). Sodium hydroxide–treated fibers were given a pretreatment with a mixture of 1% NR solution or 6% depolymerized liquid NR solution followed by 1% toluene diisocyanate (TDI) solution (Geethamma et al. 1998). The effect of chemical treatment and bonding agent on composite properties as a function of treatment time (Geethamma et al. 1995) was studied and found that rubber–coir interface bonding was improved by the addition of a resorcinol-hexamethylenetetramine dry-bonding system. When the coir fiber was treated with sodium hydroxide, the color of the fiber changed from pale yellow to brown. Also the fibers became thinner upon alkali treatment. This may be because of dissolution and leaching out of fatty acids, phenolic compounds, and their condensation products by alkali, which form the waxy cuticle layer or even the lignin component of the fiber. There was a large difference in scorch time between gum (29 minutes 48 seconds) and compounds containing chemically treated fibers and bonding agents were reported. The difference is due to the alkaline nature of the treated fiber and/or methylene donor and hexamethylenetetramine present in the bonding system. However, there is no remarkable change in scorch time with the duration of alkali treatment of fiber (Geethamma et al. 1995). The effect of chemical treatment on interfacial adhesion was also studied and reported (Geethamma et al. 1998, 1995). The interfacial adhesion of the components in the composite was improved by two means: (1) by the formation of a strong interlayer (Geethamma et al. 1998) and (2) by incorporation of an HRH bonding system (Geethamma et al. 1998, 1995). The interlayer was produced by giving a coating of either NR or depolymerized NR to the surface of the coir fibers (Geethamma et al. 1995). The tensile strength of the composite containing NaOH-treated fibers that are subjected to pretreatment with NR and TDI solution is about 120% higher than the composite containing raw coir fibers subjected to treatment with NR and TDI solution. So it is clear that even though both raw and alkali-treated coir fibers were subjected to the same chemical treatment, the latter are better at reinforcing the NR matrix. This indicates that a pretreatment of NR and TDI solution to raw coir is not sufficient to produce good interfacial adhesion with the rubber matrix because of the waxy cuticle layer on the surface of the raw coir fiber. When alkali-treated coir is given a coating of depolymerized NR and TDI solution, it exhibits certain anomalous characteristics. The tensile strength of composite containing fiber pretreated with depolymerized NR and TDI is lower than that containing NR solution-treated fibers. The moduli values at three different elongations, namely 50%, 100%, and 200%, are highest for composite having NR solution-treated fibers. This indicates the improved stiffness of the composite due to the pretreatment of fibers with NR solution. However, elongation at break values is smaller and tear strength of this composite is about 25% lower.

   The two-component (hexamethylenetetramine and resorcinol formaldehyde: HR system) (Geethamma et al. 1998, 1995) and three-component (HRH system) (Geethamma et al. 1998) bonding system is used for studying the effect of a bonding system on the adhesion of fiber to rubber matrix. The tricomponent dry-bonding system (HRH system) consists of resorcinol, hexamethylenetetramine, and hydrated silica, while in the dicomponent system (HR system), hydrated silica was absent.

The alkali treatment of the coir fiber was also found to increase wettability and the enhanced fiber to rubber adhesion in composites. The use of bonding agent also plays a key role in the adhesion between coir fiber and matrix. The swelling was found to be smaller in composites containing alkali-treated coir fiber along with the resorcinol-hexamethylenetetramine bonding agent. Analysis of stress–strain curves shows that silica is not needed in the bonding system for coir fiber–reinforced NR composites. Among the composite where raw coir fibers are used as reinforcement, the bonding system containing silica exhibits lower tensile strength than the composite without silica. Similarly, higher tensile strength is observed for the composite containing sodium carbonate-treated coir fiber vulcanized with a two-component bonding system. This indicates that silica is an unnecessary component in the bonding system. Similarly, the tensile moduli and tear strength for all three different percentage elongations, 50%, 100%, and 200%, are greater than those composites with the three-component bonding system in both longitudinal and transverse directions. The elongation at break values of mixes without silica was lower than those of mixes with silica. This confirms the better fiber–rubber adhesion in the former mixes. Thus, from the mechanical properties of the composites, it can be observed that silica is not essential in the bonding system in order to achieve good interfacial adhesion. The moduli values in the longitudinal direction are higher than those in the transverse direction. However, the stiffness across the fiber direction is somewhat similar to that of the rubber matrix. The moduli of composite containing sodium hydroxide–treated fiber pretreated with NR solution is higher than composite prepared with sodium hydroxide–treated fiber pretreated with depolymerized NR solution and untreated fiber pretreated with NR solution for all three different elongations. This is because of the higher fiber–matrix adhesion that exists in the composites containing alkali-treated coir fibers coated with NR solution. This increase in modulus is similar to that achieved by the usual chemical cross-linking reaction. But the use of short fibers leads to anisotropic behavior of the modulus, which is rather unlikely in the normal cross-linking reaction. O'Connor (1977b) compared the mechanical properties of composites reinforced with five kinds of fibers and found that their mechanical properties depend on the type, volume loading, aspect ratio, orientation, and dispersion of fiber and fiber–matrix adhesion. It was also reported that for cellulosic fibers, a dicomponent dry-bonding system consisting of hexamethylenetetramine and resorcinol is sufficient for getting good fiber–rubber adhesion, instead of the normal tricomponent dry-bonding system consisting of hexa, resorcinol, and silica.

For the entire composite, the swelling ratio, aL, in the longitudinal direction is lower than aT and aZ, which are the swelling ratios in the transverse and thickness directions, respectively. It is interesting to note that aL values of composites vulcanized with the three-component bonding system are higher than those with the two-component bonding system. This indicates that the interface is stronger in the absence of silica. This again establishes the better performance of composites that do not contain silica. The aT values (the transverse swelling ratios) of the composite without bonding agents show that, perpendicular to the preferred orientation, the swelling is independent of the nature of fibers and the value is roughly equal to that of the gum rubber. The fiber distribution is higher in the direction perpendicular to the application of pressure during molding, that is swelling is reduced in the transverse direction because of the presence of a higher amount of fibers. Hence, in most cases, the aT values are less than aZ values. Thus, it is obvious that in coir fiber–reinforced rubber composites swelling occurs in the thickness direction (Geethamma et al. 1998).

Geethamma et al. (1998) also studied the effect of chemical modification, loading, and orientation of fiber in 10 mm long short coir fiber–reinforced NR composite. The fiber was chemically treated with 5% sodium hydroxide or sodium carbonate solution for 48 hours. Sodium hydroxide–treated fibers were pretreated with a mixture of 1% NR solution or 6% depolymerized liquid NR solution followed by 1% TDI. To study the effect of fiber loading, composite is prepared with different fiber loading varying from 10 to 60 phr of NaOH-treated fiber pretreated with NR solution and TDI. In general, tensile strength of fiber-reinforced composite initially drops up to a certain level of fiber loading and then increases. This minimum volume of fiber required for improving the tensile properties of composite is known as the critical volume. The critical volume varies with the nature of fiber and matrix, fiber aspect ratio, fiber–matrix interfacial adhesion, and other factors. Critical loading of jute and silk is reported to be 40 and 7.5 phr, respectively (Setua and De 1983). The tensile properties in the longitudinal and transverse direction of the aforementioned alkali-treated coir fiber–NR composite sharply drops with increasing fiber loading up to 30 phr. The longitudinal tensile strength is slightly increased by further fiber loading, while the tensile strength in the transverse direction decreased negligibly. This decrease in tensile strength is due to the development of large stresses at low strains and the nonuniform distribution of stress below the critical fiber volume at which the fibers are not enough to restrain the matrix, large stresses will be developed at low strains and the distribution of these stresses will not be uniform. After 30 phr fiber loading, the fibers are sufficient to restrain the matrix, the stress distribution will be uniform, and therefore the fibers start reinforcing the matrix. It was reported that critical fiber volume (the minimum amount of fiber needed to restrain the matrix) is smaller when the matrix strength is higher (Dzyura 1980) in the composite. NR is a very strong rubber because of its strain-induced crystallization behavior. However, in Geethamma et al.'s (1998) study, the critical fiber loading was not obtained even at the 30 phr fiber level. A slight increase in tensile strength was observed only after 40 phr fiber loading. This may be due to the poor fiber–rubber adhesion. They observed tensile strength decreases with increase in angle θ for all fiber loadings except at 30 phr loading. This indicates that the anisotropy of composites containing 30 phr fiber is comparatively low.

The effect of treatment time of fiber with NaOH and the composite properties were reported. The tensile moduli of longitudinally oriented composites at 20% elongation and tensile strength increase with soaking time and reach a maximum at 48 hours followed by a decrease. The fiber orientation was maximum for composites containing coir fiber treated in NaOH for 48 hours. The tear strengths in both directions show maxima at 24-hour NaOH treatment. This is associated with the better interaction of the treated fiber and rubber. However, prolonged treatment up to 72 hours decreases the properties marginally. This may be due to the excessive removal of binding material, such as lignin, hemicellulose, and others, on prolonged immersion in the alkali solution. These decrease the fibrous properties of coir fibers, and as a consequence, the composite properties are decreased. This trend is true in both the transverse and longitudinal directions (Geethamma et al. 1995). In further studies, Geethamma et al. (2005) evaluated the dynamic mechanical behavior of short coir fiber–reinforced NR composites. DMA is an effective tool to determine the dynamic glass transition temperature, T_g, morphology of crystalline polymers, and damping (heat dissipation). The dynamic T_g is defined as the temperature at which maximum of the tan δ or E'' occurs or middle point of E' versus temperature curve or the region where E' increases with increasing frequency at constant temperature. For a polymer system, values should be the same at a specific frequency. In this study, the sodium hydroxide-treated coir fiber is pretreated with resorcinol–formaldehyde–latex (RFL) or TDI solution in presence of triethylamine catalyst. The TDI-treated fibers were further treated with 1% NR solution in toluene and the RFL-treated fiber is bleached in a solution of sodium silicate and hydrogen peroxide.

Dynamic mechanical properties of the coir fiber–reinforced NR composite behave quite different from gum compound. For the gum compound, the maxima in the tan δ and E'' curves and the middle point of the E' curves are almost coinciding with one another at specified frequencies. This is due to the complexity of dynamic mechanical behavior of these composites arising from the restricted movement of NR molecules in the presence of coir fibers. This behavior depends mainly on the dynamic mechanical properties of NR matrix, coir fiber, and the interface between them. It was observed that as frequency increases the values of tan δ and E'' decreased whereas the values of E' increased in the case of both gum and composites. Fiber incorporation increases the E'' of the composites at any temperature indicating the higher heat dissipation (heat build-up) of the short coir fiber–reinforced NR composites compared with that of the gum. Also, the E'' of composites exhibits a highly significant drop in the transition zone that is narrower than that of the gum compound. The position of E'' peaks of these composites also varies with fiber loading. The heat dissipation increased with fiber loading in these composites. The increase in damping at the interface with fiber loading can be explained by the fact that with a larger interfacial area, there is more energy loss. The values of E' increased with fiber incorporation in the low temperature region for the composite compared with that of gum. But this increase was less compared to that of E'' values. Also E' depends only slightly on fiber loading in the rubbery state, whereas a notable variation was observed in the glassy state. The curves of E' in the rubbery state became flatter and the rubbery range extends more to the higher temperature side when fiber loading increases. This suggests that the thermal stability of composites increases for these composites with fiber loading. A tan δ versus temperature curve of two-phase systems shows two peaks characteristic of the T_g of each component (Guo and Ashida 1993). Similarly, two prominent peaks were observed in the tan δ versus temperature curve of these composites due to the dynamic mechanical behavior in matrix and fiber at lower and higher temperature, respectively. An additional small and broad peak observed in the intermediate region represents the dynamic mechanical behavior in the interface. The magnitude of the tan δ peak is characteristic of two factors: (1) the relative concentration of the two components and (2) whether or not the phases are dispersed or continuous. Thus, the tan δ peaks in the low temperature region are larger if the component with lower T_g is the continuous phase. This is true for the coir fiber–reinforced NR composites.

To study the effect of chemical treatment of fiber, DMA is an effective method to evaluate the interfacial bonding in composites. This revealed that composite with poor interfacial bonding tends to dissipate more energy than those with good interfacial bonding. The interfacial adhesion was improved by three methods, such as surface modification of coir fiber, incorporation of HR bonding system, and the formation of an interlayer. The composite containing fibers subjected to bleaching exhibited very high tan δ values in the low temperature region, but the lowest values were observed in the high temperature region. The composite containing RFL-treated fiber showed the lowest tan δ value in the low temperature region. All the composites showed higher tan δ values in low temperature regions and lower values at higher temperature. This showed that these composites are good elastomeric compounds at higher temperatures. The untreated coir fiber has an uneven surface due to globular protrusions, which disappeared as a result of alkali treatment, leading to the formation of a large number of voids. These voids promote better mechanical anchorage between the fiber and rubber matrix. Also, the extent of interaction with NR solution during subsequent treatment will be greater owing to the increased surface area of these fibers. The activation energy, ΔE, for the glass transition of different compounds was studied using the modified form of Arrhenius equation. The increase in values of ΔE with fiber loading was observed, which indicates that mobility of the polymer chain is decreased in the presence of fibers and thereby the T_g is shifted to higher temperatures. The composite prepared with fiber treated with hydroxide, NR, and TDI solution along with two-component bonding agent (resorcinol and hexa) showed the highest ΔE and that with untreated fiber showed the least ΔE, which is comparable with gum stock. The peak temperatures of the tan δ curve of composites at low temperatures are shifted to a higher temperature range in comparison with that of gum. This shifting is due to the immobilization of the polymer matrix in the vicinity of the fiber due to interfacial bonding between the two phases. Hence, higher shifts are observed for composite with higher interfacial bonding. The highest shift in T_g for composite containing bleached fiber indicates that it has the highest interfacial bonding and the lowest shift is for composite containing untreated fiber, which is comparable with gum stock.

Geethamma et al. (1996) and Sreekala and Thomas (2003) studied the melt flow behavior of NR composites containing untreated, acetylated, and γ-irradiated coir fibers using an Instron Capillary Rheometer. Coir fibers of length 10 mm were used at different loading, namely 5, 10, 20, and 30 phr. Acetylated coir fiber was prepared by first soaking the coir fiber in 5% sodium hydroxide followed by soaking in glacial acetic acid and then in acetic anhydride containing a small amount of concentrated sulfuric acid. They analyzed the dependence of melt viscosity flow behavior index (n') and extrudate deformation of the composites with shear rate, shear stress, and fiber loading. Breaking of fibers during compound preparation and die swell of the composite also studied. Generally, reinforcing fibers were damaged during the two roll mill mixing and the subsequent fabrication stages due to the high shear force subjected by the fibers. But this damage is minimum for lignocellulosic fibers. In the case of short coir fiber–reinforced NR composites, the breakage during mill mixing was minimum. After mixing, the majority of the fibers with original length of 10 mm break down to 5–6 mm. The fiber length further breaks down to 3–5 or

1–3 mm during processing (extruded in capillary rheometer) at shear rates of 3.64 and 1224 s^{-1}, respectively.

The highest melt viscosity is reported for the gum compound compared with the compounds containing 30 phr of acylated and untreated coir fiber. This is contrary to the behavior of other natural fiber–reinforced rubber composites reported by Murthy et al. (1985) for NR short jute fiber composite and Varghese et al. (1993) for NR short sisal fiber composite; in both cases, the viscosity of the composite increases with fiber loading. They explained viscosity increase in terms of fiber–fiber interaction rather than the fiber–matrix interaction. In the case of NR coir fiber composite, the observed decrease in viscosity upon the addition of fiber is associated with the fiber migration. The melt viscosity decreases with increase in shear rate. The effect of shear stress on melt viscosity is studied with composite containing different loading of acylated or composite containing 30 phr of γ-irradiated fibers with different dosage of radiation. The melt viscosity of the composite is higher for the composite containing 5 phr fiber loading while the viscosity is lower than the gum compound upon further fiber loading. From the SEM analysis, they found the higher viscosity of the composite to be due to the nonuniform distribution of fibers, and the fibers tend to be present on the outer surface layers of the extrudate. The reduction of viscosity at higher fiber loading is explained in terms of a sheath-core structure in the extrudate where the less-viscous NR matrix encapsulates the hard coir fibers. At higher fiber loading, the majority of the fibers are concentrated in the central portion of the extrudate forming a sheath-core structure (observed in SEM images of 20 and 30 phr fiber loading). Differences in shear viscosity between the two components lead to interface distortion, which results in the encapsulation of the high viscous melt by the low viscous melt. At high shear rates, the stratification of the extrudates of coir fiber–reinforced natural composites is associated with the migration of the low-viscosity NR constituent to the surface of the extrudate forming a sheath around the more stiff coir fibers (Crowson et al. 1980; Thomas et al. 1987). The soft NR present as the sheath facilitates the easy extrusion of the composite and thus a decrease in viscosity was observed. Everage (1973) and MacLean (1973) explained the stratification of multiphase systems based on the tendency to attain a configuration with the lowest rate of viscous dissipation. The rheometric studies support the same phenomenon by observing that maximum torque is lower in the case of the fiber containing composite. The reduction in viscosity of fiber-reinforced composite may also due to the higher pseudoplasticity induced by the fibers (Dzyura et al. 1983). The melt viscosity of the irradiated fiber compounds is higher than that of the gum compound at low shear rates. This may be associated with the improved interaction between the fiber and matrix due to the γ-radiation. However, at high shear stress, this interaction disappears. At a lower shear rate, the irradiated fiber composite showed similar viscosity, which indicates that the time of γ-irradiation has little effect on melt viscosity of the composite.

The change in pseudoplasticity index with the loading of acetylated coir fibers indicated the non-Newtonian or pseudoplastic nature of a polymeric material. These composites obey the power law model for fluids. The alkali treatment of fiber improves the pseudoplastic nature of the composite. The loading of fiber first decreases the flow behavior index to a minimum then starts to increase to the maximum and thereafter decreases gradually. Incorporation of fibers enhanced the resistance to shape distortion of the extrudates. This property increases with increase in fiber loading. The die swelling of composites containing coir fibers was lower than the gum compound. The die swell values of gum rubber compound increases with increasing shear rates and it decreases with increase in fiber loading. At higher fiber loading, die swell value was found to be negligible due to the hindrance for the rubber matrix to relax.

Cellulosic microfibers were prepared from coir via a combination of chemical and mechanical treatments (Bipinbal et al. 2010). A composite of NR coir fiber (microfibers with a diameter range of 8–11 μm and length of 350–550 μm) was prepared by mixing the pulp of coir fiber treated with alkali followed by acid treatment in the concentrated NR latex. The composite solution containing different loading of coir fiber varying from 10 to 40 phr was coagulated and converted to rubber sheet. The effect of fiber loading on processing characteristics and mechanical properties of the composite are studied. The differential torque, restraining, tensile strength, modulus, and tear strength increased, whereas the scorch time, cure time, and elongation decreased with fiber loading. A sudden drop in tensile strength and tear strength is observed beyond 10 and 20 phr fiber loading, respectively, which may be due to the agglomeration effect of fibers. As expected, the observed pit formations and strations on the surface of the fibers do not effectively promote mechanical anchorage between the fiber and the rubber matrix, especially at a fiber loading more than 10 phr.

10.9   FACTORS AFFECTING THE COMPOSITE PROPERTIES

10.9.1   Mixing of Rubber Compounds

The conventional mixers like Banbury and the open two-roll mixing mill can be utilized for mixing of natural fibers with rubbers as described (Boustany and Coran 1972). The mixing procedure may be distributive or dispersive depending on the type of the fiber used. The distributive mixing increases randomness of spatial distribution of the minor constituent within the major base material without further size reduction, while dispersive mixing serves to reduce the agglomerate size.

10.9.2   Fiber Dispersion

Good dispersion of natural fibers in the rubber compounds is an essential requisite for high-performance composites. The naturally occurring cellulosic fibers tend to agglomerate during mixing due to hydrogen bonding. A pretreatment of fibers is necessary to reduce fiber–fiber interactions. Natural fibers treated with either carbon black or compositions containing latex were found to be dispersing well in the rubber matrix (Dunnom et al. 1973). Fiber length also has a small effect in facilitating better dispersion. Derringer (1971b) used commercially available fibers such as nylon, rayon, polyester, and acrylic flock cut into smaller lengths of 8–10 cm for better dispersion.

10.9.3   Fiber Breakage

The importance of fiber length and its influence on the properties of the composites were studied by several researchers (Broutman and Aggarwal 1980; Monette et al. 1993; Rosen 1965). In a composite material, fiber length is a critical factor.

It should not be too long, so as to avoid entanglement, which can cause dispersion problem. But a very small length of fiber does not offer sufficient stress transfer from the matrix to the fiber. The severity of fiber breakage mainly depends on the type of fiber and its initial aspect ratio. Fibers such as glass and carbon are brittle and they possess a lower bending strength than cellulosic fiber, which is more flexible and resistant to bending. For each type of fiber, there exists a certain aspect ratio below which no further breakage can occur depending on its resistance to bending. The aspect ratio of glass fiber is very low compared with cellulosic fibers. If the mix viscosity is high, more shear will be generated during mixing thus exceeding the critical bending stress of the fiber, which eventually results in severe breakage. O'Connar (1977a) reported fiber breakage during mixing. The lower reinforcing ability of the glass fiber has been attributed to severe reduction in their length compared with cellulosic fibers during mixing. Murthy et al. (1982) suggested that the breakage of the fiber is due to the buckling effect. Setua and De (1983, 1984) and Chakraborty et al. (1982) studied the breakage of jute and silk fibers in NR and nitrile rubber, and they found that breakage of silk fibers is less than jute fibers.

10.9.4 CRITICAL FIBER LENGTH AND ASPECT RATIO OF FIBER

In a perfectly oriented unidirectional continuous fiber-reinforced polymer composite containing fibers of uniform radius, the rate of increase of fiber stress is proportional to the interfacial shear stress and the fiber ends have very little influence on the properties of the composites. But the fiber ends in short fiber composites play a major role in the determination of the ultimate properties. The concept of critical fiber length over which the stress transfer allows the fiber to be stressed to its maximum or at which efficient fiber reinforcement can be achieved has been used to predict the strength of the composites. Vincent (2000) performed a theoretical analysis on the mechanism of stress transfer between matrix and fiber of uniform length and radius and gave the following expression for the critical fiber length (l_c).

$$\frac{l_c}{d} = \frac{sf_u}{2t_y}$$

where d is the diameter of the fiber, sf_u is the ultimate fiber strength, and t_y is the matrix yield stress in shear. The aspect ratio (the length to diameter ratio) (l/d) of fibers is a major parameter that controls the fiber dispersion, fiber–matrix adhesion that gives the optimum performance of short fiber polymer composites. If the aspect ratio of the fiber is lower than the critical aspect ratio, insufficient stress will be transferred and the reinforcement will be inefficient. Several researchers (Coran et al. 1974; Setua and De 1983; Boustany and Arnold 1976; Hu and Lim 2007) suggested that an aspect ratio in the range of 100–200 is essential for high-performance fiber–rubber composites for good mechanical properties. However, Chakraborthy et al. (1982) observed that an aspect ratio of 40 gives optimum reinforcement in the case of carboxylated nitrile rubber composite reinforced with jute fiber. Murthy et al. (1982) and Setua and De (1983) reported that aspect ratios of 15 and 32 are sufficient for reinforcement of jute fiber in NR and styrene–butadiene rubber, respectively. It was reported that for synthetic fiber such as polyester and nylon an aspect ratio of 220 and 170 gives good reinforcement in NR vulcanizates, respectively (Holbery and Houston 2006; Murthy and De 1982).

10.9.5 Fiber Orientation

Fiber orientation has a significant influence on the physicomechanical properties of fiber-reinforced rubber composites. The preferential orientation of fibers in the matrix results in the development of anisotropy in the matrix. With respect to orientation, three limits are explained as longitudinal (along machine direction), transverse (across machine direction), and random. It was observed that during mixing procedure, the lower the nip gap, the higher the anisotropy in tensile properties of the composites, implying greater orientation of fibers. This has been represented as the anisotropy index, which reduces gradually with increasing nip gap. During processing and subsequent fabrication of short fiber rubber composites, the fibers oriented preferentially in a direction depending on the nature of flow, that is convergent or divergent, as explained by Goettler et al. (1979). If the flow is convergent, the fibers align themselves in the longitudinal direction and if it is divergent they orient in the transverse direction. In longitudinally oriented composites, the effective stress transfer from the matrix to the fiber occurs in the direction of fiber alignment, and greater strength and reinforcement will be experienced by the composite. In transversely oriented composites, the stress transfer takes place in a direction perpendicular to the fiber alignment and hence fracture of the sample occurs at a lower tensile stress, which may be equal or lower than the strength of the matrix. Randomly oriented composite shows variable strength values, that is strength lies between the limits of longitudinally and transversely oriented composites. These composites are essentially isotropic in plane, that is they have desirable properties in all directions in a plane. Longitudinally oriented composites are inherently anisotropic. Many researchers have used SEM of the fracture surface to determine the fiber orientation due to the ease of sample preparation (Geethamma et al. 1995; Joseph et al. 1992; Varghese et al. 1991).

10.9.6 Fiber Concentration

Concentration of fibers in the matrix plays a crucial role in determining the mechanical properties of the fiber-reinforced rubber composites. A lower concentration of fibers gives lower mechanical strength. This has been observed not only in rubbers (Prasanthkumar and Thomas 1995b) but also in thermoplastic elastomeric matrices (Kutty and Nando 1991; Roy et al. 1992; Senapati et al. 1988, 1989). This behavior has been attributed basically to two factors: (1) dilution of the matrix, which has a significant effect at low fiber loadings and (2) reinforcement of the matrix by the fibers, which becomes of increasing importance as fiber volume fraction increases. At low fiber content, the matrix is not restrained by enough fibers and highly localized strains occur in the matrix at low strain levels causing the bond between fibers and the matrix to break, leaving the matrix diluted by nonreinforcing debonded fibers. At high fiber concentrations, the matrix is sufficiently restrained and stress is more eventually distributed, thus the reinforcement effect outweighs the dilution effect (Ku et al. 2011). As the concentration of fibers is increased to a higher level, the tensile properties gradually improve to give higher strength than that of the matrix (Ahmad et al. 2006; Lee et al. 2009; Liu et al. 2009; Lopez Manchado et al. 2003; Santos et al. 2009). The concentration of fibers beyond which the properties of the composite improve above the original matrix strength is known as optimum fiber concentration. In order to achieve improvement in mechanical properties with short fibers, the matrix is loaded beyond this volume fraction of fiber. In rubbers, this optimum fiber concentration is quite often found to lie between 25 and 35 phr. This has been observed by several researchers (Varghese et al. 1994a; Nachtigall et al. 2007; Kazayawoko and Balatinecz 1999; Arumugham and Tamaraselvy 1989; Ismail et al. 2002a) for various natural and synthetic fibers in rubbers. Quite often at concentration beyond 35–40 phr, the strength again decreases, because there is insufficient matrix material to adhere the fibers together.

10.9.7 FIBER–MATRIX ADHESION

Fiber–matrix adhesion plays a very prominent role in the reinforcement of short fibers in the rubber matrices. The fiber–matrix adhesion is important in determining the mechanical, dynamic mechanical, and rheological characteristics of the composites since the stress transfer occurs at the interface from matrix to fiber. Although the mechanism of stress transfer is not clear, it has been postulated that it takes place through shearing at the fiber–matrix interface. In composites with low fiber–matrix adhesion, Derringer (1971a) observed that a region of yielding occurs extending over a large portion of the strain range, which is accompanied by low tensile strength and high permanent set. The fiber–matrix adhesion is evaluated at the interface of the composites. Interface is an essentially bidimensional region through which material parameters such as concentration of an element, crystal structure, elastic modulus, density, coefficient of thermal expansion, and so on change from one side to another. There are two types of interface bonding in fiber-reinforced composites. They are mechanical interface bonding and chemical bonding at the interface. In the former one, a simple mechanical anchoring between the two surfaces can lead to a considerable degree of adhesion. Moreover, any contraction of the polymeric matrix on to the fiber would result in gripping of the fiber by the matrix. Chemical bonding at the interface can occur in two ways by dissolution and wettability bonding or by reaction bonding. In wettability bonding, the interaction between the fiber and the matrix occurs on an electronic scale, that is these components come into an intimate contact on an atomic scale. Hence, both surfaces should be approximately wetted to remove any impurities. In reaction bonding, transport of atoms occurs from the fiber, matrix, or both to the interface. These polar surfaces can form bonds owing to the diffusion of matrix molecules to the molecular network of the fiber, thus forming tangled molecular bonds at the interface.

Fiber–matrix adhesion has been explained by evidence such as the mechanism of dry-bonding system, fiber treatment, and determination of the adhesion level and optimization. The dry-bonding system commonly used in rubbers is the HRH system consisting of hydrated silica, resorcinol, and hexamethylenetetramine to create adhesion between fiber and rubber matrix. If the fibers are not properly bonded with the matrix, they will slide past each other under tension deforming the matrix to low strength. When the fiber–matrix interface is sufficiently strong, the load will be effectively transferred to fibers to obtain a high-performance composite. Hence, the mechanism of load transfer may take place through the shear at the interface.

O'Connar (1977a) studied the effect of three-component bonding systems such as the HRH system, and a resin bonding agent on NR composites containing various synthetic and natural fibers. To improve the adhesion between fibers and matrix, various oxidative and nonoxidative chemical treatments are available for natural and synthetic fibers (Chawla 1987; Richardson 1977). Hamed and Coran (1978) reviewed the reinforcement of elastomers with various treated short cellulosic fibers and their mechanism of reinforcement. Several researchers have investigated the use of treated short natural fibers as reinforcing elements for rubber composites (Arumugham and Tamaraselvy 1989; Das 1973; Geethamma et al. 1998; Ismail et al. 2002a, 1997b; Murthy and De 1984; Varghese et al. 1994a,c). These include jute, coir, sisal, oil palm, bamboo, and others. A good amount of adhesion is required for high-performance short fiber composites. The main problem with adhesion in short fiber–rubber composites is that it cannot be measured quantitatively. The adhesion level can be qualitatively assessed from the shapes of the stress–strain curves and the study of fracture surfaces using SEM techniques. A restricted equilibrium swelling technique can also be used to evaluate adhesion (Das 1973). But this measurement is inaccurate since the restriction may be due to the presence of fibers and the adhesion cannot be separated out. In the case of viscoelastic properties, with the increase of adhesion level a high shear will be experienced at the interface; thereby the mechanical loss associated with it also increases. At elevated temperature, the interface deteriorates and the value decreases. The studies so far reported proved that utilization of natural fibers as reinforcement in rubber composites offer economical, environmental, and qualitative advantages. By the incorporation of natural fibers along with synthetic fibers, composites with high performance can be prepared. They may find application in the automotive industry as well as the building industry.

10.10   CONCLUSION

The advantages of natural fibers as reinforcement in polymer composite and their properties were reviewed from the viewpoints of the current status and future expectations. This chapter provides a detailed overview on the structure and properties of sisal and coir fiber, the effect of chemical treatment or surface modification of the fiber on the composite properties, and the physical and mechanical properties of their composites based on NR. It can be concluded that natural fibers are promising reinforcement for NR composites on account of their economic aspect, low density, high specific strength, and high modulus. There are no health hazards like asbestos or glass fiber composites and they are environmentally friendly. These composites offer advantages of about 20%–25% reduction in processing temperature and cycle time, and a weight reduction of about 30%. In addition, minimal retooling costs due to the negligible processing equipment abrasion make these composites more attractive. Recently, there has been an increased interest in commercialization of natural fiber composites and their use, especially in decorative interior paneling due to their aesthetic appearance and low cost. However, the areas of cellulosic fiber-based green composites realize wider applications. Due to the low density and high specific properties of natural fibers, it may have good implications in the engineering and automotive industries. More efforts are needed to improve the adhesion between the fiber–matrix interface and thermal stability to open up new applications for cellulosic fiber–reinforced composites, which is still an indispensible task for scientists. A systematic and persistent research may be a good scope and better future for cellulosic fiber–reinforced green composites and thereby they can contribute to the environmental protection for a green future.

REFERENCES

Abdelmouleh, M., Boufis, S., Belgacem, M.N. and Dufresne, A. 2007. Short natural-fiber reinforced polyethylene and natural rubber composites: Effect of silane coupling agents and fibers loading. *Composites Science and Technology* 67(7–8): 1627–1639.

Abdelmouleh, M., Boufi, S., Belgacem, M.N., Duarte, A.P., Ben Salah, A. and Gandini, A. 2004. Modification of cellulosic fibers with functionalized silanes: Development of surface properties. *International Journal of Adhesion* 24: 43–54.

Ahmad, I., Baharum, A. and Abdullah, I. 2006. Effect of extrusion rate and fiber loading on mechanical properties of twaron fiber-thermoplastic natural rubber (TPNR) composites. *Journal of Reinforced and Plastics and Composites* 25: 957–965.

Alvarez, P., Blanco, C., Santamaria, R. and Granda, M. 2005. Lignocellulose/pitch based composites. *Composite Part A* 36: 649–657.

Alvarez, V.A. and Vazquez, A. 2004. Thermal degradation of cellulose derivatives/starch blends and sisal fiber biocomposites. *Polymer Degradation and Stability* 84: 13–21.

Amash, A. and Zugenmaier, P. 1998. Study on cellulose and xylan filled polypropylene composites. *Polymer Bulletin* 40: 251–258.

Anderson, M. and Tillman, A.M. 1989. Acetylation of jute: Effects on strength rot resistance, and hydrophobicity. *Journal of Applied Polymer Science* 37: 3437–3447.

Ansell, M.P., Eichhom, S.J. and Baillie, C.A. 2001. Review: Current international research into cellulosic fibers and composites. *Journal of Materials Science* 36: 2107–2131.

Arumugham, N. and Tamaraselvy, K. 1989. Coconut-fiber-reinforced rubber composites. *Journal of Applied Polymer Science* 37: 2645–2659.

Aziz, S.H. and Ansell, M.P. 2004. The effect of alkylation and fiber alignment on the mechanical and thermal properties of kenaf and hemp bast fiber composites: Part I: Polyester resin matrix. *Composite Science and Technology* 64: 1219–1230.

Baiardo, M., Frisoni, G. and Scandola, M. 2002. Surface chemical modification of natural cellulose fibers. *Journal of Applied Polymer Science* 83: 38–45.

Beall, F.C. 1986. Thermal degradation of wood. In: *Encyclopedia of Materials Science and Engineering,*Vol. 7, lst edition. Ed. Bever, M.B. Oxford: Pergamon Press.

Behera, S., Sahu, S., Patel, S. and Mishra, B.K. 2012. Sisal fiber-a potential raw material for manmade paper. *Journal of Indian Pulp and Paper Technical Association* 24: 37–43.

Bhagavan, S.S., Tripathy, D.K. and De, S.K. 1987. Stress relaxation in short jute fiber-reinforced nitrile rubber composites. *Journal of Applied Polymer Science* 33: 1623–1639.

Bipinbal, P.K., Joseph, M.J. and Kutty, S.K.N. 2010. *Cellulose microfiber-Natural rubber composites prepared by latex masterbatching: Processing characteristics and mechanical properties.* International Conference on Advances in Polymer Technology. India, 300–308.

Bisanda, E.T.N. 1993. The manufacture of roofing panels from sisal fiber reinforced composites. *Journal of Materials Processing Technology* 38: 369–380.

Bisanda, E.T.N. and Ansell, M.P. 1991. The effect of silane treatment on the mechanical and
physical properties of sisal-epoxy composites. *Composite Science and Technology* 41:
165–178.

Biswas, S., Kindo, S. and Patnaik, A. 2011. Effect of fiber length on mechanical behavior of
coir fiber reinforced epoxy composites. *Fibers and Polymers* 12: 73–78.

Bledzki, A.K. and Gassan, J. 1996. Properties and modification methods for vegetable fibers
for natural fiber composites. *Journal of Applied Polymer Science* 59: 1329–1336.

Bledzki, A.K. and Gassan, J. 1999. Composites reinforced with cellulose based fibers. *Progress
in Polymer Science* 4: 221–274.

Bledzki, A.K., Mamun, A.A., Lucka-Gabor, M. and Gutowski, V.S. 2008. The effects of acetylation on properties of flax fiber and its polypropylene composites. *Express Polymer
Letters* 2(6): 413–422.

Boustany, K. and Arnold, R.L. 1976. Short fibers rubber composites: the comparative properties of
treated and discontinuous cellulose fibers. *Journal of Elastomers and Plastics* 8: 160–176.

Boustany, K. and Coran, A.Y. 1972. US Patent No. 33679364.

Brahmakumar, M., Pavithran, C. and Pillai, R.M. 2005. Coconut fiber reinforced polyethylene
composites: Effect of natural waxy surface layer of the fiber on fiber/matrix interfacial
bonding and strength of composites. *Composite Science and Technology* 65: 563–569.

Broutman, L.J. and Aggarwal, B.D. 1980. *Analysis and Performance of Fiber Composites,
Soc. Plast. Ind.*, New York: John Wiley and Sons.

Calado, V., Barreto, D.W. and Almeida, J.R.M. 2003. Effect of a two step fiber treatment on
the flexural mechanical properties of sisal-polyester composites. *Polymer and Polymer
Composites* 11: 31–36.

Chakraborty, S.K., Setua, D.K. and De, S.K. 1982. Short jute fiber reinforced carboxylated
nitrile rubber. *Rubber Chemistry and Technology* 55: 1286–1307.

Chand, N. and Hashmi, S.A.R. 1993. Mechanical properties of sisal fiber at elevated temperatures. *Journal of Material Science* 28: 6724–6728.

Chand, N., Sood, S., Rohatgi, P.K. and Sayanarayana, K.G. 1984. Resources, structure, properties and uses of natural fibers of Madhya Pradesh. *Journal of Science and Industrial
Research* 43: 489–99.

Chand, N., Tiwary, R.K. and Rohatgi, P.K. 1988. Bibliography resource structure properties
of natural cellulosic fibers - an annotated bibliography. *Journal of Material Science* 23:
381–387.

Chawla, K.K. 1987. *Composite Materials*. New York: Springer Verlag.

Cheng, T.H., Jones, F.R. and Wang, D. 1993. Effect of fiber conditioning on the interfacial
shear strength of glass-fiber composites. *Composite Science and Technology* 48: 89–96.

Cook, J.G. 1968. *Handbook of Textile Fibers and Natural Fibers*, 4th ed. England: Morrow
Publishing.

Coran, A.Y., Boustany, K. and Hamed, P. 1974. Short-fiber–rubber composites: The properties
of oriented cellulose-fiber–elastomer composites. *Rubber Chemistry and Technology*
47: 396–410.

Crocker, J. 2008. Natural materials innovative natural composites. *Materials Technology* 2/3:
174–178.

Crowson, R.J., Folkes, M.J. and Bright, P.F. 1980. Rheology of short glass fiber-reinforced
thermoplastics and its application to injection molding I. Fiber motion and viscosity
measurement *Polymer Engineering and Science* 20: 925–933.

Cyras, V.P., Vallo, C., Kenny, J.M. and Vazquez, A. 2004. Effect of chemical treatment on
the mechanical properties of starch-based blends reinforced with sisal fiber. *Journal of
Composite Materials* 38(16): 1387–1399.

Das, B. 1973. Restricted equilibrium swelling—A true measure of adhesion between short fibers and rubber. *Journal of Applied Polymer Science* 17: 1019–1030.

Derringer, D.C. 1971a. Short fiber-elastomer composites. *Journal of Elastomers and Plastics* 3: 230–248.

Derringer, D.C. 1971b. *Rubber World.* 45: 165.

Drazal, L.T. and Madukar, M. 1993. Fiber–matrix adhesion and its relationship to composite mechanical properties. *Journal of Material Science* 28: 569–610.

Dunnom, D.D., Wagner, M.P. and Derringer, G.C. 1973. Chemical Division, PPG Industries Inc., *US Patent No. 3746669.*

Dzyura, E.A. 1980. Tensile strength and ultimate elongation of rubber-fibrous composites. *International Journal of Polymeric Materials* 8: 165–173.

Dzyura, E.A., Serebo, A.L. and Kiryushina, N.D. 1983. Degradation of components of rubber-fiber composites during processing. *Kaučukirezina* 12: 19–22.

Everage, Jr. A.E., 1973. Theory of bicomponent flow of polymer melts, I. equilibrium Newtonian tuve flow. *Transactions of the Society of Rheology* 17: 629–646.

Felix, G.M. and Gatenholm, P. 1991. The nature of adhesion in composites of modified cellulose fibers and polypropylene. *Journal of Applied Polymer Science* 42: 609–620.

Felix, G.M. and Gatenholm, P. 1993. Formation of entanglements at brushlike interfaces in cellulose-polymer composites. *Journal of Applied Polymer Science* 50: 699–708.

Feughelman, M. and Nordan, P.J. 1962. Some mechanical changes during sorption of water by dry keratin fibers in atmospheres near saturation. *Journal of Applied Polymer Science* 6: 670–673.

Gassan, J. and Bledzki, A.K. 1999a. Effect of cyclic moisture absorption desorption on the mechanical properties of silanized jute-epoxy composites. *Polymer Composite* 20: 604–611.

Gassan, J. and Bledzki, A.K. 1999b. Possibilities for improving the mechanical properties of jute/epoxy composites by alkali treatment. *Composite Science and Technology* 59: 1303–1309.

Gassan, J. and Bledzki, A.K. 2001. Thermal degradation of flax and jute fibers. *Journal of Applied Polymer Science* 82; 1417–1422.

Ge, X.C., Li, X.H. and Meng, Y.Z. 2004. Tensile properties, morphology, and thermal behavior of PVC composites containing pine flour and bamboo flour. *Journal of Applied Polymer Science* 93: 1804–1811.

Geethamma, V.G., Joseph, R. and Thomas, S. 1995. Short coir fiber-reinforced natural rubber composites: Effects of fiber length, orientation and alkali treatment. *Journal of Applied Polymer Science* 55: 583–594.

Geethamma, V.G., Joseph, R. and Thomas, S. 1998. Composite of short coir fibers and natural rubber: effect of chemical modification, loading and orientation of fiber. *Polymer* 39: 1483–1491.

Geethamma, V.G., Kalaprasad, G., Groeninckx, G.I. and Thomas, S. 2005. Dynamic mechanical behavior of short coir fiber reinforced natural rubber composites. *Composites Part A* 36: 1499–1506.

Geethamma, V.G., Ramamurthy, K., Janardhan, R. and Thomas, S. 1996. Melt flow behavior of short coir fiber reinforced natural rubber composites. *International Journal of Polymer Materials* 32: 147–161.

George, J., Bhagawan, S.S. and Thomas, S. 1998a. Effects of the environment on the properties of low-density polyethylene composites reinforced with pineapple-leaf fiber. *Composite Science Technology* 58: 1471–1485.

George, J., Bhagawan, S.S. and Thomas, S. 1998b. Improved interactions in chemically modified pineapple leaf fiber reinforced polyetheylene composites. *Composite Interfaces* 5: 201–223.

George, J., Bhagawan, S.S. and Thomas, S. 1998c. Stress relaxation behaviour of pineapple fiber reinforced low density polyethylene composites. *Journal of Reinforced Plastics and Composites* 17: 651–672.

Goettler, L.A., Lambright, A.J. and Leib, R.I. 1979. Short fiber reinforced hose – A new concept in production and performance. *Rubber Chemistry and Technology* 52: 838–863.

Gonzalez, P.I. and Yong, R.J. 1998. Crack bridging and fiber pull-out in polyethylene fiber reinforced epoxy resins. *Journal of Materials Science* 33: 5715–5729.

Gore, A. 2006. *An Inconvenient Truth: The Planetary Emergency of Global Warming and What We Can do About it*, 1st ed. Emmanaus: Rodale Press.

Guo, W. and Ashida, M. 1993. Dynamic viscoelasticities for short fiber thermoplastic elastomer composites. *Journal of Applied Polymer Science* 5 0: 1435–1443.

Hamed, P. and Coran, A.Y. (1978). Reinforcement of polymers through short cellulose fibers, page 29–50. In: *Additives for Plastics*, Vol. I. Ed. Seymour, R.B. New York: State of the art. Academic Press.

Harikumar, K.R., Joseph, K. and Thomas, S. 1999. Jute sack cloth reinforced polypropylene composites: Mechanical and sorption studies. *Journal of Reinforced Plastics and Composites* 18: 346–372.

Harper, D. and Wolcott, M. 2004. Interaction between coupling agent and lubricants in wood-polypropylene composites. *Composite Part A* 35: 385–394.

Hatakeyama, H., Nakayachi, A. and Hatakeyama, T. 2005. Thermal and mechanical properties of polyurethane-based geocomposites derived from lignin and molasses. *Composite. Part A* 36: 698–704.

Hednberg, P. and Gateholm, P. 1995. Conversion of plastic/cellulose waste into composites. 1. Model of the interphase. *Journal of Applied Polymer Science* 56: 641–651.

Hill C.A.S. and Khalil H.P.S.A. 2000a. Effect of fiber treatment on mechanical properties of coir or oil palm fiber reinforced polyester composite. *Journal of Applied Polymer Science* 78: 1685–697.

Hill, C.A.S. and Khalil, H.P.S.A. 2000b. The effect of environmental exposure upon the mechanical properties of coir or palm fiber reinforced composites. *Journal of Applied Polymer Science* 77: 1322–1330.

Hill, S. 1997. Cars that grow on trees. *New Scientist* 153(2067): 36–39.

Holbery, J. and Houston, D. 2006. Natural-fiber-reinforced polymer composites in automotive application. *Journal of the Minerals, Metals and Materials Society* 58(11): 80–86.

Hristov, V.N., Lach, R. and Grellmann, W. 2004. Impact fracture behavior of modified polypropylene/wood fiber composites. *Polymer Testing* 23: 581–589.

Hu, R. and Lim, J.K. 2007. Fabrication and mechanical properties of completely biodegradable hemp fiber reinforced polylactic acid composites. *Journal of Composite Materials* 41: 1655–1669.

Ismail, H., Edyham, M.R. and Wirjosentono, B. 2002a. Bamboo fiber filled natural rubber composites: the effect of filler loading and bonding agent. *Polymer Testing* 21: 139–144.

Ismail, H., Rosnah, N. and Rozman, H.D. 1997a. Curing characteristics and mechanical properties of short oil palm fiber reinforced rubber composites. *Polymer* 38(16): 4059–4064.

Ismail H., Rozman H.D. and Ishiaku U.S. 1997b. Oil palm fiber-reinforced rubber composite: Effects of concentration and modification of fiber surface. *Polymer International* 43: 223–230.

Ismail, H., Shuhelmy, S. and Edyham, M.R. 2002b. The effects of silane coupling agent on curing characteristics and mechanical properties of bamboo fiber filled natural rubber composites. *European Polymer Journal* 38: 39–47.

Jacob, M., Francis, B., Thomas, S. and Varughese, K.T. 2004a. Natural rubber composites reinforced with sisal/oil palm hybrid fibers: Tensile and Cure Characteristics. *Journal of Applied Polymer Science* 93: 2305–2312.

Jacob, M., Thomas, S. and Varughese, K.T. 2004b. Mechanical properties of sisal/oil palm hybrid fiber reinforced natural rubber composites. *Composite Science Technology* 64: 955–965.

Jacob, M., Francis, B., Thomas, S. and Varughese, K.T. 2006. Dynamical mechanical analysis of sisal/oil palm hybrid fiber-reinforced natural rubber composites. *Polymer Composite* 27: 671–680.

John, M.J. and Anandjiwala, R.D. 2008. Recent developments in chemical modification and characterization of natural fiber-reinforced composites. *Polymer Composites* 29(2): 187–207.

Joly, C., Gauthier, R. and Coubles, E.S. 1996. Partial masking of cellulosic fiber hydrophilicity for composite applications. Water sorption by chemically modified fibers. *Journal of Applied Polymer Science* 61: 57–69.

Joseph, K., Thomas, S. and Pavithran, C. 1992. Viscoelastic properties of short sisal-fiber-filled low density polyethylene composites: effect of fiber length and orientation. *Materials Letters* 15: 224–228.

Joseph, K., Thomas, S. and Pavithran, C. 1996. Effect of chemical treatment on the tensile properties of short sisal fiber-reinforced polyethylene composites. *Polymer* 37: 5139–5149.

Joseph, P.V., Joseph, K. and Thomas, S. 2002. Environmental effects on the degradation behavior of sisal fiber reinforced polypropylene composites. *Composite Science and Technology* 62: 1357–1372.

Kazayawoko, M. and Balatinecz, J.J. 1999. Surface modification and adhesion mechanisms in wood fiber-polypropylene composites. *Journal of Material Science* 34: 6189–6199.

Khalil, H.P., Ismail, H. and Rozman, H.D. 2001. The effect of acetylation on interfacial shear strength between plant fibers and various matrices. *European Polymer Journal* 37: 1037–1045.

Kishore, V.S. 1998. Impact behavior of glass-epoxy composites containing foam material. *Journal of Applied Polymer Science* 67: 1565–1571.

Ku, H., Wang, H., Pattarachaiyakoop, N. and Trada, M. 2011. A review on the tensile properties of natural fiber reinforced polymer composites. *Composites: Part B* 42: 856–873.

Kumar, R.P. and Thomas, S. 1995a. Tear and processing behaviour of short sisal fiber reinforced styrene butadiene rubber composites. *Polymer International* 38: 173–182.

Kumar, R.P. and Thomas, S. 1995b. Short fiber elastomer composites: effect of fiber length, orientation, loading and bonding agent. *Bulletin of Material Science* 18: 1021–1029.

Kutty, S.K.N. and Nando, G.B. 1991. Short Kevlar fiber-thermoplastic polyurethane composite. *Journal of Applied Polymer Science* 43: 1913–1923.

Lee, B.H., Kim, H.J. and Yu, W.R. 2009. Fabrication of long and discontinuous natural fiber reinforced polypropylene biocomposites and their mechanical properties. *Fiber and Polymers* 10: 83–90.

Li, H., Zadorecki, P. and Flodin, P. 1987. Cellulose fiber-polyester composites with reduced water sensitivity. (1) Chemical treatment and mechanical properties. *Polymer Composites* 8: 199–207.

Li, Y., Mai, Y.W. and Ye, L. 2000. Sisal fiber and its composites: a review of recent developments. *Composites Science and Technology* 60: 2037–2055.

Liao, B., Huang, Y. and Cong, G. 1997. Influence of modified wood fibers on the mechanical properties of wood-fiber reinforced polyethylene. *Journal of Applied Polymer Science* 66: 1561–1568.

Liu, L., Yu, J., Cheng, L. and Qu, W. 2009. Mechanical properties of poly(butylenes succinate) (PBS) biocomposites reinforced with surface modified jute fiber. *Composite Part A* 40: 669–674.

Lopez Manchado, M.A., Arroya, M., Biagiotti, J. and Kenny, J.M. 2003. Enhancement of mechanical properties and interfacial adhesion of PP/EPDM/flax fiber composites using maleic anhydride as a compatibilizer. *Journal of Applied Polymer Science* 90: 2170–2178.

Lu, J.Z., Wu, Q. and Negulescu, I.I. 2005. Wood-fiber/high-density-polyethylene composites: Coupling agent performance. *Journal of Applied Polymer Science* 96: 93–102.

MacLean, D.C. 1973. A theoretical analysis of bicomponent flow and the problem of interface shape. *Transactions of the Society of Rheology* 17: 385–399.

Maldas, D. and Kokta, B.V. 1990. Effect of recycling on the mechanical properties of wood fiber-polystyrene composites. Part 1: Chemithermomechanical pulp as a reinforcing pillar. *Polymer-Plastics Technology and Engineering* 29: 419–454.

Maldas, D. and Kokta, B.V. 1991. Effect of fiber treatment on the mechanical properties of hybrid fiber reinforced polystyrene composites: IV. Use of glass fiber and sawdust as hybrid fiber. *Journal of Composite Materials* 25: 375–390.

Malkapuram, R., Kumar, V. and Yuvraj, S.N. 2008. Recent development in natural fiber reinforced polypropylene composites. *Journal of Reinforced Plastics and Composites* 28: 1169–1189.

Mallick, P.K. 1988. *Fiber Reinforced Composite Materials, Manufacturing and Design*. Ch. 1., p.18. New York: Marcel Dekker, Inc.

Mani, P. and Satyanarayan, K.G. 1990. Effects of the surface treatments of lignocellulosic fibers on their debonding stress. *Journal of Adhesion Science and Technology* 4: 17–24.

Mannan-Kh, M. and Latifa, B.L. 1980. Effects of grafted methyl methacrylate on the microstructure of jute fibers. *Polymer* 21: 777–780.

Mohanty, A.K. and Khan, M.A. 2000. Surface modification of jute and its influence on performance of biodegradable jute-fabric/biopol composites. *Composite Science and Technology* 60: 1115–1124.

Mohanty, A.K. and Singh, B.C. 1987. Redox-initiated graft copolymerization onto modified jute fibers. *Journal of Applied Polymer Science* 34: 1325–1327.

Mohanty, A.K., Mishra, M. and Drzal, L.T. (Eds.), 2006. *Natural Fibers, Biopolymers and Biocomposites*, Serial Publications, Boca Raton: CRC Press.

Mohanty, A.K., Misra, M. and Drzal, L.T. 2002. Sustainable bio-composites from renewable resources: Opportunities and challenges in the green materials world. *Journal of Polymers and Environment* 10: 19276–19277.

Mohanty, A.K., Misra, M. and Hinrichsen, G. 2000. Biofibers, biodegradable polymers and biocomposites: An overview. *Macromolecular Materials and Engineering* 276/277: 1–24.

Mohd Ishak, Z.A., Aminullah, A., IsmailH. and Rozman, H.D. 1998. Effect of silane-based coupling agents and acrylic acid based compatibilizers on mechanical properties of oil palm empty fruit bunch filled high-density polyethylene composites. *Journal of Applied Polymer Science* 68: 2189–2203.

Monette, L., Anderson, M.P. and Grest, G.S. 1993. The meaning of the critical length concept in composites: study of matrix viscosity and strain rate on the average fiber fragmentation length in short-fiber polymer composites. *Polymer Composite* 14: 101–115.

Monteiro, S.N., De Deus, J.F., Aquino, R.C.M.P. and D'almeida, J.R.M. 2006a. Pullout tests of coir fibers to evaluate the interface strength in polyester composites. In: *Proceedings of Characterization the TMS Conference*, San Antonio, TX, USA, 1–8.

Monteiro, S.N., Lopes, F.P.D., Barbosa, A.P., Bevitori, A.B., Silva, I.L.A. and Costa, L.L. 2011. Natural lignocellulosic fibers as engineering materials: An overview. *Metallurgical and Materials Transactions A* 42: 2963–2974.

Monteiro, S.N., Lopes, F.P.D., Ferreira, A.S., and Naseimento, D.C.O. 2009a. Natural-fiber polymer-matrix composites: Cheaper, tougher, and environmentally friendly. *JOM* 61: 17–22.

Monteiro, S.N., Santafe, H.P.G.Jr. and Costa, L.L. 2009b. Mechanical behavior of polyester composites reinforced whit alkali treated coir fibers. In: *Proceedings of Characterization of Minerals, Metals & Materials - TMS Conference*, San Francisco, CA, 1–7.

Monteiro, S.N., Terrones, L.A.H., Carvalho, E.A. and D'almeida, J.R.M. 2006b. Effect of fiber/matrix interface on the strength of coir fiber reinforced polymeric composites. *Revista Matéria*, 11: 395–402.

Monteiro, S.N., Terrones, L.A.H., Lopes, F.P.D. and D'almeida, J.R.M. 2005. Mechanical strength of polyester matrix composites reinforced with coconut fiber wastes. *Revista Matéria* 10: 571–576.

Mukherjee, P.S. and Satyanarayana, K.G. 1984. Structure and properties of some vegetable fibers, Part 1. Sisal fiber. *Journal of Material Science* 19: 3925–3934.

Murthy, V.M. and De, S.K. 1982. Effect of particulate fillers on short jute fiber-reinforced natural rubber composites. *Journal of Applied Polymer Science* 27: 4611–4622.

Murthy, V.M. and De, S.K. 1984. Short-fiber-reinforced styrene butadiene rubber composites. *Journal of Applied Polymer Science* 29: 1355–1368.

Murthy, V.M., Bhowmick, A.K. and De, S.K. 1982. Scanning electron microscopy studies of failure surfaces of short glass fiber-rubber composites. *Journal of Material Science* 17: 709–716.

Murthy, V.M., Gupta, B.R. and De, S.K. 1985. Rheological behaviour of natural rubber filled with short jute fibers. *Plastics Rubber and Composites Processing and Applications* 5: 307–311.

Nachtigall, S.M.B., Cerveira, G.S. and Rosa, S.M.L. 2007. New polymeric-coupling agent for polypropylene/wood-flour composites. *Polymer Testing* 26: 619–628.

Netravali, A.N. and Chabba, S. 2003. Composites get greener. *Materials Today* 6: 22–29.

Nielsen, L.E. 1974. *Mechanical Properties of Polymers and Composites*, Volume 1. New York: Marcel Dekker, Inc.

O'Connor, J.E. 1977a. Short-fiber-reinforced elastomer composites. *Rubber Chemistry and Technology* 50: 945–958.

O'Conner, J.K. 1977b. Fiber reinforced natural rubber composites: physical and mechanical properties. *Rubber Chemistry and Technology* 50: 945–955.

Oya, N. and Hamada, H. 1998. Axial compressive behavior of reinforcing fibers and inter-phase in glass/epoxy composite materials. *Journal of Materials Science* 33: 3407–3417.

Payen A. 1838. Mémoire sur la composition du tissu propre des plantes et du ligneux, *Comptes Rendus Hebd Seances Acad Sci* 7: 1052–1056.

Pedro, J., Herrera-Franco, M. and Aguilar-Vega, J. 1997. Effect of fiber treatment on the mechanical properties of LDPE-henequen cellulosic fiber composites. *Journal of Applied Polymer Science* 65: 197–207.

Pothen, L.A., Neelakandan, N.R. and Thomas, S. 1997. Short banana fiber reinforced polyester composites: Mechanical, failure, and ageing characteristics. *Journal of Reinforced Plastics and Composites* 16(8): 744–765.

Pothen, L.A., Oommen, Z., George, J., Oommen, Z. and Thomas, S. 1999. Polyester composites of short banana fiber glass fibers. The tensile impact properties, *Polymery. Nr* 44: 750–752.

Prasad, S.V., Pavithran, C. and Rohatgi, P.K. 1983. Alkali treatment of coir fibers for coir-polyester composites. *Journal of Material Science* 18(5): 1443–1454.

Prasant, K.R. and Thomas, S. 1995a. Short fiber elastomer composites: Effect of fiber length, orientation, loading and bonding agent. *Bulletin of Material Science* 18: 1021–1029.

Prasant, K.R. and Thomas, S. 1995b. Tear and processing behavior of short sisal fiber reinforced styrene butadiene rubber composites. *Polymer International* 38: 173–187.

Prasanth, R.P., Kumar, M.L., Amma, G. and Thomas, S. 1995. Short sisal fiber reinforced styrene-butadiene rubber composites. *Journal of Applied Polymer Science* 58: 597–612.

Prasanthkumar, R. and Thomas, S. 1995a. Short fiber elastomer composites: effect of fibre length, orientation, loading and bonding agent. *Bulletin of Material Science* 18(8): 1021–1030.

Prasanthkumar, R. and Thomas, S. 1995b. Short sisal fiber reinforced styrene-butadiene rubber composites. *Journal of Applied Polymer Science* 58: 597–612.

Prasanthkumar, R. and Thomas, S. 1995c. Tear and processing behavior of short sisal fiber reinforced styrene butadiene rubber composites. *Polymer International* 38(2): 173–182.

Raghavan, J. and Wool, R.P. 1999. Interfaces in repair, recycling, joining and manufacturing of polymers and polymer composites. *Journal of Applied Polymer Science* 71: 775–785.

Ray, D., Sarkar, B.K., Rana, A.K. and Bose, N.R. 2001. Effect of alkali treated jute fibers on composite properties. *Bulletin of Material Science* 24(2): 129–138.

Richardson, M.O.W. 1977. *Polymer Engineering Composites*. London: Applied Science Publishers.

Richardson, T. 1987. *Composites: A Design Guide*. New York: Industrial Press Inc.

Rosen, B.W. 1965. *Fiber Composite Materials*, Metal Park, Ohio: American Society for metals.

Rout, J., Misra, M., Tripathy, S.S., Nayak, S.K. and Mohanty, A.K. 2001a. Novel ecofriendly biodegradable coir-polyester amide biocomposites: fabrication and properties evaluation. *Polymer Composite* 22: 770–778.

Rout, J., Misra, M., Tripathy, S.S., Nayak, S.K. and Mohanty, A.K. 2003. SEM observations of the fractured surfaces of coir composites. *Journal of Reinforced Plastics and Composites* 22: 1083–1100.

Rout, J., Tripathy, S.S., Nayak, S.K., Misra, M. and Mohanty, A.K. 2001b. Scanning electron microscopy study of chemically modified coir fibers. *Journal of Applied Polymer Science* 79: 1169–1177.

Rowell, R.M. 1985. *The Chemistry of Solid Wood, Adv. Chem. Ser 207*. Washington, DC: American Chemical Society.

Rowell, R.M., Schultz, T.P. and Narayan, R. 1992. Emerging technologies for materials and chemicals for biomass. *ACS Symposium Series* 476: 12–13.

Roy, D., Bhowmick, A.K. and De, S.K. 1992. Anisotropy in mechanical and dynamic properties of composites based on carbon fiber filled thermoplastic elastomeric blends of natural rubber and high density polyethylene. *Polymer Engineering and Science* 32: 971–979.

Rozman, H.D., Kwnar, R.N. and Ismail, H. 2000. The effect of lignin as a compatibilizer on the physical properties of coconut fiber-polypropylene composites. *European Polymer Journal* 36: 1483–1494.

Sabu, T. and Pothan, L. 2009. *Cellulose Fiber Reinforced Polymer Composites*. Philadelpha: Old City Publishing.

Saheb, D.N. and Jog, J.P. 1999. Natural fiber polymer composites: A review. *Advances in Polymer Technology* 18: 351–363.

Sahoo, P.K., Samantaray, H.S. and Samal, R.K. 1986. Graft copolymerization with new class of acidic peroxo salts as initiators. I. Grafting of acrylamide onto cotton-cellulose using potassium monopersulfate, catalyzed by Co(II). *Journal of Applied Polymer Science* 32: 5693–5703.

Samal, R.K., Rout, S.K. and Mohanty, M. 1995. Effect of chemical modification on FTIR spectra: physical and chemical behavior of coir-II. *Journal of Polymer Materials* 12: 229–233.

Samal, R.K., Samantaray, H.S. and Samal, R.N. 1989. Graft copolymerization with a new class of acidic peroxo salts as initiator. V. Grafting of methyl methacrylate onto jute fiber using potassium monopersulfate catalyzed by Fe(II). *Journal of Applied Polymer Science* 37: 3085–3096.

Santafé Júnior, H.P.G., Lopes, F.P.D., Costa, L.L. and Monteiro, S.N. 2010. Mechanical properties of tensile tested coir fiber reinforced polyester composites. *Revista Matéria* 15(2): 113–118.

Santos, E.F., Mauler, R.S. and Nachtigall, S.M.B. 2009. Effectiveness of maleated- and silanized-PP for coir fiber-filled composites. *Journal of Reinforced Plastics and Composites* 28: 2119–2129.

Satyanarayana, K.G. and Pillai, S.G.K. 1990. In: *Handbook of Ceramics and Composite,* Vol 1, edition. Ed. Cheremisinoff, N.P. New York: Marcel Dekker.

Satyanarayana, K.G., Guimaraes, J.L. and Wypych, F. 2007. Studies on lignocellulosic fibers of Brazil. Part 1: Source, production, morphology, properties and applications. *Composite Part A* 38: 1694–1709.

Satyanarayna, K.G., Kulkarni, A.G. and Rohatgi, P.K. 1981. Potential natural fibers as a resource for industrial material in Kerala. *Journal of Scientific and Industrial Research* 40: 222–237.

Satyanarayana, K.G., Pillai, C.K.S., Sukumaran, K., Pillai, S.G.K., Vijayn, K. and Rohatgi, P.K. 1982. Structure and properties of fibers from various parts of coconut palm. *Journal of Material Science* 17: 2453–2462.

Satyanarayana, K.G., Sukumaran, K., Kulkarni, A.G., Pillai, S.G.K. and Rohatgi, P.K. 1986. Fabrication and properties of natural fiber-reinforced polyester composites. *Journal of Composites* 17: 329–333.

Schniewind, A.P. 1986. Wood and fire. In: *Encyclopedia of Materials Science and Engineering,* Vol. 7, 1st edition. Ed. Bever, M.B. Oxford: Pergamon Press.

Schwartz, M.M. 1984. *Composite Materials Handbook,* New York: McGraw Hill.

Seki, Y. 2009. Innovative multifunctional siloxane treatment of jute fiber surface and its effect on the mechanical properties of jute/thermoset composites. *Materials Sciences and Engineering. A*, 508(1–2): 247–252.

Senapati, A.K., Nando, G.B. and Pradhan, B. 1988. Characterization of short nylon fiber reinforced natural rubber composites. *International Journal of Polymeric Materials* 12: 73–92.

Senapati, A.K., Nando, G.B., Kutty, S.K.N. and Pradhan, B. 1989. Short polyester fiber reinforced styrene-butadiene rubber compounds. *International Journal of Polymeric Materials* 12: 203–224.

Setua, D.K. and De, S.K. 1983. Short silk fiber reinforced natural rubber composites, *Rubber Chemistry and Technology* 56: 808–826.

Setua, D.K. and De, S.K. 1984. Short silk fiber reinforced nitrile rubber composites. *Journal of Material Science* 19: 983–999.

Shukry, N. and Girgis, B.S. 1992. Acetosolv lignins from bagasse: Characterization by TG and DTA. *Polymer-Plastics Technology and Engineering* 31: 541–551.

Sreekala, M.S. and Thomas, S. 2003. Effect of fiber surface modification on water-sorption characteristics of oil palm fibers. *Composite Science and Technology* 63: 861–869.

Sreekala, M.S., Joseph, R. and Thomas, S. 2000. Oil palm fiber reinforced phenol formaldehyde composites: Influence of fiber surface modifications on the mechanical performance. *Applied Composite Materials* 7: 295–329.

Sreekala, M.S., Thomas, S. and Neelakandan, N.R. 1996. Utilization of short oil palm empty fruit bunch fiber (OPEFB) as a reinforcement in phenol-formaldehyde resins: Studies on mechanical properties. *Journal of Polymer Engineering* 16: 265–294.

Srivastava, V.K. and Hogg, P.J. 1998. Damage performance of particles filled quasi-isotropic glass-fiber reinforced polyester resin composites. *Journal of Materials Science* 33: 1119–1128.

The Wealth of India—A Dictionary of Indian Raw Materials and Industrial Products, Vol. 5. 1959. New Delhi: CSIR, 27–29.

Thielemans, W. and Wool, R.P. 2004. Butyrated kraft lignin as compatibilizing agent for natural fiber reinforced thermoset composites. *Composite Part A.* 35: 327–338.

Thomas, M.P. and Winstone, M.R. 1998. Transverse tensile behavior of fiber reinforced titanium metal matrix composites. *Journal of Materials Science* 33: 5499–5508.

Thomas, S., Gupta, B.R. and De, S.K. 1987. Extrudate morphology of blends of plasticized poly(vinyl chloride) and thermoplastic copolyester elastomer. *Journal of Applied Polymer Science* 34: 2053–2061.

Tinh, N., Eugene, Z. and Edward, M.B. 1981a. Thermal analysis of lignocellulosic materials. Part I. Unmodified materials. *Journal of Macromolecular Science* C 20: 1–65.

Tinh, N., Eugene, Z. and Edward, M.B. 1981b. Thermal analysis of lignocellulosic materials. Part II. Modified materials. *Journal of Macromolecular Science* C 21: 1–60.

Tjong, S.C. and Meng, Y.Z. 1999. Mechanical and thermal properties of polycarbonate composites reinforced with potassium titanate whiskers. *Journal of Applied Polymer Science* 72: 501–508.

Tserki, V., Zafeiropoulos, N.E., Simon, F. and Panayiotou, C. 2005. A study of the effect of acetylation and propionylation surface treatments on natural fibers. *Composites Part A – Applied Science and Manufacturing* 36(8): 1110–1118.

Vajrasthira, C., Amomsakchai, T. and Bualek-Limcharoen, S. 2003. Fiber-matrix interactions in aramid-short-fiber-reinforced thermoplastic polyurethane composites. *Journal of Applied Polymer Science* 87: 1059–1067.

Varghese, S., Kuriakose, B. and Thomas, S. 1992. Mechanical properties of short sisal fiber reinforced natural rubber composites. *Indian Journal of Natural Rubber Research* 5: 55–62.

Varghese, S., Kuriakose, B. and Thomas, S. 1994a. Mechanical and viscoelastic properties of short fiber reinforced natural rubber composites: effects of interfacial adhesion, fiber loading, and orientation. *Journal of Adhesion Science and Technology* 8: 235–248.

Varghese, S., Kuriakose, B. and Thomas, S. 1994b. Short sisal fiber reinforced natural rubber composites: high-energy radiation, thermal and ozone degradation. *Polymer Degradation and Stability* 44: 55–61.

Varghese, S., Kuriakose, B. and Thomas, S. 1994c. Stress relaxation in short sisal-fiber-reinforced natural rubber composites. *Journal of Applied Polymer Science* 53: 1051–1060.

Varghese, S., Kuriakose, B., Thomas, S. and Joseph, K. 1995. Effect of adhesion on the equilibrium swelling of short sisal fiber reinforced natural rubber composites. *Rubber Chemistry and Technology* 68: 37–50.

Varghese, S., Kuriakose, B., Thomas, S. and Koshy A.T. 1991. Studies on natural rubber short sisal fiber composites. *Indian Journal of Natural Rubber Research* 4: 55–60.

Varghese, S., Kuriakose, B., Thomas, S. and Premalatha, C.K. 1993. Rheological behaviour of short sisal fiber reinforced natural rubber composites. *Plastics Rubber and Composites Processing and Applications* 20: 93–99.

Vasquez, A., Riccieri, J. and Carvalho, L. 1999. Interfacial properties and initial step of water sorption in unidirectional unsaturated polyester/vegetable fiber composites. *Polymer Composite* 20: 29–37.

Vincent, J.F.V. 2000. A unified nomenclature for plant fibers for industrial use. *Applied Composite Materials* 7: 269–271.

Wambua, P., Ivens, J. and Verpoest, I. 2003. Natural fibers: can they replace glass in fiber reinforced plastics? *Composites Science and Technology* 63: 1259–1264.

Wang, B., Panigrahi, S., Tabil, L. and Crerar, W. 2007. Pre-treatment of flax fibers for use in rotationally molded biocomposites. *Journal of Reinforced Plastics and Composites* 26(5): 447–463.

Warner, S. B., 1995. *Fiber Science*. New Jersey: Prentice Hall.

Wilson, P.I. 1971. Sisal, vol. II. In: *Hard Fibers Research Series*, no. 8, Rome: FAO.

Wong, S., Shanks, R. and Hodzic, A. 2004. Interfacial improvements in poly(3-hydroxybutyrate)-flax fiber composites with hydrogen bonding additives. *Composite Science and Technology* 64: 1321–1330.

Wright, J.R. and Mathias, L.J. 1993. New lightweight materials: Balsa wood-polymer composites based on ethyl α-(hydroxymethyl)acrylate. *Journal of Applied Polymer Science* 48: 2241–2247.

Xiao, Z., Zhao, L.B., Xie, Y. and Wang, Q.W. 2003. Review for development of wood plastic composites. *Journal of Northeast Forestry University* 31: 89–93.

Xie, Y., Hill, C.A.S., Xiao, Z., Militz, H. and Mai, C. 2010. Silane coupling agents used for natural fiber/polymer composites: A review. *Composites: Part A* 41: 806–819.

Zheng, Q., Song, Y.H. and Yi, X.S. 1999. Piezoresistive properties of HDPE/graphite composites. *Journal of Materials Science Letters* 18: 35–37.

11 Weathering Study of Biofiber-Based Green Composites

Vijay K.Thakur, Manju K.Thakur, and Raju K.Gupta

CONTENTS

11.1 INTRODUCTION

Biofibers are hastily emerging as a potential reinforcement in green composite materials because of their distinct advantages such as their renewable nature, biodegradability, cost-effectiveness, nonabrasiveness, high toughness, low density, enhanced energy recovery, and recyclability (Hagstrand and Oksman 2001; Kabir et al. 2012a,b; Ouajai and Shanks 2009a,b; Singha and Thakur 2009a,b). The substitution of conventional materials with biofiber-based materials is gaining immense attractiveness as a result of growing environmental awareness throughout the world (Averous 2004; Bledzki and Gassan 1999; Bledzki et al. 1996, 1998, 2010; Liu et al. 2011; Poletto et al. 2012). The biofiber-based green composite materials are up and coming as one of the most realistic alternatives to the existing pure synthetic composites prepared using established reinforcement such as carbon fibers, glass fibers, talc, and so on

(Lue and Zhang 2010; Ouajai and Shanks 2006; Panthapulakkal et al. 2006; Singha and Thakur 2010; Singha et al. 2010; Thakur et al. 2012a–c; Wambua et al. 2003). The use of biofiber as potential alternative to traditional synthetic materials for use in various fields (Debapriya et al. 2004; Hagstrand and Oksman 2001; Le Moigne and Navard 2004; Singha and Thakur 2008b; Singha et al. 2008) provides a number of advantages. Similar to biofiber, the main incentives for the fabrication of green composites are their easy availability, light weight, processability, low toxicity, low cost, and last but not the least their eco-friendly nature (Cipriani et al. 2010; Shanks et al. 2006; Shibata et al. 2013; Wong et al. 2007). The green composite materials are now in enormous demand because of the aforementioned properties especially in the automotive market (Lu and Drzal 2010; Pandey et al. 2013; Ten et al. 2010). The green composite–based materials can also be effectively used to make low-cost housing units, roofing, buildings, door panels, seat frames, and so on (Panthapulakkal et al. 2006; Singha and Thakur 2010; Singha et al. 2010; Thakur et al. 2012a–c). Nowadays, due to simultaneous increase in awareness on environment and energy, increasing attention is being paid to biofibers with a view to conserving energy and protecting the environment (Amash and Zugenmaier 1998; Panthapulakkal et al. 2006; Thakur and Singha 2010a–c). A number of research works are presently being done in the direction of using biofibers in place of synthetic fibers in various applications (Dhakal et al. 2007; Thakur et al. 2012a–c).

Saccaharum cilliare fibers are abundantly growing in hilly areas of India especially in the Himalayan region (Singha and Thakur 2008a, 2009b). These fibers are amongst the most interesting household products because of their lowest thermal conductivity and bulk density (Singha et al. 2009). Addition of these fibers can reduce the thermal conductivity of the green composite specimens and yield a suitable lightweight product. In different kinds of natural biofiber, cellulose is the main component, making them the most abundant natural polymer (Thakur and Singha 2010a–c; Thakur et al. 2010; Zain et al. 2011). It is well reported in the existing literature that the elementary unit of a cellulose macromolecule is anhydro-D-glucose, containing three hydroxyl (-OH) groups (Ouajai and Shanks 2009a,b). These hydroxyls have been found to form hydrogen bonds inside the macromolecule itself and with hydroxyl groups from moist air, making the natural fibers highly hydrophilic in nature (Thakur and Singha 2011a,b; Thakur et al. 2011). As a result of the hydrophilic groups, these natural fibers attract more moisture, making the resulting green composite materials unstable, which further promotes biological degradation (Dhakal et al. 2007; Singha et al. 2009).

Thus, the main objective of the study reported in this chapter was to further swot the physicochemical behavior of green composites prepared earlier using *S. cilliare* fibers as renewable reinforcement.

11.2 FABRICATION OF BIOFIBER-BASED GREEN COMPOSITES FOR WEATHERING STUDY

11.2.1 Materials, Methods, and Processing

The cellulosic biofibrous material used as the reinforcing material in the green biocomposites were *S. cilliare* fibers collected from the local resources of the Himalayan region in India. After proper purification, *S. cilliare* fibers were completely dried in a hot air oven and were subsequently converted into short *S. cilliare* fibers as per the standard method (Singha and Thakur 2009c). For the thermosetting polymer matrix, *Resorcinol*, formaldehyde solution, and sodium hydroxide were supplied by Qualigens Chemical Limited. The thermosetting polymer resin was synthesized by the method reported in the literature by Singha and Thakur (2008a, 2009b). Dried *S. cilliare* fibers that were chopped into short fiber dimensions were carefully mixed with thermosetting polymer resin using a mechanical stirrer with suitable loadings ranging from 10% to 40%. The fiber loading was done in terms of weight of fiber taken. For the easy removal of the green composites, surfaces of the molds that are used to prepare the composites were lubricated with oleic acid on the inner side (Thakur and Singha 2010b). This green composite mixture was then poured into specially made molds. The polymer mixture was uniformly spread on the surface of the molds. Composite sheets of size 150 mm × 150 mm × 5.0 mm were prepared by the compression molding technique (Singha and Thakur 2009a, 2010; Thakur and Singha 2010b,c). Green composite sheets were prepared by hot pressing the mold at the requisite temperature for 30 min (Singha and Thakur 2008a, 2009b). The pressure applied ranges from 3 to 4 MPa depending on the loading of *S. cilliare* fibers (Thakur and Singha 2010b).

11.3 PHYSICOCHEMICAL AND THERMAL CHARACTERIZATION OF GREEN POLYMER COMPOSITES

From the existing literature on green composite materials, it is quite clear that during the production of green composites, particular interest lies in their behavior against weathering conditions for long-term durability (Islam et al. 2005; Singha and Thakur 2009a; Thakur and Singha 2010b). A number of researchers have studied the effect of different weathering conditions on the green composites. Indeed the weathering study of different products prepared from green composites has been well reported and has been a subject of much debate (Dhakal et al. 2007; Islam et al. 2005; Singha and Thakur 2009a). In order to successfully commercialize the green composite materials, the saleable viability of the green fiber–reinforced polymer composites lies in their physical and chemical properties (Dhakal et al. 2007). Hence, keeping in mind the commercial practicability of the synthesized green composites, in this chapter a comprehensive study on swelling behavior of the green composites in different solvents, moisture absorbance at different humidity levels, and chemical resistance behavior against 1 N HCl and 1 N NaOH has been carried out. Along with physicochemical study, the thermal behavior of the optimized composite loading was also assessed (Islam et al. 2005; Singha and Thakur 2009a; Thakur and Singha 2010b). The sole aim of this study was to assess the potential of the *S. cilliare* fibers as reinforcement in a number of engineering applications.

11.3.1 Swelling Behavior

Swelling behavior of the *S. cilliare* fiber–reinforced green polymer composites was evaluated by studying the swelling in different solvents (Singha and Thakur 2009a; Thakur and Singha 2010b). Carbon tetrachloride, isobutanol, methanol, and water were used as the reference solvent for this study. This study was also carried out with the aim of learning the effect of the polar and nonpolar nature of various solvents on the swelling behavior of green polymer composites in order to assess their properties in different outdoor applications (Singha and Thakur 2009a; Thakur and Singha 2010b). Before carrying out the test, different specimens were dried in an oven for a specified time at a particular temperature and then cooled in a desiccator. A known weight (W_i) of the initial dried green polymer composite samples were immersed in 100 ml of different solvents at room temperature for 15 days. After the completion of the desired time period, the samples were filtered. The excess solvents were removed with the help of filter paper and then patted dried with the help of a lint free cloth. Finally the weight (W_f) of the composite samples was noted (Singha and Thakur 2009a; Thakur and Singha 2010b). The increase in initial weight was used to calculate the percent swelling in the following manner:

$$\text{Percent swelling } (P_S) = \frac{W_f - W_i}{W_i} \times 100$$

11.3.2   MOISTURE ABSORBANCE

Green composite samples were also subjected to moisture absorbance studies. The studies were carried out at different humidity levels in the range of 20%–90% (Singha and Thakur 2009a; Thakur and Singha 2010b). The moisture absorbance was studied by placing the known weight (W_i) of dry green composite samples in a humidity chamber set at a particular humidity level for about 12 hours. After this time, the final weight (W_f) of the green composite samples exposed at a particular relative humidity (RH) was taken (Singha and Thakur 2009a; Thakur and Singha 2010b), and the percent moisture absorbance was calculated in the following manner:

$$\% \text{ Moisture absorbance } (\%M_{\text{abs}}) = \frac{W_f - W_i}{W_i} \times 100$$

11.3.3   CHEMICAL RESISTANCE

The chemical resistance studies of the different green composites samples were carried out under the acidic and basic solutions of different molar concentrations. During the chemical resistance test, the dried green composite specimens were immersed in 100 ml of 1 N NaOH and 1 N HCl for different periods of time (24–144 hours). After the specific time periods, these green composite samples were filtered out, dried, and weighed (Singha and Thakur 2009a; Thakur and Singha 2010b), followed by the calculation of the percent chemical resistance (P_{cr}) in terms of weight loss in the following manner:

$$\text{Percent chemical resistance } (P_{\text{cr}}) = \frac{T_i - W_{\text{aci}}}{T_i} \times 100$$

where T_i = initial weight and W_{aci} = weight after a certain interval.

11.3.4 Thermal Behavior of the Green Composites

Thermal investigation of the green composite materials provides us basic information regarding thermal stability of these materials (Singha and Thakur 2009a; Thakur and Singha 2010b). Thermogravimetric analysis (TGA) of the optimized (30%) green composite samples after the swelling studies was carried out in nitrogen atmosphere on a thermal analyzer (Perkin Elmer) at a heating rate of 10°C/min. TGA is used to characterize the decomposition of green composite materials under various conditions (Singha and Thakur 2009a; Thakur and Singha 2010b). In TGA analysis a change in thermal stability is examined in terms of percentage weight loss as a function of temperature.

11.4   RESULTS AND DISCUSSIONS

Physicochemical and thermal studies have been found to be one of the most trusted tools in determining the overall behavior of green polymer composites. In this context we have studied the physicochemical properties of *S. cilliare* fiber–reinforced green polymer composites.

11.4.1 Physicochemical Behavior of Polymer Composites

S. cilliare fiber–reinforced green polymer composites show different swelling behavior in different solvents (Figure 11.1).

The swelling behavior of *S. cilliare* fiber–reinforced green polymer composites in different solvents follows the trend: $H_2O > CH_3OH > C_4H_9OH > CCl_4$. The swelling behavior of the green polymer composites increases with increase in fiber loading due to greater affinity of water for the higher number of OH groups present in the fiber-reinforced green polymer composites.

Figure 11.2 demonstrates the moisture absorbance behavior of the green polymer composites at different humidity levels as a function of fiber loading.

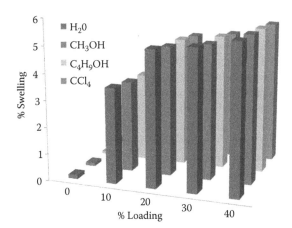

FIGURE 11.1   Swelling behavior of green polymer composites in different solvents.

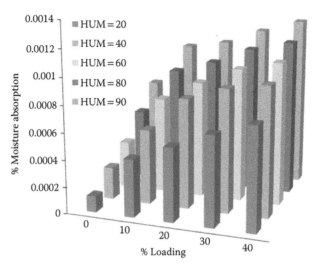

FIGURE 11.2   Moisture absorption behaviors of green polymer composites at different relative humidity levels.

From the figure it is quite clear that moisture absorbance (M_{abs}) increases with increase in humidity level ranging from 20% to 90% with increase in fiber loading.

From the chemical resistance behavior of the green composite samples, it is quite clear that resistance toward chemicals decreases with the increase in fiber loading (Figure 11.3a and b).

This may be due to the increase in the number of hydrophilic OH groups in fiber matrix bonding due to which composites with a higher number of OH groups with higher loading are vulnerable to chemical attack, resulting in decreased resistance toward the chemicals.

11.4.2   THERMAL BEHAVIOR OF GREEN POLYMER COMPOSITES

The thermal behavior of the *S. cilliare* fiber–reinforced green polymer composites was studied using TGA analysis. Figure 11.4 shows the TGA results of *S. cilliare* fiber–reinforced composite with optimum loading after the swelling study.

The green polymer composite sample has been found to exhibit intermediate behavior between the *S. cilliare* fibers and the matrix but inferior than the green composite sample without the weathering study (Singha and Thakur 2009a, 2010; Thakur and Singha 2010b). The decrease in the thermal stability of the weathered green composite sample can be attributed to the fact that during the swelling study the bonding between the polymer matrix and the reinforcement decreases, which decreases the thermal stability. Furthermore, the incorporation of the water content in the composite sample also initiates the degradation process of the green composite sample leading to a decrease in thermal stability.

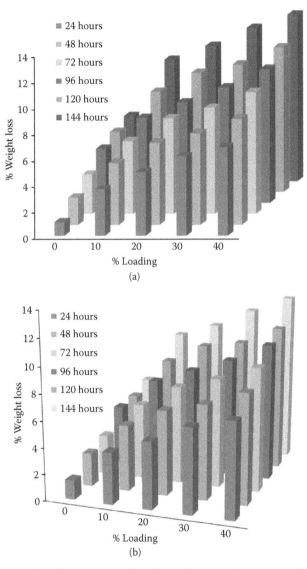

FIGURE 11.3 (a) Chemical resistance behaviors of green polymer composites at 1 N HCl concentration. (b) Chemical resistance behaviors of green polymer composites at 1 N NaOH concentration.

From the above study it is quite clear that green polymer composites are sensitive to weathering conditions. Different physical and chemical properties have been influenced by the fiber loading. Most importantly, in water absorption behavior, it has been observed that the extent of water absorption of *S. cilliare* fiber–reinforced green polymer composites depends on the RH of the environment, different fiber content, and type of polymer used.

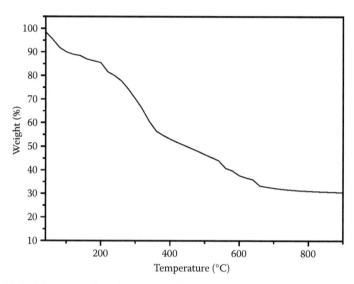

FIGURE 11.4 Thermogravimetric analysis of optimized green polymer composites sample.

11.5 CONCLUSIONS

Statistical test methods were adapted for physicochemical characterization of *S. cilliare* fiber–reinforced green polymer composites. The results of swelling, moisture, and chemical resistance behavior of green polymer composites were reported, and these could be used as a determinant for the end applications of these composites in everyday life. Physicochemical studies of the green polymer composites show that these composites are sensitive to swelling, moisture, and chemical resistance conditions due to the hydrophilic nature of the lignocellulosic fiber. Despite these limitations, the *S. cilliare* fibers can be an appropriate alternative to synthetic fibers as a reinforcing material for the preparation of various polymer matrix–based green composites. Thus, from this study, we concluded that cost-effective and environmentally friendly value-added composite materials can be obtained from *S. cilliare* biofibers.

ACKNOWLEDGMENTS

The authors thank their parental institutes for providing the necessary facilities to accomplish this research work.

REFERENCES

Amash, A. and Zugenmaier, P. 1998. Study on cellulose and xylan filled polypropylene composites. *Polymer Bulletin* 40: 251–258.
Averous, L. 2004. Biodegradable multiphase systems based on plasticized starch: A review. *Journal of Macromolecular Science. Polymer Reviews C* 44(3): 231–274.
Bledzki, A.K. and Gassan, J. 1999. Composites reinforced with cellulose based fibers. *Progress in Polymer Science* 24: 221–274.

Bledzki, A.K., Mamun, A. and Volk, J. 2010. Barley husk and coconut shell reinforced polypropylene composites: The effect of fibre physical, chemical and surface properties. *Composites Science and Technology* 70: 840–846.

Bledzki, A.K., Reihmane, S. and Gassan, J. 1996. Properties and modification methods for vegetable fibers for natural fiber composites. *Journal of Applied Polymer Science* 59(8): 1329–1336.

Bledzki, A.K., Reihmane, S. and Gassan, J. 1998. Thermoplastics reinforced with wood fillers: A literature review. *Polymer Plastics Technology and Engineering* 37(4): 451–468.

Cipriani, G, Salvini, A., Baglioni, P. and Bucciarelli, E. 2010. Cellulose as a renewable resource for the synthesis of wood consolidants. *Journal of Applied Polymer Science* 118: 2939–2950.

Debapriya, De., Debasish, De. and Basudam A. 2004. The effect of grass fiber filler on curing characteristics and mechanical properties of natural rubber. *Polymers for Advanced Technologies* 15(12): 708–715.

Dhakal, H.N., Zhang, Z.Y. and Richardson, M.O.W. 2007. Effect of water absorption on the mechanical properties of hemp fibre reinforced unsaturated polyester composites. *Composites Science and Technology* 67: 1674–1683.

Hagstrand, P.O. and Oksman, K. 2001. Mechanical properties and morphology of flax fiber reinforced melamine-formaldehyde composites. *Polymer Composites* 22(4): 568–578.

Islam, M.N., Khan, M.A. and Zaman, M.A. 2005. Study of the defects and water absorption behavior in jute-reinforced polymer composites using film neutron radiography. *Journal of Reinforced Plastics and Composites* 24: 1697–1703.

Kabir, M.M., Wang, H., Lau, K.T. and Cardona, F 2012a. Chemical treatments on plant-based natural fibre reinforced polymer composites: An overview. *Composites: Part B* 43: 2883–2892.

Kabir, M.M., Wang, H., Lau K.T., Cardona, F. and Aravinthan, T. 2012b. Mechanical properties of chemically-treated hemp fibre reinforced sandwich composites, *Composites: Part B* 43: 159–169.

Le Moigne, N. and Navard, P. 2004. Physics of cellulose xanthate dissolution in sodium hydroxide–water mixtures: A rheo-optical study. *Cellulose Chemistry and Technology* 44: 217–221.

Liu, A., Walther, A., Ikkala, O., Belova, L. and Berglund. L.A. 2011. Clay nanopaper with tough cellulose nanofiber matrix for fire retardancy and gas barrier functions. *Biomacromolecules* 12: 633–641.

Lu, J. and Drzal, L.T. 2010. Microfibrillated cellulose/cellulose acetate composites: Effect of surface treatment. *Journal of Polymer Science Part B: Polymer Physics* 48: 153–161.

Lue, A. and Zhang, L. 2010. Effects of carbon nanotubes on rheological behavior in cellulose solution dissolved at low temperature. *Polymer* 51: 2748–2754.

Ouajai, S. and Shanks, R.A. 2006. Solvent and enzyme induced recrystallization of mechanically degraded hemp cellulose. *Cellulose* 13: 31–44.

Ouajai, S. and Shanks, R.A. 2009a. Biocomposites of cellulose acetate butyrate with modified hemp cellulose fibres. *Macromolecular Materials and Engineering* 294: 213–221.

Ouajai, S. and Shanks, R.A. 2009b. Preparation, structure and mechanical properties of all-hemp cellulose biocomposites. *Composites Science and Technology* 69: 2119–2126.

Pandey, J.K., Nakagaito, A.N. and Takagi, H. 2013. Fabrication and applications of cellulose nanoparticle-based polymer composites. *Polymer Engineering and Science* 53: 1–8.

Panthapulakkal, S., Zereshkian, A. and Sain, M. 2006. Preparation and characterization of wheat straw fibers for reinforcing application in injection molded thermoplastic composites *Bioresource Technology* 97: 265–272.

Poletto, M., Zattera, A.J. and Santana, R.M.C. 2012. Structural differences between wood species: Evidence from chemical composition, FTIR spectroscopy, and thermogravimetric analysis. *Journal of Applied Polymer Science* 126: E337–E344.

Shanks, R.A., Hodzic, A. and Ridderhof, D. 2006. Composites of poly (lactic acid) with flax fibers modified by interstitial polymerization. *Journal of Applied Polymer Science* 99: 2305–2313.

Shibata, M., Yamazoe, K., Kuribayashi, M. and Okuyama, Y. 2013. All-wood biocomposites by partial dissolution of wood flour in 1-butyl-3-methylimidazolium chloride. *Journal of Applied Polymer Science* 127: 4802–4808.

Singha, A.S., Shama, A. and Thakur, V.K. 2008. Pressure induced graft co-polymerization of acrylonitrile onto *Saccharum cilliare* fiber and evaluation of some properties of grafted fibers. *Bulletin of Material Science* 31(1): 1–7.

Singha, A.S., Shama, A. and Thakur, V.K. 2009. Graft copolymerization of acrylonitrile onto *Saccaharum cilliare* fiber. *E-Polymers* 105: 1–12.

Singha, A.S. and Thakur, V.K. 2008a. Evaluation of mechanical properties of natural fiber reinforced polymer composites. *International Journal of Plastic Technology* 12: 913–923.

Singha, A.S. and Thakur, V.K. 2008b. Mechanical properties of natural fiber reinforced polymer composites. *Bulletin of Material Science* 31(5): 791–799.

Singha, A.S. and Thakur, V.K. 2009a. Chemical resistance, mechanical and physical properties of biofiber based polymer composites. *Polymer Plastics Technology and Engineering* 48(7): 736–744.

Singha, A.S. and Thakur, V.K. 2009b. Fabrication and characterization of *S. cilliare* fiber reinforced polymer composites. *Bulletin of Material Science* 32(1): 49–58.

Singha, A.S. and Thakur, V.K. 2009c. Mechanical, morphological, and thermal characterization of compression-molded polymer biocomposites. *International Journal of Polymer Analysis and Characterization* 15(2): 87–97.

Singha, A.S. and Thakur, V.K. 2010. Synthesis, characterization and study of *Saccaharum cilliare* reinforced polymer matrix based composites. *Journal of Reinforced Plastics and Composites* 29(5): 700–709.

Singha, A.S., Thakur, V.K. and Mehta, I.K. 2010. Renewable resources based green polymer composites: Analysis and characterization. *International Journal of Polymer Analysis and Characterization* 15(3): 127–146.

Ten, E., Turtle, J., Bahr, D., Jiang, L. and Wolcott, M. 2010. Thermal and mechanical properties of poly (3-hydroxybutyrate-co-3-hydroxyvalerate)/cellulose nanowhiskers composites. *Polymer* 51: 2652–2660.

Thakur, V.K. and Singha, A.S. 2010a. KPS-Initiated graft copolymerization onto modified cellulosic biofibers. *International Journal of Polymer Analysis and Characterization* 15(8): 471–485.

Thakur, V.K. and Singha, A.S. 2010b. Mechanical and water absorption properties of natural fibers/polymer biocomposites. *Polymer Plastics Technology and Engineering* 49(7): 694–700.

Thakur, V.K. and Singha, A.S. 2010c. Natural fibres-based polymers: Part I—Mechanical analysis of *Saccaharum cilliare* in forced biocomposites. *Bulletin of Material Science* 33(3): 257–264.

Thakur, V.K. and Singha, A.S. 2011a. Physico-chemical and mechanical behavior of cellulosic *Saccaharum cilliare* based biocomposites. *International Journal of Polymer Analysis and Characterization* 16(6): 390–398.

Thakur, V.K. and Singha, A.S. 2011b. Rapid synthesis, characterization, and physicochemical analysis of biopolymer-based graft copolymers. *International Journal of Polymer Analysis and Characterization* 16(3): 153–164.

Thakur, V.K., Singha, A.S., Kaur, I., Nagarajarao, R.P. and Yang L.P. 2010. Surface modified *Hibiscus sabdariffa* fibers: Physico-chemical, thermal and morphological properties evaluation. *International Journal of Polymer Analysis and Characterization* 15(7): 397–414.

Thakur, V.K., Singha, A.S. and Misra, B.N. 2011. Graft copolymerization of methyl methacrylate onto cellulosic biofibers. *Journal of Applied Polymer Science* 122(1): 532–544.

Thakur, V.K., Singha, A.S. and Thakur, M.K. 2012a. Graft copolymerization of methyl acrylate onto cellulosic biofibers: Synthesis, characterization and applications. *Journal of Polymers and the Environment* 20(1): 164–174.

Thakur, V.K., Singha, A.S. and Thakur, M.K. 2012b. In-air graft copolymerization of ethyl acrylate onto natural cellulosic polymers. *International Journal of Polymers Analysis and Characterization* 17(1): 48–60.

Thakur, V.K., Singha, A.S. and Thakur, M.K. 2012c. Surface modification of natural polymers to impart low water absorbency. *International Journal of Polymer Analysis and Characterization* 17(2): 133–143.

Wambua, P., Ivens, J. and Verpoest, I. 2003. Natural fibres: Can they replace glass in fibre reinforced plastics? *Composites Science and Technology* 63: 1259–1264.

Wong, S., Shanks, R.A. and Hodzic, A. 2007. Effect of additives on the interfacial strength of poly (l-lactic acid) and poly (3-hydroxy butyric acid)-flax fibre composites. *Composites Science and Technology* 67: 2478–2484.

Zain, M.F.M., Islam, M.N., Mahmud, F. and Jamil, M. 2011. Production of rice husk ash for use in concrete as a supplementary cementious material. *Construction and Building Materials* 25: 798–805.

12 Machining Behavior of Green Composites
A Comparison with Conventional Composites

Inderdeep Singh and Pramendra K. Bajpai

CONTENTS

12.1 INTRODUCTION

There has been a noticeable change in choice of materials for various engineering applications in the last few decades. Composite materials, especially polymer composites, have replaced metals and alloys in a wide variety of fields such as automobiles, civil engineering, aerospace, medical, electronics, agricultural, and sports equipment. Conventional polymer matrix composites (PMCs) consist of synthetic fibers such as glass, carbon, and aramid as reinforcement and petroleum-based synthetic polymers as matrix material. These polymers are either thermoplastics (such as polypropylene [PP], nylon, and polyvinyl chloride) or thermosets (such as epoxy, phenolic, and polyamide).

In spite of having very good mechanical properties, now these conventional PMCs are being discouraged where possible due to stringent environmental rules and regulations. Conventional PMCs are nonbiodegradable when disposed to the environment, which creates major ecological unbalance and hence creates a burden on the environment. Green composites (GCs) from natural resources are being looked at as a possible solution to environmental issues and a replacement to conventional polymer composites. Natural fiber–reinforced polymer (NFRP) composites are receiving widespread attention due to their environmentally friendly characteristics and competitive mechanical properties. NFRP composites are either partially biodegradable or fully biodegradable (green). Partially biodegradable composites incorporate natural fibers with synthetic polymers. In GCs, natural fibers are reinforced with biopolymers (such as polylactic acid [PLA]). The classification of polymer composites with their examples is shown in Figure 12.1. Different types of natural fibers (like sisal, nettle, ramie, flax, jute, cotton, hemp, coir, and banana) have been attempted as reinforcement in polymers to develop composites with improved overall properties (Bajpai et al. 2012). Natural fibers possess many properties, such as low specific weight, low cost, biodegradable nature, low energy consumption, renewability, and no health hazards, which are superior to synthetic fibers (Wambua et al. 2003).

Natural fiber–based composites have found their place in many application areas, especially in automobiles. Many interior and exterior parts of automobiles incorporate natural fiber–based thermoplastic polymer composites due to the advancements in polymer industry and low-cost lightweight thermoplastic composites (Holbery and Houston 2006).

Generally, simple polymer composite parts are produced to near net shape. But some machining operations are still required to give the composite products a final finish. If the composite product design involves a certain degree of intricacy, then it requires a joining process to assemble individually processed simpler parts into a complex composite product. Mechanical fastening, especially bolted joints, are the most commonly used joining technique. Hole generation is a prerequisite for

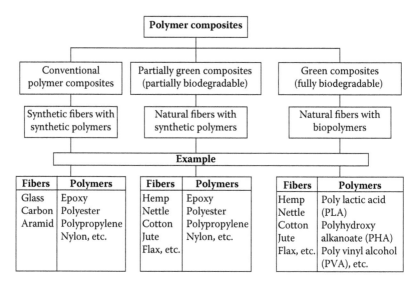

FIGURE 12.1   Classification of polymer composites.

bolted joints, which requires a drilling operation in polymer composites. The drilling operation is almost unavoidable and is one of the most frequently used machining operations in polymer composites. Drilling of polymer composites is entirely differ- ent from drilling of metals and alloys. During the drilling operation, both fibers and matrix are cut simultaneously. The mechanism of material removal is well established for metals, but due to the inherent inhomogeneity and anisotropy of PMCs, a compre- hensive investigation is still required in the field of machining of polymer composites. Drilling in PMCs results in drilling-induced damage (delamination, matrix burning, fiber pull-out, etc.), which results in poor-quality holes and parts rejection.

This chapter focuses on the drilling aspects of GCs in detail. A comparative study has been made between the drilling behavior of GCs and conventional polymer composites on the basis of process variables, drilling forces, damage, and the cutting mechanism.

12.2 DRILLING OF GREEN COMPOSITES

Though various techniques are available for hole making, conventional drilling is one of the most widely and frequently used techniques for hole generation. Conventional drilling has generated acceptable quality of drilled holes in GCs and conventional polymer composites. The drilling operation can be carried out using a radial drill- ing setup or computer numerical control (CNC) machining center. The schematic diagram of a radial drilling machine is shown in Figure 12.2. The machine has pro- visions for speed and feed change for the drilling operation. There is an option to change different drill bits in the tool holder as shown in the figure. The composite specimen is clamped in the fixture, and the machine is set at a predetermined cut- ting speed and feed rate. Generally, the cutting speed is set in RPM and feed rate is in millimeter per revolution and a sufficient range (low to high value) of variations is available in a radial drilling machine. After fixing the drill bit in the tool holder,

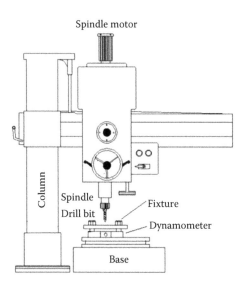

FIGURE 12.2 Radial drilling setup.

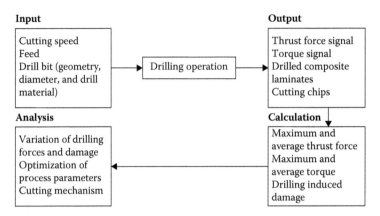

FIGURE 12.3 Drilling process in composite laminates.

the machine is turned on and the drilling action takes place within a few seconds depending upon the feed rate and the thickness of the composite laminate.

The signals of the drilling force (thrust force and torque) are observed during the drilling operation. These signals are recorded with the help of a drill dynamometer mounted above the base table of the drilling machine as shown in Figure 12.2. These signals are routed to a charge amplifier attached to a dynamometer and are recorded in a computer system through an analog/digital card. The signals are processed by a data acquisition system, and numerical values of the maximum and average thrust force and torque can be obtained. The drilling process of composite laminates is significantly governed and influenced by a number of input process parameters. Figure 12.3 shows the complete drilling process with main input and output parameters including their analysis. These parameters will be discussed in detail in the following section.

12.3 INPUT PARAMETERS

12.3.1 CUTTING SPEED AND FEED RATE

During drilling of polymer composite laminates, there is generally a similar effect of cutting speed and feed rate on the drilling of conventional and GCs. A common trend has been observed that variation of feed rate has a significant effect on drilling behavior of both types of polymer composites as compared to the effect of cutting speed. During drilling of resin-less bamboo composites, it has been found that with an increase in feed rate, the maximum thrust force increases under the cutting speed of 15 m/min with the metal drill. The specific cutting force (cutting force per unit length) decreases as the feed rate increases from 0.006 to 0.06 mm/rev at different cutting speeds ranging from 10 to 25 m/min. The specific cutting force has been found constant at the feed rate of 0.03 mm/rev and above (Mizobuchi et al. 2008). Drilling of coir/glass-reinforced polyester hybrid composites using a twist drill on a CNC machining center has shown that feed rate plays a major role in the process than the spindle speed and drill bit diameter. As the cutting speed increases, the thrust force remains almost constant, but as the feed rate increases, the thrust force first decreases and then increases with an increase in feed rate. The feed rate is also the most significant parameter for torque analysis. The spindle speed of 1503 rpm, feed rate of 0.2 mm/rev, and drill bit diameter of 8 mm have been found as the optimal parameters that produce minimum drilling forces (thrust force and torque) and tool wear during drilling of coir/glass-reinforced polyester hybrid composites (Jayabal et al. 2011). The effect of high cutting speed was studied on the thrust force, torque, tool wear, and hole quality for multifaceted and twist drill in drilling of carbon fiber–reinforced polymer (CFRP) composites. It was observed that at higher cutting speed, the tool wears out quickly, which resulted in an increase in the thrust force for both type of drills. As the cutting speed increased, the torque was found to increase for multifaceted drill points and decreased with the twist drill. It was concluded that an acceptable hole quality was observed, as the feed rates used were significantly low (Lin and Chen 1996). The influence of feed rate, cutting speed, and drill diameter has been analyzed for damage-free drilling of glass fiber–reinforced polymer (GFRP) composites. A nonlinear relationship has been observed to exist between the drilling forces (thrust force and torque) and drilling parameters such as drill diameter and feed rate (Mohan et al. 2005). The effect of drilling parameters such as cutting speed and feed rate on the cutting forces and torque has been investigated in drilling of chopped GFRP composites with different fiber volume fractions. Fiber volume fraction has been found to be directly proportional to thrust force and torque, which decrease with an increase in the cutting speed. The cutting speed has no clear effect on the delamination size, but with increase in feed rate, the delamination increases (Khashaba et al. 2007).

12.3.2 Drill Point Geometry

The selection of type of drill point geometry is a crucial decision in drilling of green and conventional composites because drilling forces and drilling-induced workpiece damage depend to a great extent upon the type of drill geometry. The machining properties of short agave fiber (both untreated and alkali treated)–reinforced epoxy composites have been studied by a drilling operation using 4, 6, 8.5, and 12.5 mm diameter high-speed steel (HSS) drill bits at a cutting speed of 260 rpm. The drill bit of 12.5 mm diameter recorded the minimum thrust force among the four drill bits used. Higher values of thrust force have been observed with composites of alkali-treated fibers, which confirm the good chemical bonding of the fiber and matrix (Mylsamy and Rajendran 2011). Though the diameter of the drill bit affects the drilling performance of polymer composites, in practical applications, generally, the diameter of the drill bit is fixed as per the design requirement. Therefore, factors that need optimization to achieve minimum drilling forces and drilling-induced damage in polymer composites are drill geometry and drill bit material. Different profiles for the drill bits (such as twist drill, trepanning tool, Jo drill, parabolic drill, C-shape drill, brad, and spur drill) or modification in various dimensions of the same drill geometry (such as change in helix angle, rake angle, web thickness, point angle) have a significant effect on drilling behavior of both green and conventional polymer composites. Some of the different types of drill geometries used for drilling of conventional and green polymer composites are shown in Figure 12.4.

The selection of drill bit material in the drilling operation of polymer composites is also an important decision. For GCs, normally, a drill bit of HSS material has

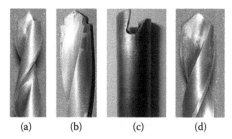

(a)   (b)   (c)   (d)

FIGURE 12.4 Different types of drill point geometries: (a) parabolic; (b) Jo drill; (c) trepanning; (d) twist drill.

been used, which effectively serve the purpose. For conventional polymer composites using glass and carbon as reinforcements, drill bits of different materials such as HSS, carbide drills, and high-speed steel–cobalt (HSS-Co) have been attempted to produce a large number of damage-free holes. During drilling of resin-less bamboo GCs with a wood drill (carbon steel) and metal drill (HSS tools), it has been found that the wood drill is inadequate to drill GC because of the larger thrust force (about four times larger) and poor hole quality as compared to the metal drill (Mizobuchi et al. 2008). The effect of various tool geometry parameters (point angle, helix angle, and chisel edge rake angle and web thickness) has been studied in the drilling process of conventional polymer composites. It has been found that the torque decreases with an increase in the point angle, whereas the thrust force increases. For larger helix and chisel edge rake angles, both the thrust force and torque decrease. The thrust force and torque also increase with the increasing web thickness and chisel edge length (Chen 1997). During drilling of graphite bismaleimide/titanium composites, it has been established that the highest temperature occurs at higher cutting speed and lower feed rates while drilling with HSS and HSS-Co drills. Carbide drills can be used to drill large number of holes to failure. The HSS and HSS-Co drills have shown damage in the form of a discoloration zone on the composite laminate, and carbide drills have resulted in minimal damage (Ramulu et al. 2001). An analytical approach has been proposed to predict the critical thrust force for different drill point geometries namely, saw drill, candlestick drill, core drill, and step drill. The analytical models presented were based on the axial loading only; the effect of torque was not considered and the type of load exerted by different drill points on the laminate to be drilled was presumed to be different (Hocheng and Tsao 2003).

Drill point geometry has a substantial effect on drilling forces and drilling-induced damage while drilling polymer composite laminates. Drilling of glass fiber–reinforced plastics with solid (twist drill and Jo drill) and hollow (trepanning and U-shape) drill point geometries has shown that the cutting mechanism of these drill geometries is substantially different and that there is a strong relationship between the drill point geometry and the drilling-induced damage. It has been shown that the twist drill is not recommended for the drilling of GFRP laminates because the thrust force generated with the twist drill is 2.5 times higher than the U-shape (which recorded minimum drilling forces and minimum delamination) drill during the drilling process (Rakesh et al. 2012).

12.4 OUTPUT PARAMETERS DURING DRILLING OPERATION

12.4.1 DRILLING FORCES

Drilling forces in terms of thrust force and torque have a direct influence on the drilling behavior of polymer composites. Different thrust force and torque signals are generated during drilling of green and conventional polymer composites. Typical thrust force signals recorded for nettle fiber–reinforced PP (nettle/PP) biocomposite with different drill geometries (twist, Jo drill, and parabolic) at a cutting speed of 900 rpm and feed rate of 0.05 mm/rev are shown in Figure 12.5. As the drilling process begins, drilling forces start to increase and then there is a zone of almost constant magnitudes. In this zone, the drill bit is fully engaged and finally the drill bit comes out of the composite laminate showing a continuous decrease in force. The drilling behavior of polymer composites has been critically analyzed with the help of drilling forces and has been correlated with the drilling-induced damage of the workpiece.

Drilling of sisal-, banana-, roselle-reinforced epoxy composites show that the cutting speed has an insignificant effect on the thrust force when drilling at low feed values. At high feed values, the thrust force and torque decreases with an increased cutting speed. Both thrust force and torque increase with the drill diameter and feed rate. Drilling forces also increase with the increase in the fiber volume fraction. The study has shown that sisal and roselle hybrid composite has shown lower fiber pull-in/pull-out during the drilling process, and it can be used for internal fixation (Chandramohan and Marimuthu 2010). Mathematical models for thrust force, torque, and damage have been proposed for the drilling of unidirectional glass/epoxy (UD-GE) conventional composites for various drill point geometries, which were in

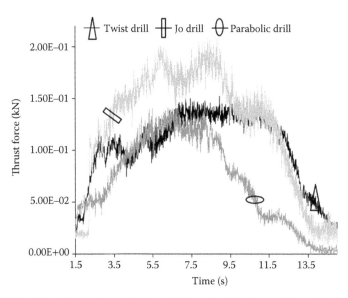

FIGURE 12.5   Thrust force signals during drilling of nettle fiber–reinforced polypropylene composite laminate.

close agreement with the experiments. It has been established that higher thrust force and torque values are recorded for the four-facet drill point geometry, and therefore, the four-facet drill is not recommended for drilling of laminated UD-GFRP composites. The thrust force and torque responses are better for eight-facet and Jo drills than the other geometries. Drilling-induced damage shows an elliptical zone with the major axis of the ellipse lying in the direction of the fibers (Singh and Bhatnagar 2006). The effect of cutting speed has been found insignificant on cutting forces for the same drill material. The cutting force has been found to be lower at lower feed rates. To improve the hole quality at exit, the feed rate at the exit should be decreased to a certain level during the drilling process (Chen 1997). It has been investigated that the torque cycle is delayed by a few seconds than the thrust force cycle during drilling of GFRP composite laminates. The drilling forces decrease with an increase in the cutting speed and increase with increase in drill diameter and fiber volume fraction. The drilling of GFRP laminates at lower feed rates has higher surface roughness as compared to higher feed rates (El-Sonbaty et al. 2004).

12.4.2 DELAMINATION

In drilling of both conventional and green polymer composites, various types of damage occur around the drilled hole and within the composite plate, which is known as the drilling-induced workpiece damage. The drilling-induced damage deteriorates the aesthetics of the composite part and the performance of the composite as well. Sometimes, due to internal damage during drilling, sudden failure of the composite part takes place. Therefore, it becomes necessary to minimize the drilling-induced damage to make the composite part safe and risk-free. Different types of drilling-induced damage such as delamination, fiber pull-out, fiber fracture, matrix cracking, matrix burning, plastic deformation, and debonding occur around the drilled hole during drilling of polymer composites. Among these damages, delamination is the most common type of damage that occurs in polymer composite laminates. Delamination at the entry of drilled hole is known as peel-up delamination and at the exit side of drilled hole is known as push-down delamination as depicted in Figure 12.6 (Hocheng and Dharan 1990). Push-down delamination is more severe than peel-up delamination. This damage can be minimized by proper optimization of process variables (cutting speed, feed rate, and drill point geometry) and sometimes using special techniques (e.g., vibration-assisted drilling) during the drilling operation. Use of a backing plate during drilling can reduce push-down delamination in composite plates.

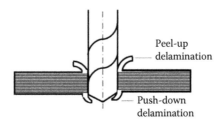

FIGURE 12.6 Type of delamination.

There are a number of techniques to measure drilling-induced damage around the drilled holes quantitatively. A nondestructive dye penetration technique has been found effective to locate the damage in conventional polymer composites. After dye penetration, the damaged area can be measured using image processing software. Other techniques such as X-ray nondestructive inspection, microscopic techniques, computerized vision inspection, ultrasonic C-scans, and radiographic techniques have been successfully used in the quantification of drilling-induced workpiece damage.

A digital scanning technique and digital radiographic technique were used in damage evaluation around drilled holes in sisal/epoxy biocomposite laminate and glass/epoxy conventional polymer composite laminate. It has been found that for large feed rate, delamination is more extended. The variation due to feed rate was more evident for a twist drill in sisal/epoxy plates and for a brad drill in glass/epoxy plates. It has been established that different drill geometry is suitable for different types of composite laminates (Durao et al. 2011). A computerized vision inspection has been proposed for damage evaluation in drilled holes in graphite/epoxy composite laminates with two different drill point geometries (dagger bit and twist drill). It has been established that the failure modes are related to the feed rate. The fraying was more substantial when low feed rates were used, and the excessive entrance delamination was associated with high feed rates (Hough et al. 1988). An experimental study has been conducted on drilling behavior of woven GFRP composites that correlated the width of the damage zone to the ratio between the cutting speed and the feed rate. As the ratio increased, the hole quality improved and the damage zone was found to decrease to the minimum value for a definite speed/feed ratio (Tagliaferri et al. 1990). To overcome the delamination in drilling of graphite/epoxy conventional composite laminates, a critical thrust force controller has been developed. Visual measurements of the drilled holes were done to compare the neural network–controlled drilling process and the conventional constant feed rate drilling process. It was proposed to incorporate an acoustic emission system in the neural network control scheme to monitor delamination (Stone and Krishnamurthy 1996). Oscillatory vibration–assisted drilling has been proposed for machining of GFRP composites to minimize drilling-induced damage. It was concluded that superimposed drilling is an effective way to perform damage-free drilling in GFRP composites (Ramkumar et al. 2004). The improvement in machinability in terms of the drilling behavior of hybrid biocomposites has been noticed when alkali-treated roselle and sisal fibers were reinforced into thermoset polyester as compared with machinability of untreated fiber composites. A profile projector and machine vision inspection system were used to analyze the profile and quality of the drilled holes (Athijayamani et al. 2010a,b). The drilling process of sisal fiber–reinforced PP (sisal/PP) biocomposites has been evaluated in terms of the drilling forces. A twist drill (solid) and trepanning tool (hollow) has been used for drilling at different cutting speed and feed rates. The variation of thrust force with feed rate is shown in Figure 12.7. It has been found that thrust force generated with the trepanning tool is quite low as compared to a twist drill during drilling of sisal/PP biocomposite, and the cutting mechanism of a hollow drill is suitable for drilling of natural fiber–reinforced thermoplastic composites (Bajpai et al. 2012).

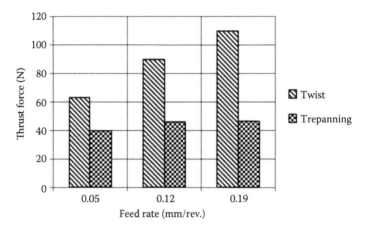

FIGURE 12.7   Variation in thrust force with cutting speed and feed rate.

12.5   APPLICATION OF PREDICTIVE TOOLS IN DRILLING OPERATION

The drilling process of green and conventional polymer composites requires proper selection of input process parameters (spindle speed, feed rate, and drill point geometry) to produce the drilled holes with minimum or no damage. There are various predictive tools and techniques, which have been used to predict the drilling behavior and to optimize the input process variables to make damage-free holes. Commonly used techniques to predict drilling behavior of polymer composites are Taguchi's experimental design, response surface methodology (RSM), genetic algorithm (GA), artificial neural network (ANN), and so on. Though the developed models using these techniques cannot be generalized, they are valid within the experimental domain.

The nonlinear regression equations of thrust force, torque, and tool wear have been developed to correlate the important machining parameters (drill bit diameter, spindle speed, and feed rate) with output responses in drilling of coir fiber–reinforced polyester composites using the Taguchi approach. The proposed models were capable of predicting the responses for drilling of coir fiber–reinforced composites within the range of variables studied and of optimizing machining parameters to reduce tool wear (Jayabal and Natarajan 2011). Mathematical models for the drilling behavior of coir/polyester composites were also developed and analyzed using the Box–Behnken design, Nelder–Mead, and GA methods to predict the main effects and interaction effects of various influential combinations of drilling parameters (Jayabal and Natarajan 2010). Nettle fiber–reinforced PLA GC laminates have been developed to analyze their machining behavior. The effect of cutting speed, feed rate, and drill point geometry on drilling forces (thrust force and torque) was statistically analyzed using RSM. The regression equations developed for thrust force and torque accurately predict the responses within the experimental domain. The variation of thrust force with feed rate and cutting speed as predicted with a Jo drill is shown in the three-dimensional surface plot in Figure 12.8.

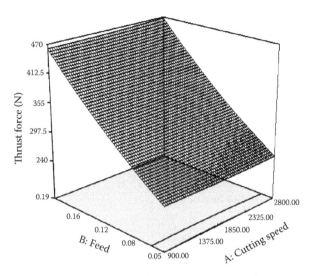

FIGURE 12.8   Effect of cutting speed and feed rate on the thrust force (Jo drill).

The delamination in drilling of GFRP composites has been analyzed using Taguchi's method and multiple regression analysis. It has been established that feed rate and drill diameter are the most influential parameters affecting drilling-induced damage, and the models developed with these techniques are suitable for predicting and optimizing the delamination factor (Lat ha et al. 2011). The microgenetic algorithm technique, a multiobjective optimization that is based on the *a posteriori* approach, can be used to optimize the drilling process. The proposed optimization technique increases the flexibility of selecting the optimal cutting parameters for drilling of polymer composite materials (Sardinas et al. 2006). RSM has been used to develop a mathematical relation with drilling parameters such as spindle speed, feed rate, and drill diameter for predicting delamination in drilling of CFRP composites. Feed rate has been found to be the main parameter that influences the delamination followed by drill diameter. It has been suggested that instead of using one bigger hole, two small holes are preferred for reducing the delamination by using proper tool geometry and drill size (Krishnamoorthy et al. 2009). Prediction of drilling forces in drilling of roselle/sisal polyester hybrid biocomposite has been carried out using an ANN and regression models (RMs). A comparison has been made on the predicted values of the thrust force and torque by both the techniques with the experimental values. The results indicate that the ANN model is more effective in prediction of drilling forces than the RM model in drilling of natural fiber hybrid composites (Athijayamani et al. 2010a,b).

12.6   GENERAL COMPARISON

In this chapter, a general analysis of results for drilling of conventional and GCs has been carried out in all the aspects including both experimental and predictive investigations. Both conventional and GCs can be machined using the same processing techniques and parameters but with different optimal settings of variables.

During the comparative study of drilling processes for conventional and GCs, it has been found that the cutting mechanism is different in both the cases that affect the quality of drilled holes; that is, drilling-induced workpiece damage. Therefore, the requirements of drill geometry and other process variables are different for both types of composites, which have also been found in some cases in the present study. Although a lot of research facts have been established in the field of hole-making in conventional composites, still a lot remains to be done in the area of making damage-free holes in GCs. The nature and type of natural fibers, the behavior of the biopolymers, and the combined effect of the two often makes the machinability of the GCs a technical challenge. The research efforts in the field of drilling of GCs are in the stage of infancy. The present chapter has reported the initial research findings in the broad area of hole-making in GCs.

12.7   CONCLUDING REMARKS

GCs derived from natural resources are acquiring their place rapidly in many application areas due to the enormous advantages associated with them. The prime advantage of GCs is their complete biodegradation characteristic, which makes these materials very desirable in today's scenario where safety of the environment and ecological balance is the main concern. The development and characterization of GCs have reached a level from where they can be molded into usable products and assemblies. Assembly requires a machining operation in GCs. The present chapter highlights the issues and challenges in machining of GCs and compares those with the drilling operation of conventional polymer composites. A complete discussion on the drilling operation in the context of conventional and GCs in terms of drilling forces and drilling-induced damage has been presented in this chapter.

REFERENCES

Athijayamani, A., Natarajan, U. and Thiruchitrambalam, M. 2010a. Prediction and comparison of thrust force and torque in drilling of natural fibre hybrid composite using regression and artificial neural network modelling. *International Journal of Machining and Machinability of Materials* 8: 131–145.

Athijayamani, A., Thiruchitrambalam, M., Natarajan, U. and Pazhanivel, B. 2010b. Influence of alkali-treated fibers on the mechanical properties and machinability of roselle and sisal fiber hybrid polyester composite. *Polymer Composites* 31: 723–731.

Bajpai, P.K., Singh, I. and Madaan, J. 2012. Development and characterization of PLA based 'green' composites: A review. *Journal of Thermoplastic Composite Materials*, doi: 10.1177/0892705712439571.

Bajpai, P.K., Singh, I. and Madaan, J. 2012. Secondary processing of natural fiber reinforced thermoplastic composite laminates. In: *Proceedings of the 8th Asian-Australasian Conference on Composite Materials* (ACCM-8), November 6–8, 2012, Kuala Lumpur, Malaysia, Paper no.: O-NAT–365.

Chandramohan, D. and Marimuthu, K. 2010. Thrust force and torque in drilling the natural fiber reinforced polymer composite materials and evaluation of delamination factor for bone graft substitutes-a work of fiction approach. *International Journal of Engineering Science and Technology* 2: 6437–6451.

Chen, W.C. 1997. Some experimental investigations in the drilling of carbon fiber-reinforced plastic (CFRP) composite laminates. *International Journal of Machine Tools and Manufacture* 37: 1097–1108.

Durao, L.M., Gonçalves, D.J.S., Tavares, J.M.R.S., de Albuquerque, V.H.C., Panzera, T.H., Silva, L.J., Vieira, A.A. and Baptista, A.M. 2011. Drilling delamination outcomes on glass and sisal reinforced plastics. In: *Proceedings of VI International Materials Symposium MATERIALS 2011,* Guimaraes, Portugal.

El-Sonbaty, I., Khashaba, U.A. and Machaly, T. 2004. Factors affecting the machinability of GFR/epoxy composites. *Composite Structures* 63: 329–338.

Hocheng, H. and Dharan, C.K.H. 1990. Delamination during drilling in composite laminates. *Transactions of the ASME, Journal of Engineering for Industry* 112: 236–239.

Hocheng, H. and Tsao, C.C. 2003. Comprehensive analysis of delamination in drilling of composite materials with various drill bits. *Journal of Materials Processing Technology* 140: 335–339.

Holbery, J. and Houston, D. 2006. Natural-fiber-reinforced polymer composites in automotive applications. *JOM The Journal of The Minerals, Metals & Materials Society* 58: 80–86.

Hough, C.L., Lednicky, T.E. and Griswold, N. 1988. Establishing criteria for a computerized vision inspection of holes drilled in carbon fiber composites. *Journal of Testing and Evaluation* 16: 139–145.

Jayabal, S. and Natarajan, U. 2010. Optimization of thrust force, torque, and tool wear in drilling of coir fiber-reinforced composites using Nelder–Mead and genetic algorithm methods. *International Journal of Advanced Manufacturing Technology* 51: 371–381.

Jayabal, S. and Natarajan, U. 2011. Drilling analysis of coir–fibre-reinforced polyester composites. *Bulletin of Material Science* 34: 1563–1567.

Jayabal, S., Natarajan, U. and Sekar, U. 2011. Regression modeling and optimization of machinability behavior of glass-coir-polyester hybrid composite using factorial design methodology. *International Journal of Advanced Manufacturing Technology* 55: 263–273.

Khashaba, U.A., Selmy, A.I., El-Sonbaty, I.A. and Megahed, M. 2007. Behavior of notched and unnotched [0/±30/±60/90]s GFR/epoxy composites under static and fatigue loads. *Composite Structures* 81: 606–613.

Krishnamoorthy, A., Boopathy, S.R. and Palanikumar, K. 2009. Delamination analysis in drilling of CFRP composites using response surface methodology. *Journal of Composite Materials* 43: 2885–2902.

Latha, B., Senthilkumar, V.S. and Palanikumar, K. 2011. Modeling and optimization of process parameters for delamination in drilling glass fiber reinforced plastic (GFRP) composites. *Machining Science and Technology* 15: 172–191.

Lin, S.C. and Chen, I.K. 1996. Drilling carbon fiber-reinforced composite material at high speed. *Wear* 194: 156–162.

Mizobuchi, A., Takagi, H., Sato, T. and Hino, J. 2008. Drilling machinability of resin-less "green" composites reinforced by bamboo fiber. In: *High Performance Structures and Materials IV.* Eds. de Wilde, W.P. and Brebbia, C.A., 185–194. Southampton, UK: WIT Press.

Mohan, N.S., Ramachandra, A. and Kulkarni, S.M. 2005. Machining of fiber-reinforced thermoplastics: Influence of feed and drill size on thrust force and torque during drilling. *Journal of Reinforced Plastics and Composites* 24: 1247–1257.

Mylsamy, K. and Rajendran, I. 2011. Influence of alkali treatment and fibre length on mechanical properties of short agave fibre reinforced epoxy composites. *Materials and Design* 32: 4629–4640.

Rakesh, P.K., Singh, I. and Kumar, D. 2012. Drilling of composite laminates with solid and hollow drill point geometries. *Journal of Composite Materials*. Published online 15 February 2012, doi: 10.1177/0021998312436997.

Ramkumar, J., Aravindan, S., Malhotra, S.K. and Krishnamurthy, R. 2004. An enhancement of the machining performance of GFRP by oscillatory assisted drilling. *International Journal of Advanced Manufacturing Technology* 23: 240–244.

Ramulu, M., Branson, T. and Kim, D. 2001. A study on drilling of composite and titanium stacks. *Composite Structures* 54: 67–77.

Sardinas, R.Q., Reis, P. and Davim, J.P. 2006. Multi-objective optimization of cutting parameters for drilling laminate composite materials by using genetic algorithms. *Composites Science and Technology* 66: 3083–3088.

Singh, I. and Bhatnagar, N. 2006. Drilling of uni-directional glass fiber reinforced plastic (UD-GFRP) composite laminates. *International Journal of Advanced Manufacturing Technology* 27: 870–876.

Stone, R. and Krishnamurthy, K. 1996. A neural network thrust force controller to minimize delamination during drilling of graphite epoxy laminates. *International Journal of Machine Tools and Manufacture* 36: 985–1003.

Tagliaferri, V., Caprino, G. and Diterlizzi, A. 1990. Effect of drilling parameters on the finish and mechanical properties of GFRP composites. *International Journal of Machine Tools and Manufacture* 30: 77–84.

Wambua, P., Ivens, J. and Verpoest, I. 2003. Natural fibres: Can they replace glass in fibre reinforced plastics? *Composites Science and Technology* 63: 1259–1264.

13 Potential Biomedical Applications of Renewable Nanocellulose

Sivoney F. de Souza, Bibin M. Cherian, Alcides L. Leão, Marcelo Telascrea, Marcia R. M. Chaves, Suresh S. Narine, and Mohini Sain

CONTENTS

Nanocellulose is observed to be a promising biomaterial for use as implants and scaffolds in tissue engineering. Due to their availability, high strength, low weight, and biodegradability, nanocellulose fibers can be used in several applications, including medicine and biomaterials such as scaffolds in tissue engineering, artificial skin and cartilage, and wound healing. Researchers are using nanocellulose to develop cartilage to create artificial outer ears. Designed studies were performed to apply nanocellulose as a blood vessel implant in humans and animals. Different strategies were developed to produce materials that use nanocellulose to apply to bone tissue engineering. Recently, the ability to cultivate nanocellulose on nerve cells is observed to be a vital achievement for many uses. Nanocellulose as an attractive fiber-reinforcing agent can present a viable material for the different kind of applications in the medical field, besides the relative low cost and ease of fabrication. With these inherent advantages of cellulose derivatives, some applications as hydrogels reinforced with cellulose appeared similar to natural tissue having hydrogel-like characteristics, which has the potential to replace materials for articular cartilage, mainly due to network structure, mechanical, viscoelastic, and swelling properties. Other applications of nanocellulose as scaffolds play a significant role in integrating the overall tissue constructs. Biocompatible nanocellulose material could potentially constitute an acceptable candidate in scaffolding of the tissue-engineered vessel due to the three-dimensional structure, controlled porosity and enhanced mechanical properties, surface adhesion capability, and increased extent of biodegradability. This chapter deals with the promising advancements made in the utilization of nanocellulose for the three main categories of biomaterials: inert, bioactive, and biodegradable materials. There is also an analysis of in vivo and in vitro research.

13.1 INTRODUCTION

Cellulose is an organic polymer of low cost, the main component of plant fibers, distinguished by properties such as hydrophilicity, chirality, large capacity for morphological and chemical modification of semicrystalline fibers, biodegradability, and strong social acceptance (Klemm et al. 2005; Szczesna-Antczak et al. 2012). This polymer is about 1.5×1012 tonnes of the total annual production of biomass and is considered an almost inexhaustible source of raw material for the increasing demand for bioproducts (Kaplan 1998). The structure of cellulose is a homopolysaccharide linear and unbranched, consisting of 10,000–15,000 D-glucose with β configuration. The glucose residues in the pulp are connected by glycosidic linkages (β1 → 4). Cellulose is composed of pyranose subunits as shown in Figure 13.1 with a strong tendency to form inter- and intramolecular hydrogen bonds.

Cellulose is found not to be uniformly crystalline. However, there are extensive regions arranged throughout the material, known as crystalline regions. The wire emerges from the linear combination of these components called microfibril (Thomas et al. 2011). Within these cellulose fibrils, there are regions where the cellulose chains are arranged in a highly ordered (crystalline) structure, and regions that are disordered (amorphous like). The crystalline regions contained within the cellulose microfibrils are extracted, resulting in cellulose nanocrystals (CNCs).

CNCs are produced by the breakdown of cellulose fibers and isolation of crystalline regions, by means of chemical processes (Oke 2010). Different names for the nanofibers are often used in the literature. These include "nanowhiskers,"

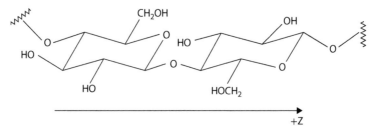

FIGURE 13.1 Fragment (repeating unit) of a cellulose chain. (From O'Sullivan, A.C. *Cellulose*, 4, 173–207, 1997. With permission.)

"nanocrystals," or "crystals" (Kvien and Oksman 2007). These crystals also have been frequently reported throughout the literature as "microfibril" or "microcrystal," although its dimensions are nanoscale. The term "whisker" is used to designate crystalline nanoparticles that are rod shaped, as can be seen from atomic force microscopy image shown in Figure 13.2, whereas the designation "nanofibrils" shall be used to refer to particles that are long and flexible, comprising alternating crystalline and amorphous sequences.

Nanocellulose is usually identified and recognized as "microfibrilated cellulose—MFC" or "nanofibrilated cellulose—NFC" and presents specific morphology where it is possible to identify one network structure. This material is obtained using special equipment called homogenizers with high pressure or grinder where the cellulose pulp is opened, spreading the microfibrils. Figure 13.3 shows the network

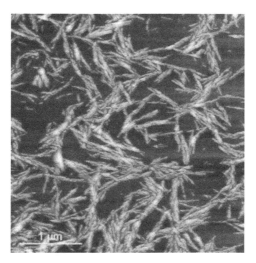

FIGURE 13.2   Atomic force microscopy image of whiskers from microcrystalline cellulose. (From Kvien, I. and Oksman, K. *Appl. Phys. A Mater. Sci. Process.*, 87, 641–643, 2007. With permission.)

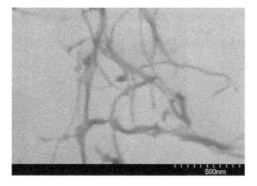

FIGURE 13.3   Morphology of microcrystalline cellulose from curaua fibers using an environmental scanning electron microscope. (From Souza, S.F., Leão, A.L., Cai, J.H., Wu, C., Sain, M. and Cherian, B.M. *Mol. Crystals Liquid Crystals*, 522, 42–52, 2010. With permission.)

structure of nanocellulose, obtained from the study by Souza et al. (2010), where curaua fibers were used as the source for NFC extraction.

Nanocellulose is currently used as a potential reinforcement material in biocomposites, and the properties are observed to be higher than those of the microfiber, due to its high surface area, which is much superior than that found in conventional microreinforcements. The benefit of nanocomposites is the ability to diminish defects found at macroscale matrix and reinforcement. In addition, cellulose fibers with diameters less than 100 nm do not affect the transparency of the polymers, since they have uniform morphology. These properties, coupled with the low density, abundance, and renewability, have triggered the attention of researchers and entrepreneurs in developing nanocomposite materials (Habibi et al. 2010; Seydibeyoglu and Oksman 2008). The composites reinforced with nanocellulose materials are quite versatile and exhibit a wide range of biomedical applications.

Biopolymers obtained from cellulose are mainly focused for intense research and development in industry and research centers (Klemm et al. 2006). The cellulose has distinctive characteristics and properties that differ from other natural and synthetic polymers. Its high availability, ease of biodegradation, easy mechanical manipulation, molding, relative strength, high stiffness, low density, and low cost makes it one of the most fascinating materials for the development of new materials with low environmental impact. As it is renewable and biodegradable, cellulose has been widely used as reinforcement material for the preparation of new biomaterials (Cherian et al. 2010; Cherian et al. 2011a). Figure 13.4 shows the thinner films made of nanocellulose especially for biomedical applications (Gatenholm and Klemm 2010).

Nanocellulose has some important properties, such as biocompatibility, low toxicity, and biodegradability together with excellent mechanical properties and thermal stability. These characteristics make it a unique candidate for medical applications, especially artificial implants, dressings, vehicles for drugs, and medical devices (Cherian et al. 2011b; Mathew et al. 2012). The bacterial nanocellulose (BNC), mainly obtained from *Acetobacter xylinum*, has been widely used in medicine. In addition to the advantages described above, the biomaterial may also be used in implantation sites without causing inflammatory reactions.

Studies and development of biomaterials and their medical applications have shown the importance of various nanocelluloses in the development of novel and

(a)   (b)

FIGURE 13.4 (a) Typical nanocellulose fleece formed in a reactor of circular profile, diameter 3 cm, thickness 2 cm. (b) Thin film of bacterial cellulose. (From Gatenholm, P. and Klemm, D. *MRS Bull.*, 3, 208–213, 2010.)

innovative products that have medical and veterinary applications. However, for medical applications, several studies have to be addressed to assess whether the new material may be applied to the body. The product obtained from Gengiflex® nanocellulose, for example, has dental applications, with the goal of assisting the recovery of the periodontal tissue (Novaes and Novaes 1997). This material allows the complete restoration of bone defects, including reestablishing normal function and aesthetics of the mouth (Novaes et al. 2003).

13.2 BIODEGRADABLE MATERIALS

Hydrolytic degradation of cellulose can be catalyzed either chemically (e.g., by the action of acids) or enzymatically. Cellulose-degrading enzymes, called cellulases, are common in certain species of fungi and bacteria and enable them to convert cellulose to glucose monomers. The human body does not contain enzymes capable of catalyzing the hydrolysis of the β-1,4 linkages of cellulose chains, rendering plant cellulose or BNC biostable and nonresorbable, at least inside the human body. In vivo and in vitro investigations have been performed to evaluate the degradability of BNC with respect to medical applications. Because glucose, as the potential degradation product, is not detectable in vivo, recent studies have investigated the macroscopic and physical composition of the material after implantation in animals. The main modification in the structure of BNC membranes after implantation into Swiss albino mice was the degradation that was detected after 90 days (Mendes et al. 2009). Souza et al. (2011) described that nanocellulose membranes have been used to evaluate the medialization, tissue response, and healing of rabbit vocal folds, after the implantation. The authors confirmed that nanocellulose is a useful material for laryngeal medialization, showing no signs of rejection or absorption (Figure 13.5).

The nanocellulose can be used as potential biodegradable platforms for controlled drug delivery, which are able to provide a drug over a long period of time and thereby reduce the number of doses of the drug during treatment. The materials containing NFC have been produced as spray dried particles, hydrogels, aerogels, and films

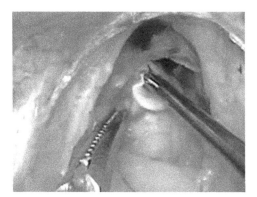

FIGURE 13.5 The endoscopic view of the nanocellulose-implanted larynx. (From Souza, F.C., Olival-Costa, H., Silva, L., Pontes, P.A. and Lancellotti, C.L.P. *J. Voice*, 25, 765–769, 2011. With permission.)

(Siro and Plackett 2010). Valo et al. (2011) showed the use of NFC for the stabilization of drug nanoparticles in suspension. These nanoparticles must be stabilized properly in such way that they aggregate it. Thus, nanoparticles could be stored for more than 10 months without major changes in their morphology.

Kolakovic et al. (2012) showed that NFC microparticles produced by a spray drying method can be used as fillers in tablet production by direct compression, as well as in wet granulation. In the spray drying process, NFC produces porous microparticles, and at the same time, produces encapsulated drugs inside the NFC carrier. These particles become enveloped by the NFC fiber network forming a matrix system structure, limiting the diffusion of the drug, and capable of controllably releasing it for a long period of time. The microparticles produced from amorphous drugs were about 5 mm in diameter. The inert matrices comprised of insoluble polymers originate porous structures in which the drug is dispersed by maintaining the apparent surface (solid–liquid interface to dissolve) throughout the dissolution step. A portion of the drug that was dried separately or located at the particle surface affected the final delivery of the drug. The drug release studies demonstrated a sustained-release profile of drugs for a period of two months.

Tablets prepared with inert polymers form systems that do not change throughout the gastrointestinal tract and eliminated virtually intact.

Another versatile product, Biofill®, is used as a bandage, which can be applied to cases of second- and third-degree burns. Biofill® is an innovative product that temporarily replaces the human skin. One of the major limiting factors is its elasticity when applied in high mobility (Cherian et al. 2011b; Fontana et al. 1990). Nevertheless, the material promotes immediate pain relief after application, reducing discomfort after surgery and the rate of infection. Biofill® reduces time and cost of treatment (Fontana et al. 1990).

The medical use of nanocellulose has vast potential in the area of burns. Because burns are complex lesions and generally result in destruction of the skin tissues, these can be classified and treated differently according to the damage (Latarjet 1995). Third-degree burns are those that cause complete destruction of all epidermis and dermis, extending to the subcutaneous tissue. This requires that the treatment must be more complex and requires the consideration of skin grafts. The healing process involves the regeneration and repair of the epidermis from the dermis, which results in scar tissue formation (Balasubramani et al. 2001).

The use of biomaterials allows burns to be cured more rapidly during the healing treatment and significantly reduces the pain and the chance of infection (Demling and De Santi 1999; Gallin and Hepperle 1998; Jones et al. 2002; Prasanna et al. 2004). Several studies indicate that nanocellulose is extremely efficient in reducing scars when applied topically on wounds or burns (Czaja et al. 2007a,b). Burns covered with cellulose membranes heal faster when compared with normal healing. The membrane adheres to all contours of the body, which facilitates the entire process of regeneration of the skin, reduces pain, and leaves the wounded area without scar formation.

Scarring is a very important aesthetic factor, since many of the burns leaves deep marks, and these have an important psychological impact on patients (Cherian et al. 2011b). The never-dried microbial cellulose membrane is a nonpyrogenic and highly biocompatible biomaterial with high mechanical strength. Figures 13.6 and 13.7 show the nanocellulose dressing applied on wounded areas of the body (Czaja et al. 2006).

Basmaji (2010) showed the regeneration ability of the product NanoskinV® (nanocellulose produced by Innovatec—Biotechnology Research and Development). This material has proven to be very effective in cleansing wounds (Figure 13.8). NanoskinV® is a film made of a highly hydrated random set of fibers that assembles into a ribbon of less than 100 nm width. Its application inhibits bacterial growth in wounds thereby accelerating the healing process.

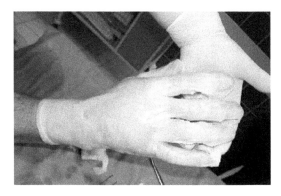

FIGURE 13.6 Bacterial cellulose dressing applied on a wounded hand. (From Czaja, W., Krystynowicz, A., Bielecki, S. and Brown, R.J. *Biomaterials,* 27, 145–151, 2006. With permission.)

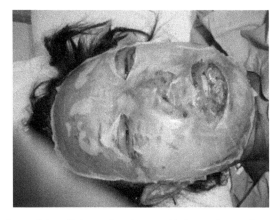

FIGURE 13.7 Bacterial cellulose dressing applied on face. (From Muangman, P., Opasanon, S., Suwanchot, S. and Thangthed, O. *J. Coll. Certified Wound Specialists,* 3, 16–19, 2011. With permission.)

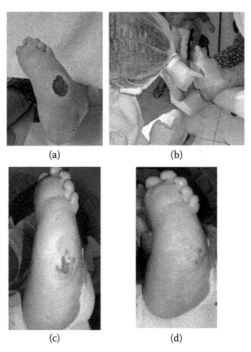

(a)   (b)

(c)   (d)

FIGURE 13.8   Diabetic varicose ulcer treatment using NanoskinV®. Complete healing was observed after three months (d). (From Basmaji, P. 2010. Nanoskin for medical application. 3rd International Seminar on Nanosciences and Nanotechnologies, Sep. 7–9, 2010, Havana, Cuba.)

13.3 BIOACTIVE BIOMATERIALS

Bioactivity means that a biomaterial has a direct influence on the physiology and morphology of living cells by controlling their adhesion, migration, proliferation, differentiation, and release of extracellular matrix (ECM) molecules leading to the formation of new tissue. Several factors, such as structural and surface properties, the inclusion of composite partners, and the ability to release incorporated substances (e.g., growth factors, cytokines), contribute to the bioactivity of biomaterials.

The bones are the largest and most remarkable nanocomposite of our body. Despite their importance, many health problems are related to bone defects and traumatic processes and require that grafts are performed to improve the quality of life (Murugan and Ramakrishna 2005). According to data, millions of fractures occur each year around the world and many of them require grafts. The most common fractures are of the hip, ankle, tibia, and fibula.

Several nanocomposites are being researched in developing prosthetics that are more accessible and more efficient. One, hydroxyapatite (HA), based on collagen, has been used as a base material for bone graft for replacement. Its main characteristics are related to its composition and structural similarity with the bones and functional properties (high surface area and mechanical resistance). Several studies suggest that the biocomposite HA/collagen graft can be replaced by promoting more effective bone regeneration (Murugan and Ramakrishna 2005). Figure 13.9 shows the design strategy of a tissue-engineered nanocomposite bone graft. Its application in bone grafts is extremely promising, where the material is implanted in bone defects or broken parts and accelerates the regeneration of new bone cells. These types of biocomposites have been extensively applied in orthopedic and dental applications (Bodin et al. 2007a; Cherian et al. 2011b; Helenius et al. 2006; Murugan and Ramakrishna 2005; Silva 2009).

There are various fields in which nanocellulose is used with HA for bone regeneration. Nanocomposites prepared with HA covered by nanocellulose were applied in the dental cavities of cats (*Felis catus*) by Silva (2009). It was found that nanocellulose associated with HAP endorse faster bone regeneration when compared with the control group eight days after the procedure and after a delay of 30 days, although after 50 days they had identical tissues (Silva 2009).

In nose surgery (rhinoplasty), carved or crushed cartilage used as a graft has some disadvantages; chiefly it may be perceptible throughout the nasal skin after tissue resolution is complete, and to overcome these problems and to obtain a smoother surface, researches initiated the use of Surgicel®-wrapped diced cartilage. Surgicel® is a hemostatic agent (blood clot inducing material) made of an oxidized cellulose polymer (the unit is polyanhydroglucuronic acid). The favorable results obtained by this technique have led the authors to use Surgicel®-wrapped diced cartilage routinely in all types of rhinoplasty (Erol 2000). Postoperative appearance after one year showed that overcorrection persisted and needed to be revised. When the excised tissue was carefully observed, numerous pieces of diced cartilage were noticeable. They

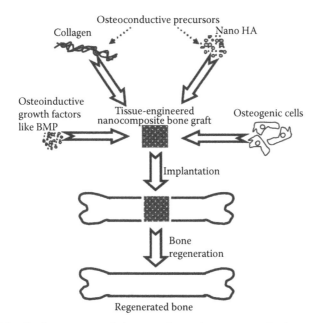

FIGURE 13.9 Design strategy of tissue-engineered nanocomposite bone graft. (From Murugan, R. and Ramakrishna, S. *Compos. Sci. Technol.*, 65, 2385–2406, 2005. With permission.)

were obviously surrounded with fibrous tissue. Histologic study of the same specimen showed mosaic-type alignment of cartilage islands surrounded with mature fibrous tissue.

13.4 INERT BIOMATERIALS

Biomaterials can be classified as "nearly inert" or bioactive, depending on the reaction at the interface between the implant and the tissue around natural. The inert biomaterials cause minimal or no response from the host tissue, which is kept unchanged.

The biocompatibility of the materials can be influenced by inflammation. However, when aggregates such as biomaterials are used in implants interface with surrounding tissue, the implant is not chemically or biologically active. Thus, almost all inert biomaterials can promote any bone repair and are almost exclusively used as bone replacement devices (Barone et al. 2011).

Cellulose nanofibers were also used in the load-bearing component of ligament/ tendon substitutes. Mathew and Oksman (2010) showed that characteristics such as low toxicity, biocompatibility, and biodegradability together with excellent mechanical properties of nanocellulose-based nanocomposites make them an excellent load-bearing component for biomedical applications.

Although natural tendons and ligaments are tough, injuries such as rupture of the anterior cruciate ligament can occur and cellulose nanofibers are important in stabilizing the knee (Fu et al. 1999). Various prostheses have been developed (Bolton and Bruchman 1983; Chazal et al. 1985; De Santis et al. 2004; Gissefalt et al. 2002; Jenkins et al. 1977; Legnani et al. 2010; McCartney et al. 1993; Meyers et al. 2008; Migliaresi and Nicolais 1980; Nachemson and Evans 1968). However, the biggest challenge is that these materials have no biomechanical properties in their original structure. Secondary studies reported that many polymers had progressive degradation of generating mechanical pain along with other problems (McCartney et al. 1993).

De Santis et al. (2004) studied replacement ligaments and tendons with carbon–fiber composites and reported a maximum strength of 28 MPa and a pressure of 30% compared with the original ligament and a maximum resistance pressure of 38 MPa and 18%, compared with the original tendon.

Mathew et al. (2012) showed that nanocellulose can be employed as a candidate for tendon and ligament development that offers mechanical properties similar to or better than the natural origin tendons or ligaments. When developing artificial tendons or ligaments, biocompatibility must take into account that the body temperature is 37°C and humidity is high (>60%); therefore, all testing related to the mechanical properties of the biomaterial must be evaluated under these conditions (Hukins et al. 1999). Various studies (Mathew et al. 2012) showed that nanoreinforced cellulose presented excellent stability, good strength properties and pressure (25–30 MPa and 20%–28%), low toxicity, and excellent compatibility. All these properties are required for development of biomaterials with the good mechanical properties and excellent cytocompatibility required for medical applications.

Despite the significant progress made toward the understanding of anatomy, composition, and biomechanical healing of tendons and ligaments, there are actually few materials that are ideal replacements for ligaments and tendons (Mathew and Oksman 2010). Natural tendons and ligaments are able to withstand high voltages. This factor is limiting the development of new materials.

Synthetic polymers evaluated clinically for reconstruction of ligaments and tendons, including polytetrafluoroethylene, polyethylene terephthalate, braided carbon fibers, and polypropylene, were developed and studied, and these showed none of biomechanical properties of native structures (Bolton and Bruchman 1983; Jenkins et al. 1977; Mathew and Oksman 2010; Mccartney et al. 1993). The nanocellulose isolated from wood fibers, resulting in a homogeneous and uniform product, has allowed the development of materials with mechanical properties similar to or better than the natural ligaments or tendons.

Processing and properties of nanocomposites, especially in body conditions of 37°C and high humidity (98%), were studied by Mathew and Oksman (2010). The researchers evaluated the microstructural morphology of the nanocomposites developed. The nanofiber networks were tested for performance traction and sterilization, using gamma rays, in simulated body conditions. The results showed that, despite a reduction in tensile strength, the nanomaterials showed greater deformation compatible to natural ligaments or tendons.

A study with adhesion proteins proposes to improve the adhesion of biomaterials of various organic structures (Bodin et al. 2007b). Many cellulose derivatives of polymers are applied in systems for purifying blood clotting systems and plasma expansions in aqueous systems. For applications in this area, it is essential that the new material is hemocompatible, has high purity, and allows molding. The possibility of developing hydrogels, mainly from BNC, expands the possibilities for use in biomedical and pharmaceutical areas, because it has high structural similarity with that of the ECM of many tissues (Bodin et al. 2007b). Some studies report the use of specific peptides to improve the fixation and cell proliferation in bacterial cellulose, promoting development of new biomaterials for medical applications in blood vessels (Brumer et al. 2004; Zhou et al. 2006a,b).

Advances in the development of prosthetic heart valves and techniques for their implementation have allowed extension of the survival time of patients improving their quality of life. The development of durable, biocompatible prosthetic heart valves is already a reality (Cherian et al. 2011a). In the development of new biomaterials for the medical field, one key factor is biocompatibility. The body that receives these new materials is not rejected by host tissues. Therefore, it is important as the physical mechanics of the new material is to evaluate and develop the one that does not cause thrombosis and did not induce tissue mineralization (Cherian et al. 2011a).

Biopolymers, such as polyurethanes–nanocellulose materials, are potential alternatives in the development of biocompatible heart valves. Cardiovascular diseases are a major cause of mortality throughout the world. The reduction in the diameter of arteries causes a reduction in blood flow, which, among other factors, is largely responsible for sudden death due to cardiovascular disease. The possibility of using polyurethane reinforced with nanocellulose has been an option in developing innovative vascular prostheses. These biomaterials, as well as presenting a low probability of causing thrombosis, also have exceptional mechanical physical properties that allow the development of the prosthesis for adaptive blood vessels (Cherian et al. 2011a).

Studies done by Klemm et al. (2001) showed that BNC has important properties that allow its use as a material for medical purposes: high stability and dimensional control of the architecture of the biopolymer; biopolymer production in a predetermined shape and high resolution; macropores control, and others. The application of biogenerated hydrogels has been tested for the development of implants in hard and soft tissue with consistent progress.

A new material synthesized based on BNC, known as BASYC®, is basically a tube used as a substitute for a blood vessel. Tubes of .01 mm internal diameter and 5 mm length were developed and tested in mice (Figure 13.10). In the initial study, microsurgical implants were attached to BASYC® artificial defects of the carotid arteries of rats for one year. These long-term results showed the incorporation of joining pipes and active fibroblasts. Other grafts were made in pigs and were removed and analyzed after three months. The results showed that the use of BNC for tubular implants is a highly attractive approach for development of vessels in vivo (Gatenholm and Klemm 2010).

Cherian et al. (2011a) noted that the development of the polyurethane reinforced with nanocellulose allows the development of heart valves with biological characteristics compatible to the organism, good resistance to fatigue, and good hemodynamics. These results showed that the nanocellulose prosthetic heart valve and vascular prostheses showed excellent biocompatibility with the living tissues (Figures 13.11 and 13.12).

Degenerative or traumatic lesions of the meniscus may lead to osteoarthritis. Collagen has been widely used in the treatment and regeneration of meniscus tissue and BNC. The BNC was cast in dimensions similar to a meniscus. The character wedge meniscus was created by introducing a carrier of silicone (Cherian et al. 2011b).

Shortly after resection of the lateral meniscus, an implant was done and tests showed that the BNC has five times better mechanical properties than those observed for collagen (Bodin et al. 2007b). Due to the fact that it is cheaper and permits cellular migration that confers the properties described above, BNC becomes an attractive biomaterial for studies aimed at the development of prostheses that can be used for sports injuries. About 15 million people worldwide suffer trauma of the knee joint, and 225 thousand people suffer from meniscus repair.

FIGURE 13.10 BASYC® tubes sufficient for experimental microsurgical applications with different inside diameter, different wall thickness, and different length. (From Gatenholm, P. and Klemm, D. *MRS Bull.*, 3, 208–213, 2010.).

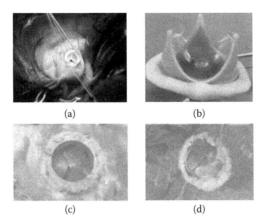

(a)   (b)

(c)   (d)

FIGURE 13.11 Nanocellulose–polyurethane prosthetic heart valve: (a) valve implant, (b) heart valve, (c) viewed in situ immediately prior to explant (inflow surface), (d) viewed in situ immediately prior to explant (outflow surface). (From Cherian, B.M., Leão, A.L., Souza, S.F., Costa, L.M.M., Olyveira, G.M., Kottaisamy, M., Nagarajan, E.R. and Thomas, S. *Carbohyd. Polym.*, 86, 1790–1798, 2011. With permission.)

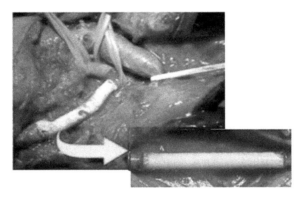

FIGURE 13.12 Vascular prostheses made of nanocellulose–polyurethane placed between the brachiocephalic trunk and the right common carotid artery in a 26-year-old man with multiple endocrine neoplasia (2B MEN 2B). (From Cherian, B.M., Leão, A.L., Souza, S.F., Costa, L.M.M., Olyveira, G.M., Kottaisamy, M., Nagarajan, E.R. and Thomas, S. *Carbohyd. Polym.*, 86, 1790–1798, 2011. With permission.)

13.5 OTHER APPLICATIONS

13.5.1 Immobilization of Enzymes

According to Anirudhan and Rejeena (2012), the immobilization of enzymes on biopolymer substrates has been considered a powerful technique in biomedical applications. Proteases are used in industrial and biomedical applications, among which trypsin (TRY) stands out. TRY is a pancreatic protease with significant activities related to the amino acids such as arginine, lysine, and ornithine. Of all the digestive endopeptidases, TRY has the highest specificity (Ghosh and Banerjee 2002). It is an important molecule, due to a wide variety of possible applications, such as hydrolysis of milk casein and predigestion of baby food (Wang and Sun 2007), the semisynthesis of human insulin (Jönsson and Johansson 2004), the process of harvesting cells, and proteomic experiments (Teke and Baysal 2007). However, the earlier mentioned applications are limited as regards the problem of the instability and rapid loss of catalytic activity during periods of work and storage. Several authors have shown the use of polymers grafted nanofibers to stabilize enzymes (Bayramoğlu et al. 2008; Krogh et al. 1999; Li et al. 2007). Various carriers have been used to immobilize TRY, for example, chitosan, silica gel coated with chitosan, glycidyl methacrylate-modified cellulose, and polyester fibers (Manrich et al. 2008; Nouaimi et al. 2001; Peng et al. 2012; Xi and Wu 2006). Nevertheless, the medical requirements greatly limit the number of carriers that can be used for the immobilization of enzymes for therapeutic purposes. Carriers should be nontoxic, biocompatible, and noncarcinogenic. Cellulose has been recommended as a suitable matrix for immobilizing drugs and hormones (Anirudhan and Rejeena 2012).

13.5.2 MEMBRANES FOR FILTRATION

The biopolymers developed with nanocellulose can also be applied as a membrane for filtration. Studies developed by Ferraz et al. (2012a) showed that nanocomposites of cellulose are used as potential candidates to generate new and innovative membranes for hemodialysis. The novel composites combine passive ultrafiltration and a large surface area (about 80 m^2/g). The new membranes have hemocompatibility and are viable for removing toxic substances. Biomembranes have greater commercial value due to their hemocompatibility. The exchange capacity of the materials was measured in pH 7.4 phosphate solution (600 ± 26 and 706 ± 31 mmol/g) and oxalate solution (523 ± 5) in 0.1 M. These results show that nanocellulose is a promising material for the development of membrane filtration (Ferraz et al. 2012a).

13.5.3 COSMETIC TISSUES

Mihranyan et al. (2012) described how cellulose, chitosan, and various types of proteinaceous polymers are used in cosmetics formulas. Generally, these materials show low toxicity and are the most commonly used additives in pharmaceuticals, cosmetics, and food. For example, Cosmospheres® are prototypes of delivery vehicles for active ingredients in cosmetic and are small beads usually based on lactose and microcrystalline cellulose (MCC) especially developed for the use in cosmetic products. When applied to the skin, Cosmospheres GT-S provides a smooth and esthetic feeling (Mihranyan et al. 2012).

The nanocellulose has high water content (95%), high purity, and mechanical stability. It may thus be employed in cosmetic applications due to its high moisture (e.g., ingredient in creams). An example of a mask designed with nanocellulose is the product called NanoMasque®, made from pure BNC. This product can be impregnated with various extracts and assets that can act directly on the skin. Human trials described that the applied nanocellulose showed no unwanted skin reactions (Cherian et al. 2011b).

13.6 IN VITRO AND IN VIVO ANALYSIS

Biocompatibility studies are the basis for any material that can be applied in the biomedical field. Nevertheless, it is important to evaluate the cytotoxicity using in vitro tests prior to the in vivo tests, mainly because of their low costs, rapid results, and easy access at the laboratory scale.

Cytotoxicological tests represent the initial phase to evaluate the biocompatibility of any material with potential for medical applications. In vitro tests are done to evaluate and simulate human behavior, analyzing the interaction between the material and cell culture. Different methods for cytotoxicological evaluation have been developed and standardized depending on the specific structure, chemical composition, degradation properties, and others from the material to be analyzed. In vitro techniques can be employed to identify possible adverse effects of materials for medical devices on cell functions (Northup 1997).

In vivo analysis is a fundamental study for any biomaterial to be used in human bodies. While the in vitro analysis gives more information about the cells and biomaterial interactions, the in vivo assessment gives some responses from tissues, organs, blood, proteins, and other elements present in living organisms. Animals are used to simulate the real environment found in a human being.

For in vivo assessment, it is desirable to delineate a protocol according to the end use and period of time inside the body, and try to approximate the study to the real application and obtain strong conclusions (Northup 1997). Some standards for the in vitro assessment specifying the cell line, the medium, and techniques for the cytotoxicity evaluation are already available. In addition, all standard protocols stipulate materials to be used as negative and positive controls. Positive control is a substance that has a cytotoxic effect in a reproducible way, and negative control is a material or substance that does not produce a cytotoxic effect. There are two types of in vitro tests: methods of direct and indirect contacts. At first, the cells are placed in contact with the test material and usually sown in the cell suspension over the material. On the other hand, indirect contact methods can be divided into two types: those in which the material to be tested is separated from the cells by a diffusion barrier (agar or agarose) and those in which substances are extracted from the material to be tested through a solvent and placed in contact with the cells (Malmonge et al. 1999).

The fact that cellulose is biodegradable, and the notion that the biosources are deemed "safe," may suggest that cellulose nanowhiskers are indeed benign (Clift et al. 2011). Meanwhile, nanocellulose as nanowhiskers, nanofibers, or mirofibrillated need to be further studied to understand their behaviors since different results and point of views have been reported, mainly because each source, production process, purification, and design of in vitro tests can significantly change the cytotoxicity.

The studies are showing different health effects depending on the procedures for the isolation of cellulose as MFC, nanowhiskers or nanofibers, sources, end treatments, purification process, even some chemical modifications where it is possible to observe totally different behaviors on the material such as its structure, surface, and composition and consequently, different results of cytotoxicity and biocompatibility.

The cytotoxicity of MFC was performed by Vartiainen et al. (2011) from elementally chlorine-free bleached birch kraft pulp and complementary in vitro analysis evaluating cell viability. This test was done using two different cells: mouse macrophages and human macrophages derived from blood human cells. The cell viability was done 24 hours after the cells were exposed to the MFC. Nonexposed cells acted as negative control (CTR), and all the cells were compared with cells exposed to MCC from Avicel PH-101. The behavior of MFC was better when compared with that of MCC and even better than the negative control, since it reduced the rate of dead cells; the graphs are shown in Figure 13.13. Similar behavior was reported by Vartiainen et al. (2011) when used in the human macrophages cells.

Another important technique to evaluate the cytotoxicity in vitro is to analyze the immunologic response. It can be done by correlating special proteins present in the inflammatory reaction. The inflammatory reaction is accompanied by a response known as the systemic acute phase response, and it is characterized by production of several hormones, leukocytosis, and proteins. Cytokines, such as interleukin (IL)-1, IL-6, Tumor Necrosis Factor (TNF)-α, leukemia inhibitory factor, and Oncostatin M, that are produced at the site of inflammation and play a crucial role in acute phase response (Bilate 2007). Peak production of acute phase protein usually occurs between 12 and 24 hours after the onset of acute inflammatory response.

Macrophages could engulf all the particles, components, molecules, and others recognized as antigens and subsequently produce cytokines and chemokines, which attract responder cells to the site of inflammation. Accordingly, the cytokine profile was used to estimate the cytotoxicity of MFC. The cytokine levels were obtained after mouse and human macrophages were exposed to the MFC samples for 6 hours in three different sample concentrations (Vartiainen et al. 2011), and it was acquired with no significant inflammatory effects in the presence of the MFC when compared with the control; the obtained results are shown in Figure 13.14. The behavior of MFC was even better than MCC, because it was so similar to the negative control. The cells not exposed to the samples were used as a negative control and bacterial lipopolysaccharide was used as positive control at concentration of 100 μg/mL (Figure 13.14).

BNC is another important source of cellulose; it assembles into several micrometer long ribbons that form a dense reticulated network and have been studied

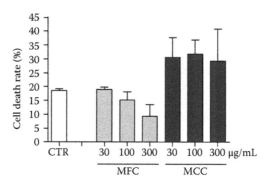

FIGURE 13.13 Cell viability of mouse macrophage cells after 24 hours of exposure to three different concentrations (30, 100, and 300 μg/mL) of microfibrillated cellulose (MFC) and microcrystalline cellulose (MCC). (From Vartiainen, J., Pöhler, T., Sirola, K., Pylkkänen, L., Alenius, H., Hokkinen, J., Tapper, U. et al. *Cellulose,* 18, 775–786, 2011. With permission.)

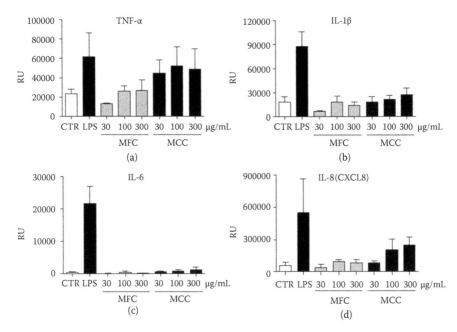

FIGURE 13.14 Ratio of mRNA expressions of (a) TNF-α, (b) IL-1β, (c) IL-6, and (d) IL-8 in human monocyte-derived macrophages after exposure to microfibrillated cellulose (MFC) and MCC at the concentrations of 30, 100, and 300 μg/mL. (From Vartiainen, J., Pöhler, T., Sirola, K., Pylkkänen, L., Alenius, H., Hokkinen, J., Tapper, U. et al. *Cellulose*, 18, 775–786, 2011. With permission.)

continuously as the scaffold for wound care and tissue engineering applications. One of the properties desired for a scaffold material is that it is bioresorbable and since human bodies do not have the ability to degrade cellulose, Hu and Catchmark (2011) studied techniques to make bacterial cellulose (BC) more bioresorbable incorporating cellulase enzymes. They used buffer components into BNC in order to retain the activity of the cellulases. The degradation was 30% without incorporation of buffer ingredients and it increased to 97% in the presence of the suboptimal pH environment of 7.4. They used simulated body fluid and simulated tissue padding, both mimicking the real wound environment, and also demonstrated some improvements in terms of material degradation. Measurements of mechanical properties of materials revealed that BC materials had tensile strength and extensibility similar to human skin, especially when hydrated with saline water prior to use.

Another significant evaluation was the period of time until almost total cellulose degradation by ranging the enzyme concentration. Hu and Catchmark (2011) believed that the degradation rate can be controlled by loading a lower amount of cellulases because it is important to ensure the integrity of materials until cells have attached completely and begun to proliferate. They confirmed the relevance of buffer ingredients in materials that helps form an optimal pH microenvironment for cellulases and consequently induces the release of high levels of glucose from BNC.

Cytotoxicity using different strategies of nanowhiskers obtained from filter paper was evaluated by Clift et al. (2011). The tests were done to determine how cotton cellulose nanowhisker interacted with the triple cell coculture and with each one isolated. The cells were human monocyte-derived macrophages, and dendritic cells (obtained by incubating monocytes with the growth factors GM-CSF and IL-4) were isolated from human whole blood and cultured at 37°C, at 5% CO_2.

The tests were done with multiwalled carbon nanotube and crocidolite asbestos fibers, besides the nanowhiskers from cotton. The carbon nanotube was dispersed in Pluronic F127 to provide a stable and well-dispersed sample with no adverse effects in vitro. The asbestos fibers were weighed out and suspended directly in cell culture media. Samples were then sonicated in a water bath for 10 minutes. Stock solutions of both carbon nanotubes and asbestos fibers were tested at low and high concentration, 0.005 and 0.03 mg/mL, respectively. They were exposed, as a suspension, to either monocultures or the triple-cell coculture for 24 hours at 37°C, 5% CO_2, and for all the described tests, nanocrystals resulted in superior and less toxic to the cell cultures.

Some trials of producing membranes for hemodialysis have been studied using composites of nanocellulose and the conducting polymer polypyrrole (PPy). Because of its potential use for active ion extraction, it would be possible to produce electrochemically controlled membranes (Ferraz et al. 2012b). They evaluated the biocompatibility of nanocellulose–PPy composites on indirect toxicity assays with fibroblasts and monocyte cell lines following ISO-10993-5 (2009) and in vivo tests in mice. The composites did not induce any cytotoxic response in vitro or in vivo. The images of an indirect test using a human dermal fibroblasts culture are displayed in Figure 13.15 and the cell viability was assessed using trypan blue staining (95%–99% viable cells); the samples were incubated with Hanks' balanced salt solution for 48 hours at room temperature and sterilized by autoclaving at 1.5 MPa for 20 minutes. Extensive rinsing

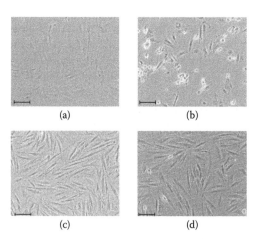

(a)   (b)

(c)   (d)

FIGURE 13.15 Representative light microscopy images of human dermal fibroblasts on tissue culture plates cultured for 24 hours in: (a) tamoxifen (TMX) extract medium (negative control), (b) 5% dimethyl sulfoxide (DMSO) in culture medium (positive control), (c) polymer polypyrrole–cellulose composite extract medium, and (d) Cladophora cellulose/microfibrillated cellulose extract medium. The scale bars represent 50 μm. (From Ferraz, N., Strømme, M., Fellström, B., Pradhan, S., Nyholm, L. and Mihranyan, A. *J. Biomed. Mater. Res. A*, 100, 2128–2138, 2012. With permission.)

and 48 hours of incubation in a biological buffer previous to the preparation of the culture medium extracts were, however, necessary to obtain a noncytotoxic composite (Ferraz et al. 2012b).

The present results show that the biocompatibility of PPy composites depends on the rinsing and pretreatment with the composite as well as the aging of the material, because it was identified that the aging of the composite had a negative effect on biocompatibility (Ferraz et al. 2012b).

Lima et al. (2012) performed genotoxic of cellulosic nanofibers using plant and animal cells. Initially, the cellulose nanowhiskers were obtained from colored cotton that was removed just as the waxes using ethanol and cyclohexane, while the curaua fibers were treated with NaOH 17% to obtain a pulp, after both materials (white, brown, green, ruby cotton, and curaua pulp) were hydrolyzed with H_2SO_4 (6.5 mol/L) were purified until pH between 6.5 and 7.0, and the results demonstrated that nanofibers derived from cotton (white, green, and brown) and curaua could cause alterations in plant cells, while the brown cotton and curaua nanofibers were genotoxic in animal cells (human lymphocytes and mouse fibroblasts). Therefore, the animal cells appeared to be less liable to genetic alterations and possessed more effective repair mechanisms. An important observation was that neither plant nor animal cells showed any genotoxic alterations when exposed to ruby cotton nanowhiskers, under the conditions employed, indicating that ruby cotton might be the material of choice for biomedical application.

Aggregation of nanowhiskers appeared to reduce toxicity and demonstrated better results in culture media (used in the experiments involving animal cells) than in water (used in the plant cell tests). Further studies will be desirable to better understand the influence of aggregation on nanofiber toxicity. It is therefore important to consider the environment in which the nanofibers will be used, since in some applications fiber aggregates are used to reinforce polymeric matrices, where the fibers can improve thermal stability, mechanical resistance, and permeability to liquids and gases, even at low fiber concentrations. In their intact state, these matrices are unlikely to release nanofibers; however, decomposition processes could result in slow release of fibers to the wider environment.

There are some important uses of biomaterials nanocellulose derivatives in veterinary medicine: for pipe insulation peripheral nerve reconstruction, wound healing of experimental bovine mammary papillae; integument experimental wound healing in horses and pigs; prophylaxis of the laminectomy membrane in dogs; and incisional healing of corneal lesions in experimental dogs (Cherian et al. 2011b).

The nanocellulose membranes were tested in dogs for wound healing (Iamaguti et al. 2008). The cellulose membranes were applied on fibrocartilaginous component getting good tissue and reintegration cartilages. The membranes were tested through an incision of 5–7 cm in lateral parapatellar skin, followed by incising the joint capsule and retinaculum until exposing the knee joint.

Falcao et al. (2008) investigated the kind of host tissue by incorporation of nanocellulose membranes produced by bacteria and expanded polytetrafluoroethylene (ePTFE) defects in the abdominal wall of rats. The cellulose membrane was sutured with Prolene® and continuous suture. Then, the skin was closed with thread of Mononylon® through an interrupted suture. Observations were made in the presence of wrapping with and without infiltration in the implants in rats of the nanocellulose and ePTFE groups. The researchers observed that incorporation composed by wrapping associated with host tissue infiltration is seen only in ePTFE implants.

Tests performed on 36 fetal sheep conducted by Oliveira et al. (2007) showed that using cellulose film is more suitable as a substitute for dura mater to cover and protect nervous tissue. The film looks promising for the correction of intrauterine myelomeningocele, preventing adhesion to neural tissue superficial planes and minimizing the deleterious effects of the intrauterine environment on the spinal cord (Figure 13.16).

13.7 CONCLUDING REMARKS

Cellulose has extensive and unlimited biological applications. One immense advantage is that this polymer is abundant, renewable, and available in nature. The development of new products for biomedical applications from nanofiber is using their great properties such as higher strength, durability, and stability. In addition, products made basically of nanocellulose have shown good reception in bodies, either human or animal.

Thus, this chapter is intended to present the various possible applications of biomaterials derived from cellulose and new technologies developed for these purposes. The development of different materials or composites needs to be studied further, and different application opportunities are necessary to resolve the problems related to the implementation of these biomaterials in medicine.

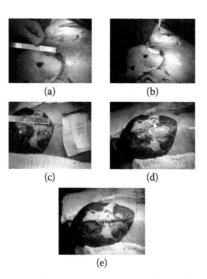

FIGURE 13.16 Stages of corrective surgery in the sheep fetus. (a) Exposed fetal back during correction surgery at 100 days of gestation. (b) Human acellular dermal matrix (black arrow) covering the defect. (c) Same aspect of another animal at 100 days of gestation, induced meningomyelocele (MMC). (d) Biosynthetic cellulose (black arrow) was used to cover the defect before it is recovered with skin. (e) Postoperative sutured skin. (From Oliveira, R.C.S., Valente, P.R., Abou-Jamra, R.C., Araújo, A., Saldiva, P.H. and Pedreira, D.A.L. *Acta Cir. Bras.*, 22, 174–181, 2007. With permission.)

Meanwhile, it is significant to highlight the relevance of the in vitro and in vivo assessment, because for each type of nanocellulose process, modification or end treatment, purification, composite, ways techniques to implant and others can change significantly and are necessary to be estimated.

REFERENCES

Anirudhan, T.S. and Rejeena, S.R. 2012. Adsorption and hydrolytic activity of trypsin on a carboxylate-functionalized cation exchanger prepared from nanocellulose. *Journal of Colloid and Interface Science* 381: 125–136.

Balasubramani, M., Kumar, T.R. and Babu, M. 2001. Skin substitutes: A review. *Burns* 27: 534–544.

Barone, D.T.J., Raquez, J.M. and Dubois, P. 2011. Bone-guided regeneration from inert biomaterials to bioactive polymer (nano) composites. *Polymer Advanced Technology* 22: 463–475.

Basmaji, P. 2010. Nanoskin for medical application. *3rd International Seminar on Nanosciences and Nanotechnologies*, Sep. 7–9, 2010, Havana, Cuba.

Bayramoğlu, G., Yilmaz, M., Senel, A.U. and Arica, M.Y. 2008. Preparation of nanofibrous polymer grafted magnetic poly (GMA-MMA)-g-MAA beads for immobilization of trypsin via adsorption. *Biochemical Engineering Journal* 40: 262–274.

Bilate, A.M.B. 2007. Inflammation, acute phase proteins, cytokines and chemokines. *Temas de Reumatologia Clínica* 8: 86–90.

Bodin, A., Bäckdahl, H., Fink, H., Gustafsson, L., Risberg, B. and Gatenholm, P. 2007a. Influence of cultivation conditions on mechanical and morphological properties of bacterial cellulose tubes. *Biotechnology and Bioengineering* 97: 425–434.

Bodin, A., Concaro, S., Brittberg, M. and Gatenholm, P. 2007b. Bacterial cellulose as a potential meniscus implant. *Journal of Tissue Engineering and Regenerative Medicine* 1: 406–408.

Bolton, C.W. and Bruchman, B. 1983. Mechanical and biological properties of the GORE-TEX expanded polytetrafluoroethylene (PTFE) prosthetic ligament. *Aktuelle Probl Chir Orthop* 26: 40–51.

Brumer, H., Zhou, Q., Baumann, M.J., Carlsson, K. and Teeri, T.T. 2004. Activation of crystalline cellulose surfaces through the chemoenzymatic modification of xyloglucan. *Journal of the American Chemical Society* 126: 5715–5721.

Chazal, J., Tanguy, A., Bourges, M., Gaurel, G., Escande, G., Guillot, M., and Vanneuville, G. 1985. Biomechanical properties of spinal ligaments and histological study of the supraspinal ligament in traction. *Journal of Biomechanics* 18: 167–176.

Cherian, B.M., Leão, A.L., Souza, S.F., Thomas, S., Pothan, L.A. and Kottaisamy, M. 2010. Isolation of nanocellulose from pineapple leaf fibres by steam explosion. *Carbohydrate Polymers* 81: 720–725.

Cherian, B.M., Leão, A.L., Souza, S.F., Costa, L.M.M., Olyveira, G.M., Kottaisamy, M., Nagarajan, E.R. and Thomas, S. 2011a. Cellulose nanocomposites with nanofibres isolated from pineapple leaf fibers for medical applications. *Carbohydrate Polymers* 86: 1790–1798.

Cherian, B.M, Leão, A.L., Souza, S.F., Thomas, S., Pothan, L.A. and Kottaisamy, M. 2011b. Cellulose nanocomposites for high-performance applications. In: *Cellulose Fibers: Bio- and Nanopolymer Composites*. Eds. Susheel, K., Kaith, B.S. and Kaur, I., 539–588. Berlin: Springer-Verlag.

Clift, M.J.D., Foster, E.J., Vanhecke, D., Studer, D., Wick, P., Gehr, P., Rutishauser, B.R. and Weder, C. 2011. Investigating the interaction of cellulose nanofibers derived from cotton with a sophisticated 3D human lung cell coculture. *Biomacromolecules* 12: 3666–3673.

Czaja, W., Krystynowicz, A., Bielecki, S. and Brown, R.J. 2006. Microbial cellulose—The natural power to heal wounds. *Biomaterials* 27: 145–151.

Czaja, W., Krystynowicz, A., Kawecki, M., Wysota, K., Sakiel, S., Wróblewski, P., Glik, J., Nowak, M. and Bielecki, S. 2007a. Biomedical applications of microbial cellulose in burn wound recovery. In: *Cellulose: Molecular and Structural Biology: Selected Articles on the Synthesis, Structure, and Applications of Cellulose*. Eds. Brown, R.M. Jr. and Saxena, I.M., 307–321. Dordrecht: Springer.

Czaja, W.K., Young, D.J., Kawecki, M. and Brown, R.M. Jr. 2007b. The future prospects of microbial cellulose in biomedical applications. *Biomacromolecules* 8: 1–12.

Demling, R.H. and De Santi, L. 1999. Management of partial thickness facial burns (comparison of topical antibiotics and bio-engineered skin substitutes). *Burns* 25: 256–261.

De Santis, R., Sarracino, F., Mollica, F., Netti, P.A., Ambrosio, L. and Nicolais, L. 2004. Continuous fiber reinforced polymers as connective tissue replacement. *Composites Science and Technology* 64: 861–871.

Erol, O.O. 2000. The Turkish delight: A pliable graft for rhinoplasty. *Plastic and Reconstructive Surgery* 105: 2229–2241.

Falcao, S.C., Neto, J.E. and Coelho, A.R.B. 2008. Incorporation by host tissue of two biomaterials used as repair of defects produced in abdominal wall of rats. *Acta Cirurgica Brasileira* 23: 78–83.

Ferraz, N., Carlsson, D.O., Hong, J., Larsson, R., Fellström, B., Nyholm, L., Strømme, M. and Mihranyan, A. 2012a. Haemocompatibility and ion exchange capability of nanocellulose polypyrrole membranes intended for blood purification. *Journal of the Royal Society Interface* 9: 1943–1955.

Ferraz, N., Strømme, M., Fellström, B., Pradhan, S., Nyholm, L. and Mihranyan, A. 2012b. In vitro and in vivo toxicity of rinsed and aged nanocellulose–polypyrrole composites. *Journal of Biomedical Materials Research. Part A* 100: 2128–2138.

Fontana, J.D., Souza, A.M., Fontana, C.K., Torriani, I.L., Moreschi, J.C., Gallotti, B.J., de Souza, S.J, Narcisco, G.P., Bichara, J.A. and Farah, L.F. 1990. Acetobacter cellulose pellicle as a temporary skin substitute. *Applied Biochemistry and Biotechnology* 24–25: 253–264.

Fu, F., Bennet, C.H., Latterman, C. and Ma, C. 1999. Current trends in ACL ligament reconstruction. *American Journal of Sports Medicine* 27: 821–830.

Gallin, W.J. and Hepperle, B. 1998. Burn healing in organ cultures of embryonic chicken skin: A model system. *Burns* 24: 613–620.

Gatenholm, P. and Klemm, D. 2010. Bacterial nanocellulose as a renewable material for biomedical applications. *MRS Bulletin* 3: 208–213.

Ghosh, S. and Banerjee, A. 2002. A multitechnique approach in protein/surfactant interaction study physicochemical aspects of sodium dodecyl sulfate in the presence of trypsin in aqueous medium. *Biomacromolecules* 3: 9–16.

Gissefalt, K., Edberg, B. and Flodin, P. 2002. Synthesis and properties of degradable poly (urethane urea)s to be used for ligament reconstructions. *Biomacromolecules* 3: 951–958.

Habibi, Y., Lucia, A.L. and Rojas, O.J. 2010. Cellulose nanocrystals chemistry, self-assembly, and applications. *Chemical Reviews* 110: 3479–3500.

Helenius, G., Bäckdahl, H., Bodin, A., Nannmark, U., Gatenholm, P. and Risberg, B. 2006. In vivo biocompatibility of bacterial cellulose. *Journal of Biomedical Materials Research. Part A* 76: 431–438.

Hu, Y. and Catchmark, J.M. 2011. In vitro biodegradability and mechanical properties of bioabsorbable bacterial cellulose incorporating cellulases. *Acta Biomaterialia* 7: 2835–2845.

Hukins, D.W.L., Leahy, J.C. and Mathias, K.J. 1999. Biomaterials defining the mechanical properties of natural tissues and selection of replacement materials. *Journal of Materials Chemistry* 9: 629–636.

Iamaguti, L.S., Brandão, C.V.S., Minto, B.W., Mamprim, M.J., Ranzani, J.J.T. and Gomes, D.C. 2008. Use of biosynthetic cellulose membrane in experimental trochleoplasty in dogs. Clinical, radiographic and macroscopic. *Vet Zootec* 15: 160–168.

ISO-10993-5. 2009. *Biological Evaluation of Medical Devices.* Part 5: Tests for in vitro cytotoxicity. ISBN: 1-57020-355-5.

Jenkins, D.H.R., Forster, I.W., Mckibbin, B. and Ralis, Z.A. 1977. Induction of tendon and ligament formation by carbon implants. *Journal of Bone and Joint Surgery-British Volume* 59: 53–57.

Jones, I., Currie, L. and Martin, R. 2002. A guide to biological skin substitutes. *British Journal of Plastic Surgery* 55: 185–193.

Jönsson, M. and Johansson, H. O. 2004. Effect of surface grafted polymers on the adsorption of different model proteins. *Colloids and Surfaces. B, Biointerfaces* 37: 71–81.

Kaplan, D. L. 1998. Introduction to biopolymers and renewable resources. In: *Biopolymers from Renewable Resources.* Ed. Kaplan, D.L., 1–29. Berlin: Springer.

Klemm, D., Heublein, B., Fink, H.P. and Bohn, A. 2005. Review: Cellulose: Fascinating biopolymer and sustainable raw material. *Angewandte Chemie (International ed. in English)* 44: 3358–3393.

Klemm, D., Schumann, D., Kramer, F., Heßler, N., Hornung, M., Schmauder, H.P. and Marsch, S. 2006. Nanocellulose as innovative polymers in research and application. *Polysaccharides* 205: 49–96.

Klemm, D., Schumann, D., Udhardt, U. and Marsch, S. 2001. Bacterial synthesized cellulose— Artificial blood vessels for microsurgery. *Progress in Polymer Science* 26: 1561–1603.

Kolakovic, R., Laaksonen, T., Peltonen, L., Laukkanen, A. and Hirvonen, J. 2012. Spray-dried nanofibrillar cellulose microparticles for sustained drug release. *International Journal of Pharmaceutics* 430: 47–55.

Krogh, T.N., Berg, T. and Højrup, P. 1999. Protein analysis using enzymes immobilized to paramagnetic beads. *Analytical Biochemistry* 274: 153–162.

Kvien, I. and Oksman, K. 2007. Orientation of cellulose nanowhiskers in polyvinyl alcohol. *Applied Physics A: Materials Science & Processing* 87: 641–643.

Latarjet, J. 1995. A simple guide to burn treatment. *Burns* 21: 221–225.

Legnani, C., Ventura, A., Terzaghi, C., Borgo, E. and Albisetti, W. 2010. Anterior cruciate ligament reconstruction with synthetic grafts. A review of literature. *International Orthopaedics* 34: 465–471.

Li, Y., Yan, B., Deng, C., Tang, J., Liu, J. and Zhang, X. 2007. On-plate digestion of proteins using novel trypsin-immobilized magnetic nanospheres for MALDI-TOF-MS analysis. *Proteomics* 7: 3661–3671.

Lima, R., Feitosa, L.O., Maruyama, C.R., Abreu, B.M., Yamawaki, P.C., Vieira, I.J., Teixeira, E.M., Corrêa, A.C., Caparelli Mattoso, L.H. and Fernandes Fraceto, L. 2012. Evaluation of the genotoxicity of cellulose nanofibers. *International Journal of Nanomedicine* 7: 3555–3565.

Malmonge, S.M., Zavaglia, C.A.C., Santos, A.R. Jr. and Wada, M.L.F. 1999. Evaluation of the cytotoxicity of hydrogels poliHEMA: An in vitro study. *Revista Brasileira de Engenharia Biomédica* 15: 49–54.

Manrich, A., Galvão, C.M.A., Jesus, C.D.F., Giordano, R.C. and Giordano, R.L.C. 2008. Immobilization of trypsin on chitosan gels use of different activation protocols and comparison with other supports. *International Journal of Biological Macromolecules* 43: 54–61.

Mathew, A.P. and Oksman, K. 2010. Cellulose nanofiber based composites for use as ligament or tendon substitute. Paper presented at International Conference on Nanotechnology for the Forest Products Industry, Sep. 27–29, 2010, Otaniemi, Espoo, Finland.

Mathew, A.P., Oksman, K., Pierron, D. and Harmand, M.F. 2012. Fibrous cellulose nanocomposite scaffolds prepared by partial dissolution for potential use as ligament or tendon substitutes. *Carbohydrate Polymers* 87: 2291–2298.

McCartney, D.M., Tolin, B.S., Schwenderman, A., Freidmen, M.J. and Woo, S.L.Y. 1993. Prosthetic replacement for anterior cruciate ligament. In: *The Anterior Cruciate Ligament: Current and Future Concepts.* Ed. Jackson, D.W., 343–356. New York: Raven Press Ltd.

Mendes, P.N., Rahal, S.C., Pereira-Junior, O.C.M., Fabris, V.E., Lenharo, S.L.R., Lima-Neto, J.F. and Cruz, L.A.F. 2009. In vivo and in vitro evaluation of an *Acetobacter xylinum* synthesized microbial cellulose membrane intended for guided tissue repair. *Acta veterinaria Scandinavica* 51: 1–8.

Meyers, M.A., Chen, P.Y., Lin, Y.M.A. and Seki, Y. 2008. Biological materials: Structure and mechanical properties. *Progress in Materials Science* 53: 1–206.

Migliaresi, C. and Nicolais, L. 1980. Composite materials for biomedical applications. *The International Journal of Artificial Organs* 3: 114–118.

Mihranyan, A., Ferraz, N. and Strømme, M. 2012. Current status and future prospects of nano-technology in cosmetics. *Progress in Materials Science* 57: 875–910.

Muangman, P., Opasanon, S., Suwanchot, S. and Thangthed, O. 2011. Efficiency of microbial cellulose dressing in partial-thickness burn wounds. *The Journal of the American College of Certified Wound Specialists* 3: 16–19.

Murugan, R. and Ramakrishna, S. 2005. Development of nanocomposites for bone grafting. *Composites Science and Technology* 65: 2385–2406.

Nachemson, A. and Evans, J. 1968. Some mechanical properties of the third human lumbar interlaminar ligament (ligamentum flavum). *Journal of Biomechanics* 1: 211–220.

Northup, S.J. 1997. In vitro Assessment of tissue compatibility. In: *Biomaterials Science: An Introduction to Materials in Medicine.* Eds. Ratner, B.D., Hoffman, A.S., Schoen, F.J. and Lemons, J.E., 215–220. California: Academic Press.

Nouaimi, M., Möschel, K. and Bisswanger, H. 2001. Immobilisation of typsin on polyester fleece via different spacers. *Enzyme and Microbial Technology* 29: 567–574.

Novaes, A.B. Jr., Marcaccini, A.M., Souza, S.L., Taba, M. Jr. and Grisi, M.F. 2003. Immediate placement of implants into periodontally infected sites in dogs: A histomorphomet-ric study of bone–implant contact. *The International Journal of Oral & Maxillofacial Implants* 18: 391–398.

Novaes, A.B. and Novaes, A.B. 1997. Soft tissue management for primary closure in guided bone regeneration: Surgical technique and case report. *The International Journal of Oral & Maxillofacial Implants* 12: 84–87.

Oke, I.W. 2010. Nanoscience in nature: Cellulose nanocrystals. *Studies Undergraduate Researches at Guelph, Winter* 3: 77–80.

Oliveira, R.C.S., Valente, P.R., Abou-Jamra, R.C., Araújo, A., Saldiva, P.H. and Pedreira, D.A.L. 2007. Biosynthetic cellulose induces the formation of a neoduramater following prenatal correction of meningomyelocele in fetal sheep. *Acta Cirurgica Brasileira* 22: 174–181.

O'sullivan, A.C. 1997. Cellulose the structure slowly unravels. *Cellulose* 4: 173–207.

Peng, G., Zhao, C., Liu, B., Ye, F. and Jiang, H. 2012. Immobilized trypsin onto chitosan modified monodisperse microspheres: A different way for improving carrier's surface biocompatibility. *Applied Surface Science* 258: 5543–5552.

Prasanna, M., Mishra, P. and Thomas, C. 2004. Delayed primary closure of the burn wounds. *Burns* 30: 169–175.

Seydibeyoglu, M.O. and Oksman, K. 2008. Novel nanocomposites based on polyurethane and micro fibrillated cellulose. *Composites Science and Technology* 68: 908–914.

Silva, E.C. 2009. *Synthetic Hydroxyapatite in Dental Alveolus After Exodontia in* Felis catus: *Clinical, Radiological and Histomorphometric Studies*, M.S. Dissertation, Universidade Federal de Viçosa, Viçosa, Brazil.

Siro, I. and Plackett, D. 2010. Microfibrillated cellulose and new nanocomposite materials: A review. *Cellulose* 17: 459–494.

Souza, F.C., Olival-Costa, H., Silva, L., Pontes, P.A. and Lancellotti, C.L.P. 2011. Bacterial cellulose as laryngeal medialization material: An experimental study. *Journal of Voice* 25: 765–769.

Souza, S.F., Leão, A.L., Cai, J.H., Wu, C., Sain, M. and Cherian, B.M. 2010. Nanocellulose from curava fibers and their nanocomposites. *Molecular Crystals and Liquid Crystals* 522: 42–52.

Szczesna-Antczak, M., Kazimierczak, J. and Antczak, T. 2012. Nanotechnology—Methods of manufacturing cellulose nanofibres. *Fibres & Textiles in Eastern Europe* 20: 8–12.

Teke, A.B. and Baysal, S.H. 2007. Immobilizantion of urease using glycidyl methacrylate grafted nylon-6-membranes. *Process Biochemistry* 42: 439–443.

Thomas, S., Paul, S.A., Pothan, L.A. and Deepa, B. 2011. Natural fibres: Structure, properties and applications. In: *Cellulose Fibers: Bio-and Nano-Polymer Composites. Green Chemistry and Technology.* Eds. Kalia, S., Kaith, B.S. and Kaur, I., 3–42. Berlin: Springer.

Valo, H., Kovalainen, M., Laaksonen, P., Häkkinen, M., Auriola, S., Peltonen, L., Linder, M., Jarvinen, K., Hirvonen, J. and Laaksonen, T. 2011. Immobilization of protein-coated drug nanoparticles in nanofibrillar cellulose matrices—enhanced stability and release. *Journal of Controlled Release* 156: 390–397.

Vartiainen, J., Pöhler, T., Sirola, K., Pylkkänen, L., Alenius, H., Hokkinen, J., Tapper, U. et al. 2011. Health and environmental safety aspects of friction grinding and spray drying of microfibrillated celulose. *Cellulose* 18: 775–786.

Wang, D.M. and Sun, Y. 2007. Fabrication of superporous cellulose beads with grafted anion-exchange polymer chains for protein chromatography. *Biochemical Engineering Journal* 37: 332–337.

Xi, F. and Wu, J. 2006. Preparation of macroporous chitosan layer coated on silica gel and its application to affinity chromatography for trypsin inhibitor purification. *Reactive and Functional Polymers* 66: 682–688.

Zhou, Q., Baumann, M.J., Brumer, H. and Teeri, T.T. 2006a. The influence of surface chemical composition on the adsorption of xyloglucan to chemical and mechanical pulps. *Carbohydrate Polymers* 63: 449–458.

Zhou, Q., Baumann, M.J., Piispanen, P.S., Teeri, T.T. and Brumer, H. 2006b. Xyloglucan and xyloglucan endo-transglycosylases (XET): Tools for ex vivo cellulose surface modification. *Biocatalysis and Biotransformation* 24: 107–120.

14 Green Composites from Functionalized Renewable Cellulosic Fibers

Vijay K.Thakur, Manju K.Thakur, and Raju K.Gupta

CONTENTS

14.1 INTRODUCTION

During the last few years, major improvements in eco-friendly green composites in material application have been achieved by using composite materials made with biomass from natural resources as reinforcement (Auras et al. 2004; Chao-Lung et al. 2011; Debapriya and Basudam 2004; Garlotta 2002; Hagstrand and Oksman 2001). Reinforcement of plastics by biomass-based materials such as natural fibers began with cellulose fiber in the early eighteenth century and was later further developed (Favaro et al. 2010; Singha and Thakur 2008; Singha et al. 2008). Nowadays, for a number of industrial applications, especially in the automotive industry, biomass-based composites are of immense importance (Averous 2004; Singha et al. 2009). These composite materials are very attractive because they allow combining properties in ways that are not found in nature in single materials (Bledzki and Gassan 1999; Bledzki et al. 1996, 2010). These materials often offer very high mechanical properties with lightweight structure suitable for a number of applications (Alawar et al. 2009; Singha and Thakur 2009a,b). Typical composite materials based on lignocellulosic fibers consist of a discontinuous reinforcement phase embedded in a continuous polymer matrix phase (Ayswarya et al. 2012; Kabir et al. 2012a,b). Biocomposites are composite materials comprising one or more phase derived from a renewable resource (Panthapulakkal et al. 2006). Biocomposites can be defined as materials made from renewable and sustainable biomasses that also include agricultural wastes, wood, or other plants (Thakur and Singha 2010a–c). The main components in renewable plant biomass are cellulose, hemicelluloses, and lignin along with minor amounts of pectin, wax, and proteins (Abdelmouleh et al. 2004; Singha and Thakur 2010a,b; Thakur et al. 2012a–c). The composition of these components can vary based on factors such as age, type, and origin of the plant. The growing demand for eco-friendly green materials based on renewable resources material is showing a promising future, and the use of biomass-based material is strongly influencing the automotive industry, aerospace, and food products, as well as material for engineering applications (Ayswarya et al. 2012; Wambua et al. 2003). Indeed, human beings have been using biomass-based natural polymers to make life possible on this earth for centuries. People used to prepare a variety of materials from cellulosic fibers, wood bark, cotton, wool, silk, animal skin, and natural rubber (Zain et al. 2011). Indeed, biomass-based materials, especially lignocellulosic fibers, have the potential to tap the markets of filler and reinforcement for synthetic polymer matrices (Thakur and Singha 2010a–c; Thakur et al. 2010). Compared to the conventional fillers such as glass fibers, nylon, and aramid, lignocellulosic materials such as cellulosic fibers and wood fibers have many advantages as they are renewable, cheaper, lighter, and less abrasive to processing equipment (Thakur and Singha 2011a,b; Thakur et al. 2011). Depending on their origin, natural cellulosic fibers can be classified into two main groups: nonwood natural fibers and wood fibers. Nonwood natural fibers are generally classified into bast, leaf, straw, seed, and fruit. Suitable examples of these natural cellulosic fibers are sisal, pineapple, flax, hibiscus sabdariffa, hemp, kenaf, cotton, and so on (Rozman et al. 2003). The properties of lignocellulosic natural fibers such as strength, stiffness, and density depend on the internal structure and the composition of the fiber components (Siriwardena et al. 2003; Yussuf et al. 2010). The cellulose content in lignocellulosic fibers determines the tensile and flexural properties of the fibers while the lignin is believed to have an influence on the structure and morphology of the fibers (Ruseckaite et al. 2007). On the other hand, the waxy substances influence the fibers' wettability and adhesion characteristics (Mahlberg et al. 1998; Ndazi et al. 2007).

Although the incorporation of lignocellulosic fibers as reinforcement in polymers brings a number of benefits such as tensile strength and relatively high modulus of elasticity, still they suffer from few drawbacks, especially decrease in properties due to weather conditions (Abdelmouleh et al. 2004; Tserki et al. 2005). Indeed, during their exterior use, biomass-based composites are readily degraded by adverse weather conditions (Seki 2009). The prime reasons for the outdoor instability of the cellulosic fibers and cellulosic fiber–based composites are their instability due to moisture absorption/desorption, breakdown of cellulosic polymers by UV light, and decay by microorganism (Huda et al. 2006, 2008).

In this chapter, we report some of our previous results on functionalization of natural cellulosic fibers and new studies on the mechanical behavior of modified cellulosic fiber–reinforced green composites.

14.2 FUNCTIONALIZED CELLULOSIC FIBER–REINFORCED GREEN COMPOSITES

14.2.1 MATERIALS, METHODS, AND INSTRUMENTS USED

Reagent grade chemicals, namely, sodium hydroxide (NaOH), ethanol, acetone, aminopropyl triethoxy silane, phenol, and formaldehyde solution, were kindly supplied by Qualigens Chemicals Ltd. These chemicals were used as received without any further purification. *Saccaharum cilliare* fibers were collected from the local resources of Himalayan region and were purified through Soxhlet extraction in acetone for 72 hours by the standard method. Figure 14.1 shows the scanning electron microscope (SEM) image of the raw *S. cilliare* fiber (Thakur et al. 2010).

Phenol formaldehyde (PF) has been used as a novel polymer matrix resin for the preparation of polymer composites. The polymer was synthesized by the standard method reported in the literature (Singha and Thakur 2009a, 2010b; Thakur et al. 2010). Weights of the samples were taken on a Shimadzu electronic balance

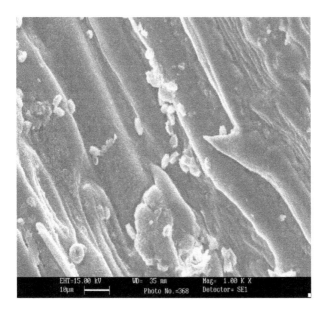

FIGURE 14.1   SEM image of raw *Saccaharum cilliare* fiber. (Reprinted from Thakur, V.K. et al., *Int. J. Polym. Anal. Charact.*, 15, 397–414, 2010. With permission. Copyright 2010 Taylor & Francis.)

(LIBROR AEG-220), thermal studies were carried out on a Thermal Analyzer (Perkin Elmer), and SEM micrographs were taken on a LEO 435VP. Polymer resin and composite samples were cured using a compression molding machine (Thakur et al. 2010). Mechanical properties such as tensile, compressive, and flexural stress and wear resistance were tested using standard testing machines.

14.2.2 FUNCTIONALIZATION OF *SACCAHARUM CILLIARE* FIBERS

Functionalization of *S. cilliare* fibers was carried out by first treating them with NaOH solution of optimized concentration (Thakur et al. 2010). Table 14.1 shows the effect of different concentrations of NaOH on *S. cilliare* fibers.

After the treatment process, the fibers were thoroughly washed and dried in an oven at 50°C to a constant weight as per standard method. Figure 14.2 shows the morphological image of the *S. cilliare* fiber after treatment with optimized NaOH concentration (Thakur et al. 2010).

Subsequently, the fibers were treated with aminopropyl triethoxy silane to carry out the silane functionalization. Mercerized *S. cilliare* fibers obtained through optimized NaOH treatment were dipped into 2% of silane solution, which was prepared by mixing aminopropyl triethoxy silanes with an ethanol/water mixture in the ratio 60:40. The pH of the solution was maintained between 3.5 and 4 using METREPAK Phydrion buffers (Thakur et al. 2010). After the stipulated time, the ethanol/water mixture was drained out and the fibers were dried in air and then in an oven at 50°C to a constant weight. The SEM image of the *S. cilliare* fibers after silane treatment is shown in Figure 14.3 (Thakur et al. 2010).

From the morphological images, it is clear that the surface of the *S. cilliare* fibers is successfully modified by silane functionalization.

TABLE 14.1

Effect of Reaction Time on Mercerization of *Saccaharum cilliare* Fibers

No.	Time (Minutes)	% Wt. Loss
1	0	0
2	45	10.51
3	90	15.34
4	135	17.12
5	180	20.20
6	225	21.37
7	270	24.46
8	315	22.63
9	360	22.89

Source: Reprinted from Thakur, V.K. et al., *Int. J. Polym Anal. Charact*, 15, 7, 397–414, 2010. With permission. Copyright 2010 Taylor & Francis.

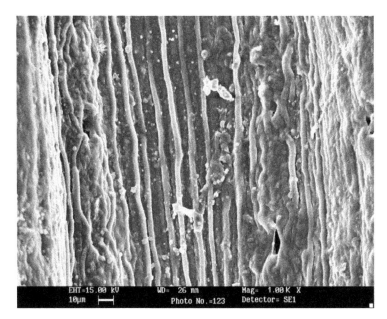

FIGURE 14.2 SEM image of mercerized *Saccaharum cilliare* fiber. (Reprinted from Thakur, V.K. et al., *Int. J. Polym. Anal. Charact.*, 15, 397–414, 2010. With permission. Copyright 2010 Taylor & Francis.)

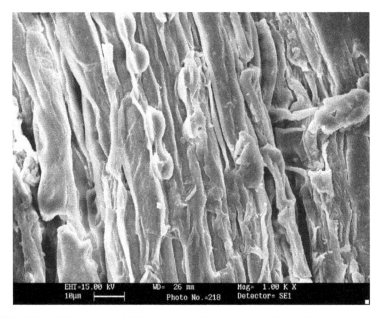

FIGURE 14.3 SEM image of silane-treated *Saccaharum cilliare* fiber. (Reprinted from Thakur, V.K. et al., *Int. J. Polym. Anal. Charact.*, 15, 397–414, 2010. With permission. Copyright 2010 Taylor & Francis.)

14.2.3 FABRICATION AND CHARACTERIZATION OF GREEN POLYMER COMPOSITES PREPARED USING FUNCTIONALIZED FIBERS

For the preparation of green polymer composites using *S. ciliare* fibers as reinforcement, polymer resin matrix, that is, –PF, had been synthesized by the standard method by using phenol and formaldehyde in different molar ratios (1.0:1.0, 1.0:1.5, 1.0:2.0, 1.0:2.5, and 1.0:3.0) in the reaction kettle (Singha and Thakur 2010b; Thakur et al. 2010).

On the other hand, the reinforcements, that is, functionalized *S. cilliare* fibers, were used in three different forms (particle, short, and long fiber reinforcements) for the fabrication of green polymer composites (Thakur et al. 2010). These functionalized fibers of different dimensions were mixed thoroughly with polymer matrix using a mechanical stirrer with suitable loading (1.0:0.1) in terms of weight (Thakur et al. 2010) for the fabrication of PF matrix–based polymer composites. Composite sheets of size 150 × 150 × 5.0 mm were prepared by the compression molding technique (Singha and Thakur 2009a; Thakur et al. 2010).

The green composites thus fabricated were subjected to evaluation of their mechanical properties such as tensile, compressive, flexural stress, and wear resistance in accordance with the ASTM D 3039, ASTM D 3410, ASTM D 790, and ASTM D 3702 methods, respectively (Singha and Thakur 2009a, 2010b; Thakur and Singha 2010b,c; Thakur et al. 2010). Surface morphology of the composite sample was observed by SEM. To achieve good electric conductivity, the optimized sample was carbon sputtered followed by sputtering a gold platinum mixture before examination (Thakur and Singha 2010b,c). The percent swelling (P_s) was calculated from the increase in initial weight as follows:

$$\text{Percent swelling}: (P_S) = \frac{W_f - W_i}{W_i} \times 100$$

where W_f = final weight and W_i = initial weight.

The percent chemical resistance (P_{cr}) was calculated in terms of weight loss as follows:

$$\text{Percent chemical resistance}: (P_{cr}) = \frac{T_i - W_{aci}}{T_i} \times 100$$

where T_i = initial weight and W_{aci} = weight after certain interval.

14.3 RESULTS AND DISCUSSION

It has been well documented that the coupling agents usually improve the degree of cross-linking in the interface region and offer a perfect bonding between the polymer matrix and the fibers. Also, in some of our earlier studies, we have reported that surface functionalization through coupling agents improves the physicochemical properties of the cellulosic fibers. Among the various coupling agents, silane coupling agents were found to be effective in modifying the natural fiber–matrix interface in a number of polymer systems. Silane modification is based on the use of reactants that bear reactive end groups, which, on one end, can react with the matrix and, on the other end, can react with the hydroxyl groups of the fiber. Generally, the alkoxy or ethoxies are the end groups that can form stable covalent bonds reacting with the hydroxyl groups of the cellulosic fibers. In our green composites, silane coupling agent has also been found to be effective in improving the interface properties. The comparative results of the silane functionalized fibers with those of nonfunctionalized fibers show that the mechanical properties of the modified fiber–reinforced composites are increased to a significant extent. During the functionalization of *S. ciliare* fibers, silanes after hydrolysis undergo condensation and bond formation stage and can form polysiloxane structures by reaction with the hydroxyl group of the fibers that during the fabrication of composites forms strong bonds with the polymer matrix resulting in higher mechanical properties. Furthermore, during the fabrication of *S. cilliare* fiber–reinforced composite, particular interest lies in the behavior of these materials against weather conditions. The effect of different weather conditions on the overall properties of the green polymer composites has been a subject of much interest and significance. The commercial viability of the green composites from functionalized *S. cilliare* fibers lies in their physical and chemical properties. Hence, keeping in mind the commercial viability, a comprehensive study on swelling behavior in different solvents and chemical resistance behavior against 1N HCl and 1N NaOH of functionalized fiber–reinforced green polymer composites has been carried out to assess the potential of functionalized fibers as reinforcement in a number of engineering applications.

14.3.1 MECHANICAL PROPERTIES OF FUNCTIONALIZED *S. CILLIARE* FIBER–REINFORCED GREEN COMPOSITES

It has been observed that functionalized *S. cilliare* fiber–reinforced polymer composites with particle reinforcement exhibit higher mechanical properties such as tensile strength, compressive strength, flexural strength, and wear resistance followed by short-fiber and long-fiber-reinforced composites (Figure 14.4a through d).

Similar trends were also observed for the mechanical properties of the composites prepared using raw *S. cilliare* fibers. The enhanced mechanical properties of the functionalized fiber–reinforced composites in particle form as compared to short and long fibers can be attributed to the higher interfacial bonding strength between the matrix and reinforcement due to higher surface area as well as adhesion due to fiber surface modification. From the foregoing results, it is clear that silane functionalization improves the interfacial bonding strength between the fibers and polymer matrix. Such bonding is highly desirable to achieve good fiber reinforcement. The interfacial bonding strength depends on the surface topology of the fibers.

14.3.2 WATER ABSORPTION BEHAVIOR OF THE FUNCTIONALIZED FIBER–REINFORCED COMPOSITES

Functionalized *S. cilliare* fiber–reinforced polymer composites with different dimensions have been found to show different absorption behaviors in different solvents (Figure 14.5).

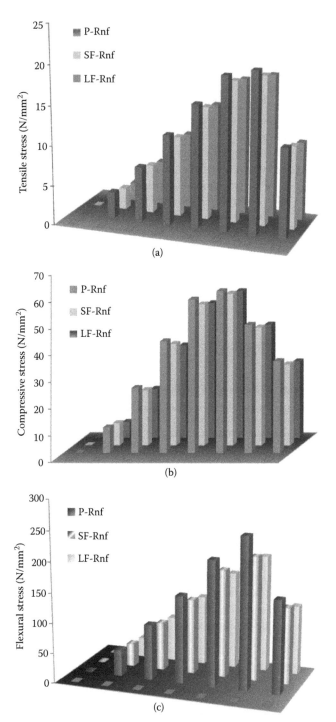

FIGURE 14.4 Tensile/compressive/flexural stress and wear resistance results of functionalized *Saccaharum cilliare* fiber–reinforced green composites (a through d).

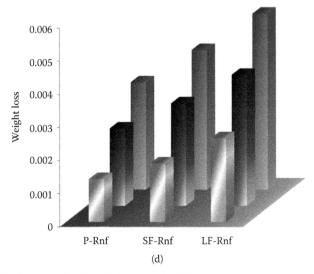

(d)

FIGURE 14.4 (*Continued*)   Tensile/compressive/flexural stress and wear resistance results of functionalized *Saccaharum cilliare* fiber–reinforced green composites (a through d).

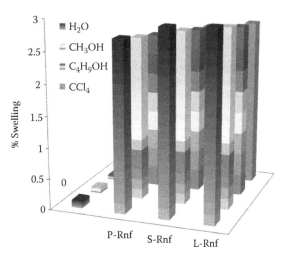

FIGURE 14.5   Absorption behaviors of silane functionalized *Saccaharum cilliare* fiber–reinforced polymer composites in different solvents.

The *S. cilliare* fiber–reinforced PF composites in different solvents follow the trend $H_2O > CH_3OH > C_4H_9OH > CCl_4$ for the absorption behavior. On comparison with raw fiber–reinforced composites, functionalized fiber–based composites showed more resistance toward swelling, again confirming the strong interfacial bonding between the polymer matrix and the functionalized fibers.

14.3.3 CHEMICAL RESISTANCE BEHAVIOR OF THE FUNCTIONALIZED FIBER–REINFORCED COMPOSITES

In the case of chemical resistance behavior, it has been observed that resistance of silane functionalized *S. cilliare* fiber–reinforced composites toward chemicals decreases with the increase in fiber dimension and is much better than those of the raw fiber–reinforced composites (Figures 14.6 and 14.7) (Thakur et al. 2010).

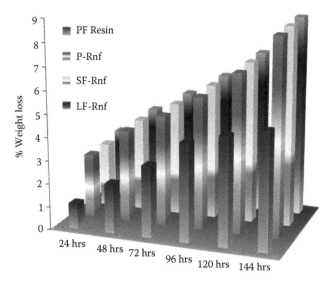

FIGURE 14.6 Chemical resistances of silane functionalized *Saccaharum cilliare* fiber–reinforced polymer composites at different time intervals against 1N HCl.

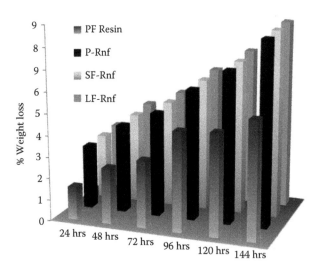

FIGURE 14.7 Chemical resistances of silane functionalized *Saccaharum cilliare* fiber–reinforced polymer composites at different time intervals against 1N NaOH.

FIGURE 14.8   SEM image of silane functionalized *Saccaharum cilliare* fiber–reinforced composites with particle reinforcement.

The superior behavior of functionalized fiber–reinforced composites can be attributed to the increase in fiber matrix bonding between the matrix and the reinforcement.

14.3.4   Morphological Analysis of Polymer Biocomposites

Figure 14.8 shows the morphological image of the silane functionalized *S. cilliare* fiber–reinforced composite with particle reinforcement.

From the image, it is obvious that better interfacial bonding takes place in the green composite prepared using silane functionalized *S. cilliare* fibers, which results in higher mechanical and physicochemical properties.

14.4   CONCLUSION

In this chapter, we have discussed the effect of silane functionalization on natural fiber–reinforced composites prepared using modified *S. cilliare* fibers. The chemical treatment of the fibers was aimed at improving the adhesion between the fibers' surface and the polymer matrix. The mechanical properties of the functionalized fiber–reinforced composites have been found to increase as compared to raw fiber–reinforced composites along with a reduction in water absorption of composites.

ACKNOWLEDGMENTS

The authors thank their parental institutes for providing the necessary facilities to accomplish the present research work.

REFERENCES

Abdelmouleh, M., Boufi, S., Belgacem, M.N., Duarte, A.P., Ben Salah, A. and Gandini, A. 2004. Modification of cellulosic fibres with functionalised silanes: Development of surface properties. *International Journal of Adhesion and Adhesives* 24(1): 43–54.

Alawar, A., Hamed, A.M. and Al-Kaabi, K. 2009. Characterization of treated date palm tree fiber as composite reinforcement. *Composites Part B: Engineering* 40(7): 601–606.

Auras, R., Harte, B. and Selke, S. 2004. An overview of polylactides as packaging materials. *Macromolecular Bioscience* 4(9): 835–864.

Averous, L. 2004. Biodegradable multiphase systems based on plasticized starch: A review. *Journal of Macromolecular Science. Polymer Reviews C* 44(3): 231–274.

Ayswarya, E.P., Vidya Francis, K.F., Renju, V.S. and Thachil, E.T. 2012. Rice husk ash—A valuable reinforcement for high density polypropylene. *Materials and Design* 41: 1–2.

Bledzki, A.K. and Gassan, J. 1999. Composites reinforced with cellulose based fibers. *Progress in Polymer Science* 24: 221–274.

Bledzki, A.K., Mamun, A. and Volk, J. 2010. Barley husk and coconut shell reinforced polypropylene composites: The effect of fibre physical, chemical and surface properties. *Composites Science and Technology* 70: 840–846.

Bledzki, A.K., Reihmane, S. and Gassan, J. 1996. Properties and modification methods for vegetable fibers for natural fiber composites. *Journal of Applied Polymer Science* 59(8): 1329–1336.

Chao-Lung, H., Le Anh-Tuan, B. and Chun-Tsun, C. 2011. Effect of rice husk ash on the strength and durability characteristics of concrete. *Construction and Building Materials* 25: 3768–3772.

Debashish, D., Debapriya, D. and Basudam, A. 2004. The effect of grass fiber filler on curing characteristics and mechanical properties of natural rubber. *Polymers for Advanced Technologies* 15(12): 708–715.

Favaro, S.L., Lopes, M.S., Vieira de Carvalho Neto, A.G., Rogerio de Santana, R. and Radovanovic, E. 2010. Chemical, morphological and mechanical analysis of rice husk/post-consumer polyethylene composites. *Composites Part A: Applied Science and Manufacturing* 41: 154–160.

Garlotta, D. 2002. A literature review of poly(lactic acid). *Journal of Polymer and the Environment* 9(2): 63–84.

Hagstrand, P.O. and Oksman, K. (2001). Mechanical properties and morphology of flax fiber reinforced melamine-formaldehyde composites. *Polymer Composites* 22(4): 568–578.

Huda, M.S., Drzal, L.T., Misra, M. and Mohanty, A.K. 2006. Wood-fiber-reinforced poly (lactic acid) composites: Evaluation of the physicomechanical and morphological properties. *Journal of Applied Polymer Science* 102(5): 4856–4869.

Huda, M.S., Drzal, L.T., Mohanty, A.K. and Misra, M. 2008. Effect of fiber surface-treatments on the properties of laminated biocomposites from poly(lactic acid) (PLA) and kenaf fibers. *Composites Science and Technology* 68(2): 424–432.

Kabir, M.M., Wang, H., Lau, K.T. and Cardona, F. 2012a. Chemical treatments on plant-based natural fibre reinforced polymer composites: An overview. *Composites Part B: Engineering* 43B: 2883–2892.

Kabir, M.M., Wang, H., Lau, K.T., Cardona, F. and Aravinthan, T. 2012b. Mechanical properties of chemically-treated hemp fibre reinforced sandwich composites, *Composites Part B: Engineering* 43: 159–169.

Mahlberg, R., Niemi, H.E.M., Denes, F. and Rowell, R.M. 1998. Effect of oxygen and hexamethyldisiloxane plasma on morphology, wettability and adhesion properties of polypropylene and lignocellulosics. *International Journal of Adhesion and Adhesives* 18: 283–297.

Ndazi, B.S., Karlsson, S., Tesha, J.V. and Nyahumwa, C.W. 2007. Chemical and physical modifications of rice husks for use as composite panels. *Composites Part A: Applied Science and Manufacturing* 38: 925–935.

Panthapulakkal, S., Zereshkian, A. and Sain, M. 2006. Preparation and characterization of wheat straw fibers for reinforcing application in injection molded thermoplastic composites. *Bioresource Technology* 97: 265–272.

Rozman, H.D., Yeo, Y.S., Tay, G.S. and Abubakar, A. 2003. The mechanical and physical properties of polyurethane composites based on rice husk and polyethylene glycol. *Polymer Testing* 22: 617–623.

Ruseckaite, R.A., Ciannamea, E., Leiva, P. and Stefani, P.M. 2007. Particleboards based on rice husk. In: *Polymer and Biopolymer Analysis and Characterization*. Eds. Zaikov, G.E. and Jimenez, A. 1–12. Nova Science Publishers: New York.

Seki, Y. 2009. Innovative multifunctional siloxane treatment of jute fiber surface and its effect on the mechanical properties of jute/thermoset composites. *Materials Science and Engineering: A* 508(1–2): 247–252.

Singha, A.S., Shama, A. and Thakur, V.K. 2008. Pressure induced graft co-polymerization of acrylonitrile onto *Saccharum cilliare* fiber and evaluation of some properties of grafted fibers. *Bulletin of Material Science* 31(1): 1–7.

Singha, A.S., Shama, A. and Thakur, V.K. 2009. Graft copolymerization of acrylonitrile onto *Saccaharum cilliare* fiber. *E-Polymers* 105: 1–12.

Singha, A.S. and Thakur, V.K. 2008. Mechanical properties of natural fiber reinforced polymer composites. *Bulletin of Material Science* 31(5): 791–799.

Singha, A.S. and Thakur, V.K. 2009a. Chemical resistance, mechanical and physical properties of biofiber based polymer composites. *Polymer-Plastics Technology and Engineering* 48(7): 736–744.

Singha, A.S. and Thakur, V.K. 2009b. Fabrication and characterization of *S. cilliare* fiber reinforced polymer composites. *Bulletin of Materials Science* 32(1): 49–58.

Singha, A.S. and Thakur, V.K. 2010a. Renewable resources based green polymer composites: Analysis and characterization. *International Journal of Polymer Analysis and Characterization* 15(3): 127–146.

Singha, A.S. and Thakur, V.K. 2010b. Synthesis characterization and study of pine needles reinforced polymer matrix based composites. *Journal of Reinforced Plastics and Composites* 29(5): 700–709.

Siriwardena, S., Ismail, H. and Ishiaku, U.S. 2003. A comparison of the mechanical properties and water absorption behavior of white rice husk ash and silica filled polypropylene composites. *Journal of Reinforced Plastics Composites* 22: 1645–1666.

Thakur, V.K. and Singha, A.S. 2010a. KPS-initiated graft co polymerization on to modified cellulosic biofibers. *International Journal of Polymer Analysis and Characterization* 15(8): 471–485.

Thakur, V.K. and Singha, A.S. 2010b. Mechanical and water absorption properties of natural fibers/polymer biocomposites. *Polymer-Plastics Technology and Engineering* 49(7): 694–700.

Thakur, V.K. and Singha, A.S. 2010c. Natural fibres-based polymers: Part I—Mechanical analysis of pine needles reinforced biocomposites. *Bulletin of Materials Science* 33(3): 257–264.

Thakur, V.K. and Singha, A.S. 2011a. Physico-chemical and mechanical behavior of cellulosic pine needles based biocomposites. *International Journal of Polymer Analysis and Characterization* 16(6): 390–398.

Thakur, V.K. and Singha, A.S. 2011b. Rapid synthesis, characterization, and physico chemical analysis of biopolymer-based graft copolymers. *International Journal of Polymer Analysis and Characterization* 16(3): 153–164.

Thakur, V.K., Singha, A.S., Kaur, I., Nagarajarao, R.P. and Yang, L.P. 2010. *Silane* function-alization of *Saccaharum cilliare* fibers: Thermal, morphological and physicochemical study. *International Journal of Polymer Analysis and Characterization* 15(7): 397–414.

Thakur, V.K., Singha, A.S. and Misra, B.N. 2011. Graft copolymerization of methyl methac-rylate onto cellulosic biofibers. *Journal of Applied Polymer Science* 122(1): 532–544.

Thakur, V.K., Singha, A.S. and Thakur, M.K. 2012a. Graft copolymerization of methyl acry-late onto cellulosic bio fibers: Synthesis, characterization and applications. *Journal of Polymers and the Environment* 20(1):164–174.

Thakur, V.K., Singha, A.S. and Thakur, M.K. 2012b. In-air graft copolymerization of ethyl acrylate onto natural cellulosic polymers. *International Journal of Polymer Analysis and Characterization* 17(1): 48–60.

Thakur, V.K., Singha, A.S. and Thakur, M.K. 2012c. Surface modification of natural poly-mers to impart low water absorbency. *International Journal of Polymer Analysis and Characterization* 17(2): 133–143.

Tserki, V., Matzinos, P., Kokkou, S. and Panayiotou, C. 2005. Novel biodegradable composites based on treated lignocellulosic waste flour as filler. Part I. Surface chemical modi-fication and characterization of waste flour. *Composites Part A: Applied Science and Manufacturing* 36(7): 965–974.

Wambua, P., Ivens, J. and Verpoest, I. 2003. Natural fibres: Can they replace glass in fibre reinforced plastics? *Composites Science and Technology* 63: 1259–1264.

Yussuf, A.A., Massoumi, I. and Hassan, A. 2010. Comparison of polylactic acid/kenaf and polylactic acid/rice husk composites: The influence of the natural fibers on the mechani-cal, thermal and biodegradability properties. *Journal of Polymers and the Environment* 18(3): 422–429.

Zain, M.F.M., Islam, M.N., Mahmud, F. and Jamil, M. 2011. Production of rice husk ash for use in concrete as a supplementary cementious material. *Construction and Building Materials* 25: 798–805.

15 Properties and Characterization of Natural Fiber–Reinforced Polymeric Composites

Hom N. Dhakal and Zhong Y. Zhang

CONTENTS

15.1 INTRODUCTION

The term "composite material" can be broadly defined as the result of combining two or more materials on a microscopic scale, each of which has their own unique properties, to produce a new material that has properties far superior than either of the base materials. The constituent that is continuous, present normally in greater quantity, in the composite is called the matrix, and the second constituent is referred to as the reinforcement (fibers) as it enhances the properties of the matrix (Hull and Clyne 1996; Matthews and Rawlings 1995).

Composites are classified according to the matrix used; polymeric matrix composite (PMC), ceramic matrix composite, and metallic matrix composite. The matrix in the composite serves three major functions:

1. Support and transfer the stresses to the fibers that carry most of the load
2. Protect the fibers against physical damage and the environment
3. Reduce propagation of cracks in the composite by virtue of the ductility and toughness of the plastic matrix

Since the main interest in this chapter is PMCs, the presented discussion on properties and characterization is focused on natural plant fiber–reinforced polymeric composites.

PMCs are used in commercial applications due to their good mechanical properties, light weight, and low cost. The use of natural fibers as reinforcement in PMCs has generated much interest in recent years due to environmental impact considerations, the need to redirect agricultural production from the food industry to other applications, as well as the urgency to find alternative, energy-saving materials (Bledzki and Gassan 1999). The use of bast fibers such as hemp (*Cannabis sativa* L.) and flax (*Linum usitatissimum*) fiber bundles as reinforcing agents in composites, for example, offers many advantages over synthetic fibers (glass, carbon, or aramid) including good specific strengths and modulus, unlimited and sustainable availability, low density, reduced tool wear, enhanced energy recovery, reduced dermal and respiratory irritation, and good biodegradability (Bolton 1995; Richardson et al. 1998).

15.2 REINFORCEMENTS FOR POLYMERIC COMPOSITES

Typical reinforcements for composites include glass and carbon fibers. Glass/carbon fiber–reinforced polymer composites have several advantages over their metal matrix counterparts, most commonly recognized as light weight, corrosion resistance, high strength, increased "strength-to-weight ratio," and the "designability" of composites. Composite materials enable the designer to choose the optimum combination of fiber reinforcement and matrix to develop a material particularly designed for a specific application as these composites can be fabricated to the required mechanical properties.

Environmental legislation as well as consumer pressure for the adaptation of a "cradle to the reincarnation" (or "cradle to the cradle," if materials are recycled) concept for material use throughout the world has triggered a paradigm shift toward using natural materials as a substitute for nonrenewable man-made fibers such as glass and carbon (Bledzki et al. 2002). Over the last decade, a number of researchers have been involved in investigating the exploitation of natural fibers as load-bearing constituents. The use of these renewable materials in composites has increased significantly due to their relative cheapness compared to conventional materials (Dhakal et al. 2009). Natural fibers have densities of 1.25–1.50 versus 2.5 g/cm³ for E-glass. This lower density (approximately 80% lighter than glass) helps to give them a higher "strength-to-weight ratio" for reinforcing composites compared to glass fiber (Gassan and Bledzki 1998).

Natural fibers are classified according to their source: plants, animals, and minerals. In general, it is the plant fibers that are used as reinforcement in polymeric composites. Many varieties of plant fibers exist and they can be divided into five main categories depending on the part of the plant from which they are extracted (Richardson et al. 1998).

1. Bast (bark) fibers (flax, hemp, jute, kenaf, ramie, etc.)
2. Leaf fibers (sisal, henequen, coir, abaca, pineapple, etc.)
3. Seed or hair fibers (cotton, kapok, etc.)
4. Fruit fibers (coconut, palm oil, etc.)
5. Wood fibers (pine, spruce, oak, etc.)

Composites that use polymer resin as the matrix and natural fiber as the reinforcement are called natural fiber–reinforced composites (NFRC). Although NFRCs are a relatively new generation of engineering materials, natural fibers including hemp, sisal, flax, and jute are already harvested in many countries around the world at an industrial scale, which is exciting and has drawn a great deal of interest as these materials can be used for demanding product applications in the automotive, building, furniture, and aerospace applications (Bolton 1995).

Many research and developments into using natural fibers as reinforcement in polymer composites have shown that the specific properties of these materials, particularly hemp, flax, sisal, and jute, are comparable with conventional reinforcing materials such as glass and carbon fibers. However, the application of these fibers is limited to interior parts and nonstructural components of automobiles (Gayer and Schuh 1996). The main reasons for this are the weak interface between the reinforcement and the matrix as a result of inherent moisture absorption behavior, and the fact that their structure and damage mechanisms are not well understood.

Natural fibers are hydrophilic and have low moisture resistance, which can lead to poor interfacial bonding between fibers and matrix. The behavior of NFRC is dominated by the arrangement and the interaction of the stiff, strong fibers with the less stiff, weaker polymer matrix. Because these materials are relatively new, not much information is available in the literature about their physical, mechanical, thermal, and surface properties, particularly the characterization of these important properties. This chapter attempts to highlight the importance of above-mentioned properties and their characterization.

15.3 NATURAL PLANT FIBERS FOR COMPOSITE REINFORCEMENTS

All plant species are built up of cells. When a cell is very long in relation to its width, it is called a fiber. For example, wood fibers are mostly 50–100 times longer than they are wide. Knowledge about fiber length and width is important for comparing different kinds of natural fibers. A high aspect ratio (length/width) is crucial in cellulose-based fiber composites as it give an indication of possible strength properties. Hemp single fiber, for example, has an aspect ratio of 1000, which is good for mechanical properties. The fiber is like a microscopic tube (i.e., wall surrounding a central void referred to as the lumen). Moreover, when the cell wall is made up mainly (85% or more) of cellulose, hemicellulose, and lignin, we talk about lignocellulosic fibers, and this includes woody species, scrubs, and most agricultural crops. Typical lignocellulosic fibers from agriculture are found, for example, in straws, flax, hemp, jute, and sisal. Nonlignocellulosic fibers are fibers that do not contain lignin and are found in potatoes, beets, and cotton among other crops (Faruk et al. 2012).

15.3.1 ADVANTAGES OF PLANT FIBERS

The work of Peijs (2000), Bledzki et al. (1997), and Eichhorn and Young (2004) highlighted that natural fibers have been used for a long time in different industries and have the potential to be used increasingly as reinforcements in polymer engineering composites because they offer several advantages compared to conventional glass fibers. The key features of plant fibers are high specific strength because of their low density, relative cheapness compared to conventional materials such as glass and carbon fibers, and their ability to recycle. Other highly desirable qualities of natural fibers are that they are nontoxic, easy to fabricate, and producible with low investment at low cost, which makes the material an interesting product for low-wage countries. Other advantages include reduced tool wear in machining operations, enhanced energy recovery, and reduced dermal and respiratory irritation.

The total energy required to produce natural fibers is less compared to glass fibers (Joshi et al. 2004). Glass fiber production requires 5–10 times more nonrenewable energy than natural fiber production. As a result, the pollutant emissions from glass fiber production are significantly higher than from natural fiber production. This strengthens natural fibers sustainable, renewable advantage over glass fibers.

15.3.2 DISADVANTAGES OF PLANT FIBERS

Natural fibers have some disadvantages, which are absorption of moisture and high moisture regain property, susceptibility to fungal and insect attack, poor interfacial adhesion, and lower strength properties. The variable quality of fibers, which depends on unpredictable influences such as weather and growing conditions, makes it difficult to predict the properties natural plant fibers.

The structure of natural fibers has other weaknesses. Natural fiber is filled with cellulose material, which acts as an insulator, thus a natural fiber composite shows much lesser thermal conductivity when compared to a glass fiber–reinforced polymer composite. Unlike synthesized fibers, natural fiber properties are dependent on the weather during the growing season and may suffer mechanical damage during mechanized harvesting reducing the mechanical advantages (Summerscalesa et al. 2010).

15.3.3 NATURAL PLANT FIBERS: COMPOSITION AND STRUCTURE

Natural plant fibers, especially bast fibers such as hemp, jute, and flax fibers, are multicelled in structure (Figure 15.1). The cell walls of fibers are made up of a number of layers: the so-called primary wall (the first layer deposited during cell development) and the secondary wall (S), which is again made up of three layers (S_1, S_2, and S_3). As in all lignocellulosic fibers, these layers mainly contain cellulose, hemicellulose, and lignin in varying amounts. The individual fiber is bonded together by a lignin-rich region known as the middle lamella. Cellulose attains the highest concentration in the S_2 layer (about 50%) and lignin is most concentrated in the middle lamella (about 90%). The S_2 layer is the thickest layer by far (32–150 laminas) and dominates the properties of the fibers.

Review of relevant natural fiber properties (Table 15.1) as well as the properties of glass fiber has shown that natural bast fibers can be an alternative to glass fibers as reinforcement in composite materials. The results presented show that glass fibers are superior to plant fibers, and moreover, the ultimate tensile strength, in particular, is larger for glass fiber than that of natural bast fibers. However, if one considers the specific modulus of hemp fibers (modulus/density), then one obtains an average value of 35.5 GPa. A similar calculation for glass fibers using a modulus of 70 GPa gives a specific modulus of 28 GPa. This indicates that the mechanical properties of hemp fiber are approaching the properties of glass fibers. A similar calculation for flax fibers using an average modulus of 70 GPa gives a specific modulus of 46 GPa. These properties make hemp and flax fibers good reinforcing

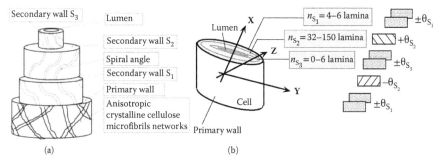

FIGURE 15.1 (a) Cell wall structure. (b) Model for illustration of one single fiber cell. Reproduced with permission from Springer Publishing: Gassan J., Chate A. and Bledzki A.K. *Journal of Materials Science*, 36, 2001, 3715, Copyright 2001.

TABLE 15.1

Physical and Mechanical Properties of Selected Natural Bast Fibers (Bundles)

Fiber Type	Density (g/cm³)	Moisture Absorption (%)	Elongation at Break (%)	Tensile Strength (MPa)	Young's Modulus (GPa)	Specific Modulus (GPa) $\left(\frac{E}{\rho}\right)$
Jute	1.3	12	1.5–1.8	393–773	10–30	7–21
Flax	1.5	7	2.7–3.2	345–1035	60–80	26–46
Hemp	1.4	8	1.6	690	30–70	21–50
E-glass	2.5	—	2.5	2000–3500	70.0	28

Source: Eichhorn S.J. et al., *J. Mat. Sci.*, 36, 2107–2131, 2001a; Wambua P.W. et al., *Comp. Sci. Technol.*, 63, 1259–1264, 2003.

 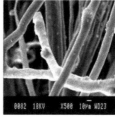 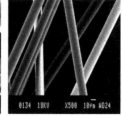

FIGURE 15.2 Nonwoven hemp mat and scanning electron microscopic image of hemp fiber and glass fiber. (Adapted from Dhakal H.N., Zhang Z.Y., Bennett N. and Reis P.N.B, Composite Structures, 94, 2756, 2012; Dhakal H.N., Zhang Z.Y., Richardson, M.O.W. and Errajhi, O.A.Z., *Composite Structures*, 81, 559, 2007b. With permission from Elsevier Ltd. Copyright 2012, 2007).

materials. In particular, hemp and flax fiber composites have the potential to become the market leader for environmentally friendly reinforcing material for composites (Dissanayake et al. 2008).

One major obstacle concerning the development and use of natural fiber–reinforced polymer composites relates to incompatibility between the hydrophilic natural fibers and generally hydrophobic thermoplastic and thermoset matrices. Synthetic fibers such as glass and carbon can be produced with a definite range of reproducible characteristics, whereas the properties of natural fibers vary depending on the fiber diameter, structure, climate conditions, age, and digestion process (Figure 15.2). As a result, compared to composites from synthetic reinforcements, natural plant fiber–reinforced composites have high variation in properties, sensitivity to moisture, and wet environments. These limitations can be disadvantageous in structural applications (Bledzki et al. 1996; De Rosa et al. 2012). It is, therefore, important that appropriate treatments are applied to enhance the adhesion between fiber and the matrix or fiber–matrix interface.

15.4 CHARACTERIZATION OF PHYSICAL PROPERTIES

15.4.1 DENSITY MEASUREMENT

As highlighted in earlier sections, the microstructure of natural fiber is extremely complex due to the hierarchical organization at different length scales and the different materials present in variable proportions. To calculate the properties of natural fiber–reinforced polymeric composites, it is important to determine accurate physical properties of natural fibers. It is not easy to attain a reliable value for diameter and density of natural fibers due to their porous nature. Sections 15.4.2 through 15.4.6 are aimed at highlighted how the density, diameter, void content, and volumetric interactions of natural fiber–reinforced polymeric composites are calculated and analyzed.

It is important to measure the densities of fiber, matrix, and the composite to measure the volume fractions in composite materials accurately. However, because of the complex nature of natural fibers, it is always difficult to measure their density accurately. Prattern (1981) has reviewed a number of methods for the measurement of density of small samples. Other techniques such as the displacement method (ASTM D 792), and "The Berman Density Balance" method (Berman 1939), can also be adapted to determine the density of substances.

The determination of fiber density by the pycnometry method, for example, is based on four gravimetric measurements: the empty pycnometer (C1), the water-filled pycnometer (C2), the pycnometer containing the dry fibers (C3), and the pycnometer filled with both fibers and water (C4). From these measurements, the following quantities are calculated:

1. $m_f = C3 - C1$
2. $m_w = C2 - C1$
3. $m_{w^*} = C4 - C3$

where m_f is mass of the dry fibers, m_w is the mass of the maximum amount of water in the pycnometer, and m_{w^*} is the mass of the reduced amount of water in the pycnometer owing to the fiber volume.

The fiber dry density, ρ_f, is then calculated by the standard equation

$$\rho_f = \frac{m_f \rho_w}{m_w - m_{w^*}} \Rightarrow \rho_f = \frac{m_f}{(m_w/\rho_w) - (m_{w^*}/\rho_w)} \tag{15.1}$$

where ρ_w is the density of water.

15.4.2 COMPOSITE VOID CONTENT

Given densities and fiber content of the composites, the void contents can be determined according to ASTM D 2734.

$$V_v = 1 - \rho_c \left(\frac{w_f}{\rho_f} + \frac{w_m}{\rho_m} \right) \tag{15.2}$$

where V_v is volume fraction of voids, ρ_c is density of composite, w_f is weight fraction of fiber, w_m is weight fraction of matrix, ρ_f is density of fiber in g/cm^3, and ρ_m is density of matrix in g/cm^3.

15.4.3 COMPOSITE VOLUMETRIC INTERACTION

The following improved model of composite volumetric interaction is based on similar models presented by Andersen and Lilholt (1999).

Composite volume fraction V_f can be calculated from the ideal fiber volume fraction with no voids (V_{f*}) and the actual void content (V_v).

$$V_f = V_f(1 - V_v) \tag{15.3}$$

$$V_f = \frac{w_f}{\rho_f} \tag{15.4}$$

$$\frac{w_f}{\rho_f} + \frac{w_m}{\rho_m}$$

15.4.4 COMPOSITE DENSITY MEASUREMENT

The density of the composite can be calculated using the following equation.

$$\rho_c = \rho_f \times V_f + \rho_m \times V_m \tag{15.5}$$

where ρ_c is density of composite, ρ_f is density of fiber, V_f is fiber volume fraction, ρ_m is density of matrix, and V_m is volume fraction of matrix.

15.4.5 COMPOSITE VOID CONTENTS

The properties of natural fiber composites depend on the properties of the individual components (i.e., fibers and matrix) as for any composite system. However, it is recognized that the inherent heterogeneous nature of NFRCs means that the interface region between fiber and matrix also plays an important role in defining the composite properties (Dhakal et al. 2009). The renewed interest in the use of natural fiber composites in primary structural applications has also resulted in attention being focused on the role of defects, in particular voids, in the reduction of the performance of these composite materials. With most synthetic composite materials, such as glass/polyester and carbon/epoxy, considerable knowledge has been built up over the years to control and optimize the fabrication process, and the void content is normally low (<2%) (Lystrup 1998). As far as void content in natural fiber composites is concerned, the fabrication techniques are not yet fully developed and the natural origin of the fiber component necessarily induces an element of variation into the composites; both factors contribute in the creation of voids making a noteworthy contribution to the overall composite volume in natural fiber composite systems (Madsen and Lilholt 2003).

15.4.6 Composite Volume Fractions

Composite volume fractions can be calculated using Equation 15.6. The volume fractions, assuming no void contents, were calculated using the following parameters:

$$V_f = \frac{w_f}{\rho_f}$$ (15.6)

$$\frac{w_f}{\rho_f} + \frac{w_m}{\rho_m} 2$$

Similarly, matrix volume fraction can be calculated using the rule of mixture equation assuming no void contents:

$$V_m = 1 - V_f$$ (15.7)

Composite volume fractions without considering void contents do not represent accurate results. Therefore, volume fractions with the consideration of void contents were calculated using Equation 15.8.

$$V_f = V_f(1 - V_v)$$ (15.8)

15.5 CHARACTERIZATION OF MOISTURE ABSORPTION

Despite the above-mentioned advantages, both synthetic (glass/carbon/aramid) and natural fiber–reinforced polymeric composite materials are susceptible to environmental exposure. Between these two, NFRCs are more vulnerable because of the hydrophilic nature of natural fibers resulting in low moisture resistance behavior. Therefore, aging is an important factor that must be considered while designing polymeric composite materials.

To understand the water sorption characteristics of natural fiber–reinforced polymeric composites, it is important to achieve an understanding of the water sorption mechanism in the reinforcing fibers themselves.

Mass transport of small molecules through polymeric materials is very complex, which is associated with the transport mechanisms related to the physical properties of the polymeric materials. The mass transport process in polymeric materials consists of three main steps as follows:

1. In the first step, small solvent molecules are first absorbed on the surface of the polymeric material.
2. The second step involves the diffusion of molecules through the polymer.
3. Finally, the solvent molecules desorb on the downstream surface of the polymer.

Moisture diffusion in polymeric composites has shown to be governed by three different mechanisms similar to polymeric matrix materials as explained earlier.

1. The first involves the diffusion of water molecules inside the microgaps between polymer chains.
2. The second involves capillary transport into the gaps and flaws at the interfaces between the fiber and the matrix.
3. The third involves transport of microcracks in the matrix arising from the swelling of fibers (particularly in the case of natural fiber composites).

Several diffusion models have been proposed for modeling the moisture diffusion behavior of natural fiber–reinforced polymeric composites. Among them, for steady-state one-dimensional (1D) diffusion, a model described by Fick's laws is frequently used, which is described as follows (Equations 15.9 and 15.10):

$$J = -D_x \frac{\partial c}{\partial x} \qquad (15.9)$$

$$\frac{\partial c}{\partial t} = -\frac{\partial J}{\partial x} \qquad (15.10)$$

In Equation 15.9 (Fick's first law), J is the flux of the material, c is the concentration of diffusion substance, D is the diffusion coefficient, x is the space coordinate along the direction of diffusion, and $\frac{\partial c}{\partial x}$ is the rate of change of the concentration of the diffusing material.

Fick's first law of diffusion is based on the following assumptions:

- The flux (J) through a unit area of material is proportional to the concentration (c) gradient measured normal to the material.
- The diffusion coefficient (D) is constant.

Case I or Fickian diffusion behavior: Water absorption is rapid and linear in the beginning, then slows, and approaches to saturation. Equilibrium is reached in relatively short time, and the sorption process is independent of the swelling kinetics (Errajhi et al. 2005; Shen and Springer 1999).

Fick's second law (Equation 15.10) is based on the following assumptions:

- The concentration (c) changes as a function of time (t) to the change in flux (J) with respect to position.
- The surface concentration attains its equilibrium value immediately upon a change in conditions and remains constant through the sorption process.

When moisture absorption curve fits the linear Fickian diffusion curve, as shown in Figure 15.4, the moisture absorption process can be considered Fickian diffusion. Subsequently, the maximum moisture content and mass diffusivity can be obtained using Equation 15.11.

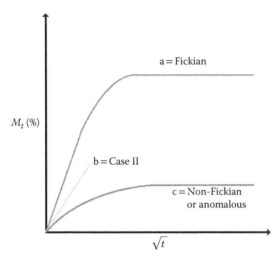

FIGURE 15.3   Different curves explaining different sorption mechanisms.

By integrating Equation 15.9 over time and space, the moisture uptake of the specimen when exposed to a constant temperature and humidity environment can be described as follows:

$$\frac{M_t}{M_\infty} = 1 - \sum_{n=0}^{\infty} \frac{8}{\left[(2n+1)\pi\right]^2} \exp\left[\frac{-D_x(2n+1)^2 \pi^2 t}{h^2}\right] \qquad (15.11)$$

where M_t = mass of absorbed moisture at time t, M_∞ is saturated mass of absorbed moisture, h is thickness of the sheet in dry condition, and D is the diffusion coefficient or diffusivity.

Case II: In some polymer systems, there will be some sharp boundaries, which move linearly with time, existing between the swollen and unswollen regions characterized as a "Case II" diffusion or sorption process. In this, the rate of diffusion and penetrant mobility are much greater compared to other relaxation processes. This sorption process is a strong function of swelling kinetics.

Non-Fickian or anomalous: In this, diffusion occurs when the penetrant mobility and segmental relaxation are comparable. Case I and Case II can be viewed as two extreme diffusion processes with anomalous diffusion lying between them. Representative curves for these three different sorption behaviors are presented in Figure 15.3.

The diffusion behavior can be classified mathematically from the shape of the initial portion of the absorption curve. If the mass uptake (M_t) can initially be represented by

$$M_t = k.t^n \qquad (15.12)$$

where t is the time, and k and n are the constants, then Fickian sorption corresponds to $n = 0.5$. Case II sorption is associated with $n = 1$. Anomalous sorption is characterized by $0.5 < n < 1$. k is the initial slope of the plot.

15.5.1 Moisture Sorption Measurements

The percentage of water absorption in the composites is calculated by the weight difference between the samples immersed in water and the dry samples using Equation 15.13.

$$M_t = \frac{W_t - W_o}{W_o} \times 100 \qquad (15.13)$$

where M_t is moisture uptake, and W_o and W_t are the mass of the specimen before and during aging, respectively.

The diffusion coefficient is an important parameter in Fick's law. Solving the diffusion equation for the weight of moisture, and rearranging in terms of the percent moisture content, the following relationship is obtained:

$$M_t = 4M_\infty \sqrt{\frac{D_x \cdot t}{\pi \cdot h^2}} \qquad (15.14)$$

Rearranging, and solving for the diffusion coefficient, D_x, or diffusivity for the initial stage of diffusion, constant slope

$$D_x = \frac{\pi}{16M_\infty^2}\left[\frac{M_t}{\sqrt{t/h}}\right]^2 \qquad (15.15)$$

where M_∞ is the equilibrium moisture content (maximum water uptake) of the sample. Using the weight-gain data of the material with respect to time, a graph of weight gain versus time is plotted.

The diffusion properties of composites described by Fick's law is evaluated by weight-gain measurements of the specimen immersed in water by considering the slope of the first part of the weight-gain curve versus the square root of time.

15.5.2 Factors Affecting Moisture Absorption Process

Moisture diffusion in natural fiber–reinforced polymer composites depends on many factors (Dhakal and Zhang 2012), which are briefly discussed in the following points:

1. Volume fraction of fiber: The work done by Dhakal et al. (2007a) has shown that as the fiber volume fraction increases, the moisture absorption percentage also increases. This behavior is more pronounced for natural fiber composites than conventional fiber–reinforced composites. The reason for this is that when composites are exposed to moisture, the hydrophilic natural fiber composite swells. As a result of swelling, microcracking occurs and the composites absorb more water.

2. Void content: During water immersion, water molecules first enter the free space of microvoids formed by cavities and cracks in the matrix, rapidly penetrate, and diffuse along the interface because of capillarity, increasing the weight of the sample.

3. Humidity and temperature: Higher temperature always accelerates the absorption process due to increased rate of diffusion. When the temperature of immersion is increased, the moisture saturation time is greatly reduced.

4. Nature of the resin system.

5. Fiber orientation.

6. Types of fibers: Prolonged immersion would induce chemical reaction between the water molecules and the glass fiber as well as the matrix causing some elements to leach or dissolve in the water; this would result in a decrease in weight.

7. Fiber-surface coating.

8. Sample thickness.

9. Interfacial bond: It is understood that water immersion for a long time will deteriorate the matrix, the reinforcing materials, and the interface leading to increased swelling and delamination.

10. Manufacturing process: The manufacturing process also has role to play in moisture absorption behavior. Void contents in composite materials largely depend on how it was manufactured.

15.5.3 MOISTURE ABSORPTION BEHAVIOR OF NATURAL FIBER–REINFORCED COMPOSITES

Cellulosic fibers are hydrophilic and absorb moisture. Lignocellulosic fibers are 3D, polymeric composites, made primarily of cellulose, hemicellulose, and lignin. Although all types of lignocellulosic fibers differ in chemical composition, within certain limits, all lignocellulosic fibers have very similar properties. That is, they all swell and shrink as the moisture content of the cell wall changes; they burn, they decay, and they all are degraded by acids, bases, and ultraviolet (UV) radiation.

The lignocellulosic structure of plant fibers changes dimensions with changing moisture content because the cell wall polymers contain hydroxyl and other oxygen-containing groups that attract moisture through hydrogen bonding. The hemicelluloses are mainly responsible for moisture sorption, but the accessible cellulose, noncrystalline cellulose, lignin, and surface of crystalline cellulose also play major roles. Moisture swells the cell wall, and the fiber expands until the cell wall is saturated with water (fiber saturation point, FSP). Beyond this saturation point, moisture exists as free water in the void structure and does not contribute to further expansion. This process is reversible, and the fiber shrinks as it loses moisture below the FSP.

Water absorption tests on maleic anhydride-modified wood fiber–reinforced composites (Marcovich et al. 2001) indicated that they were more hydrophobic than the unmodified ones. This was also shown by Saha et al. (2000) for acrylonitrile-modified jute fiber composites; water absorption was much reduced for cold and boiling water compared to unmodified jute fiber composites. Influence of water uptake on pineapple leaf–reinforced low-density polyethylene (LDPE) composites have been evaluated by George et al. (1998) in terms of fiber loading, temperature, and chemical treatment. It was found that water uptake has a major influence on the mechanical properties of the studied composite system.

15.5.4 EFFECTS OF WATER ABSORPTION ON THE PROPERTIES

In the course of their service life, PMCs are exposed to various environmental conditions. In such exposure, water diffuses through the fiber, matrix, and fiber–matrix interfaces. The diffusion of water into the composites can cause swelling, plasticization, and hydrolysis, profoundly affecting the physical, mechanical, and thermal properties as a result of weaker interfacial bonding between fibers and matrix.

All polymer composites absorb moisture in humid atmosphere and when immersed in water. When natural fibers are used as reinforcement for polymer composites, moisture can be a serious problem in terms of durability, swell, and stability of mechanical properties. Moisture causes natural fibers to swell, and after prolonged exposure to high humidity, rotting takes place through fungal infection.

Various studies on environmental effects on the properties of NFRCs have shown that moisture causes degradation and reduction in mechanical properties. Most NFRCs are very sensitive to the environment, such as acids, alkaline conditions, UV radiation, and high temperature. Both acidic and alkaline solutions can reduce the bending strength and modulus of NFRCs (Chow and Li 2003; George et al. 1998).

Rout et al. (2001) investigated the influence of fiber treatment on the performance of coir–polyester composites. In their study, coir fibers were subjected to various surface treatments including alkali treatment, bleaching, and vinyl grafting in an attempt to improve fiber–matrix adhesion. They also have reported in their study that all natural fibers are hydrophilic in nature and their moisture content can reach 3%–13%. They further suggested that fiber–matrix adhesion could be improved by modifying the surface topology of fibers by a suitable pretreatment. Their investigation claims that the extent of water absorption decreases considerably on surface modifications, which eventually lead to increased mechanical properties of the NFRCs.

Chow and Li (2003) studied the effect of moisture absorption on the mechanical properties of sisal fiber–reinforced polypropylene (PP) composites. They reported that water absorption has significantly weakened the sisal fiber/PP interface. This weak interface has contributed to lower tensile properties. They reported that impact strength of the composites was increased due to moisture uptake. However, they found that the water uptake period should be short to improve impact strength. They further outlined that longer immersion time could damage the interfacial bond due to the water absorbed, which leads to lower impact strength.

A detailed study on the effects of environmental conditions on mechanical and physical properties of flax fibers has been conducted by Stamboulis et al. (2001). They investigated the effect of upgrading treatment by comparing the moisture absorption and residual mechanical properties of upgraded Duralin and Green flax fibers. Their report concludes that Duralin fibers exhibit a somewhat higher and more uniform strength with less scatter showing the upgrading process absorbing less moisture, which retains mechanical properties, if not improved. They revealed that the average tensile strength of flax fibers changes with relative humidity as well as test duration.

15.6 CHARACTERIZATION OF MECHANICAL PROPERTIES

The structural reliability of composites depends on the interface between the matrix and the fibers. One of the most significant factors to be considered is its load-bearing capacity or mechanical properties. It is obvious that the reinforcement must be the stronger component if it is to give good mechanical properties. So, the reinforcement must have a higher elastic modulus. The bond between the matrix and the reinforcement (fibers) is critical, because good interfacial adhesion between the matrix and the fibers transfers the stress from the matrix to the fibers improving the mechanical properties of the composites (Bisanda and Ansell 1991).

15.6.1 Tensile Properties

In the tensile test, the tensile strength and the tensile modulus of the composite are estimated by interpolating between the stress/strain curves for a fiber and a matrix material. Tensile strength gives a measure of the ability of a composite material to withstand forces that pulls it apart and this determines to what extent the material stretches prior to breaking. The three main parameters investigated in standard tensile testing are the ultimate tensile strength, tangent modulus, and percentage strain to failure.

1. Ultimate tensile strength

$$\sigma = \frac{F}{b \cdot h} \tag{15.16}$$

   where
   σ = ultimate tensile strength in megapascal (MPa)
   F = maximum load in newton (N)
   b = initial average width of the specimen in millimeters
   h = initial average thickness of the specimen in millimeters

2. Tangent modulus

$$E_t = \frac{L_o}{b \times h} \frac{\Delta F_1}{\Delta Z_1} \tag{15.17}$$

   where ΔF_1 is the change in load in newton (N), E_t is the tangent modulus in megapascal (MPa), L_o is the gauge length in millimeters, b is initial average width of the specimen in millimeters, h is initial average thickness of the specimen in millimeters, and ΔZ_1 is increase in the distance between reference points corresponding to the difference in load ΔF_1 expressed in millimeters.

3. Percentage strain to failure

$$a = \frac{100.Z_r}{L_o} \tag{15.18}$$

where a is strain to failure in percent, Z_r is increase in the distance between reference point measured at failure and expressed in millimeters, and L_o is gauge length in millimeters.

15.6.2 Tensile Properties

Tensile testing of natural fiber composites is widely used to assess their properties and quality. Testing is an important part in the process of developing this type of composite material. From an engineering point of view, the three most important tensile properties are stiffness, ultimate stress, and strain to failure. These properties are measured with relatively short gauge lengths (150–250 mm, BS EN 2747: 1998, ASTM D 2256, ISO 2062).

To predict the tensile behavior of NFRCs, several models have already been developed (Davies and Bruce 1997; Gassan et al. 2001; Zeidman and Sawhney 2002). These models utilize important parameters such as cellulose content, primary and secondary walls, and spiral angle to calculate the elastic properties of natural fiber–reinforced polymeric composites. It is widely agreed that whatever the origin of the fiber is, the Young's modulus in the fiber axis decreases with the increase of spiral angle, and increases linearly with the cellulose content.

The work of Kuruvilla et al. (1993) involving sisal fiber–reinforced polyethylene composites produced by a hand-operated ram-type injection molding method indicates that the tensile properties of NFRCs depend on processing, fiber length, and fiber orientation. The tensile strength noted for approximately 20% and 30% w/w of randomly oriented sisal fiber is 12.5 and 14.7 MPa, respectively. Similarly, the flexural modulus recorded for 20% and 30% w/w of sisal fiber is 0.45 and 0.78 GPa, respectively. The tensile strength values recorded in their study are lower than that of hemp-reinforced polyester composites in this study. However, the tensile modulus obtained in their work for 30% w/w sisal fiber LDPE composite is (approximately 27% lower) in close agreement with the 28% w/w (0.15 fiber volume fraction) of hemp fiber values obtained by Dhakal et al. (2007a). Their report suggested that the hemp fiber–reinforced unsaturated polyester (HERUPE) composites show an increasing trend in tensile strength with the increase of fiber volume fraction. The introduction of 0.21 and 0.26 fiber volume fraction of hemp reinforcement into the unreinforced polyester increased the tensile strength by 207% and 241%, respectively.

The fractured surface shown by scanning electron microscopy (SEM) images confirmed that fiber matrix debonding with evidence of considerable delamination and fiber splitting and fiber breakage were the main dominant failure modes observed on these specimens in the tensile test (Figure 15.4).

It has been observed (Joseph et al. 1999a) that the tensile properties of solution-mixed randomly oriented sisal fiber–reinforced PP at 30% w/w of fiber has exhibited tensile strength of 33.84 MPa and Young's modulus of 0.94 GPa, respectively. These values are in close agreement with the tensile strength (48.65 MPa) and modulus (0.995 GPa) values obtained from 28% w/w of hemp-reinforced polyester composites by Dhakal et al. (2007a).

Studies by Chen et al. (1998) on bamboo-reinforced PP composites indicated that the tensile strength and modulus of composites increases linearly up to the

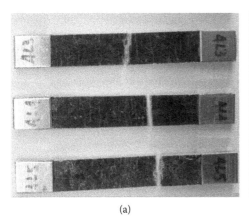
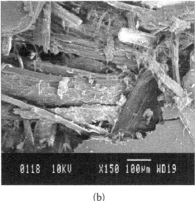

(a)   (b)

FIGURE 15.4 Tensile tested hemp/polyester composite specimen (a) and scanning electron microscopic image of fractured surface in tensile testing (b). (Adapted from Dhakal H.N., Zhang Z.Y. and Richardson M.O. 2007a. *Composites Science and Technology*, 67(7–8), with permission from Elsevier Ltd. Copyright 2007.)

fiber threshold weight fraction; above that value, the properties decrease even if the weight fraction of fiber is increased. The tensile strength and modulus for 50% w/w of bamboo/PP recorded in their study is 36 MPa and 3.6 GPa, respectively. The results are explained by the modified rule of mixtures equation, which is commonly applied in studying the tensile properties of composites with discontinuous short fibers:

$$\sigma_c = K V_f \sigma_f \frac{(1 - L_r)}{2L} + (1 - V_f)\sigma_m \qquad (15.19)$$

$$L_r = \frac{D\sigma_f}{2\tau} \qquad (15.20)$$

where σ_c is the tensile strength of the composite, K is an empirical fiber efficiency parameter, V_f is the volume fraction of the fiber, σ_f is the tensile strength of the fiber, σ_m is the tensile strength of the matrix at breaking strain of the fiber, L is the fiber length, L_r is the critical fiber length, D is the diameter of the fiber, and τ is the interface shear strength.

From Equations 15.19 and 15.20, it is apparent that the value of tensile strength increases with the increasing interface shear strength, when K, L, and D are assumed constant in all systems. This implies that the improvement of adhesion between hemp fiber and polyester matrix increases the tensile strength of the composites.

The structural reliability of composites depends on the interface between the matrix and the fibers; one of the most significant factors to be considered is its load-bearing capacity or mechanical properties. It is obvious that the reinforcement must be the stronger component if it is to give good mechanical properties. So, the reinforcement must have the higher elastic modulus. The bond between the matrix and the reinforcement (fibers) is critical, since good interfacial adhesion between the matrix and the fibers transfers the stress from the matrix to the fibers improving the mechanical properties of the composites (Bisanda and Ansell 1991).

On the basis of consideration of bond energies between atoms in the molecular structure of cellulose, the theoretical stiffness and ultimate tensile stress of crystalline cellulose loaded on the chain direction have been estimated to be in the range of 60–120 GPa and 12,000–19,000 MPa, respectively, for some plant fibers (Lilholt and Lawther 2000). These estimates can be thought of as upper limits. A number of structural aspects serve, however, to restrain the practical attainable tensile properties of plant fibers, for example, the degree of cellulose crystallinity, the microfibril angle (MFA), and the cellulose content.

The tilt angle of the cellulose fibrils with respect to the longitudinal cell axis is called MFA as shown in Figure 15.5. As can be recognized from the description of plant fiber structure, plant fibers themselves can be considered as composite materials with the stiff and strong cellulose microfibrils embedded in a hemicellulose lignin matrix. However, the composite structure in plant fiber is rather complex.

Table 15.2 presents typical reported tensile properties of important plant fibers. Tensile strength and stiffness for ramie fibers have been reported in the ranges of 345–1035 MPa and 61–128 GPa, respectively. It can be seen from Table 15.2 that the measured ultimate stress of plant fiber is much below the theoretical estimates.

This lower value for ultimate stress might be explained by the presence of fiber defects, which have been shown to affect the failure mechanisms of plant fibers (Eichhorn et al. 2001b). The measured stiffness values recorded for this particular fiber are close to the theoretical estimated values. The observed large variation in

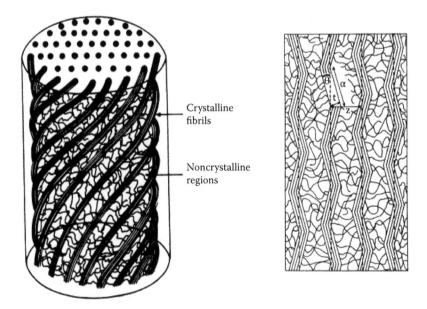

Crystalline
fibrils

Noncrystalline
regions

FIGURE 15.5 Microfibril angle. (Adapted from Bledzki A.K. and Gassan J. 1999. *Progress in Polymer Science*, with permission from Elsevier Ltd. Copyright 1999.)

TABLE 15.2
Spiral Angles, Cellular Contents, and Tensile Properties for Natural Fibers

Fiber	Spiral Angle (°)	Cellulose Content (wt.%)	Ultimate Stress (MPa)	Stiffness (GPa)
Ramie	8	69–83	400–938	61.4–128
Coir	30–49	43	175	11.0
Flax	6–10	64–71	345–1035	27.6
Hemp	6	57–77	300–800	30–60
Jute	8	61–72	393	26.5
Sisal	10–25	66–70	511–635	9.4–22.0
Softwood	3–50	40–45	100–170	10–50

Source: Data for microfibril angles are from Gassan J. et al. (*J. Mat. Sci.*, 36, 3715–3720, 2001); data for cellulose content and tensile properties are from Bledzki A.K and Gassan J. (*Prog. Polym. Sci.*, 24, 221–274, 1999).

the measured tensile properties is a typical trait for materials with natural origin, but some variation is also added by the experimental testing procedure. The measurement of tensile properties of single plant fiber is not a simple task, and problems are especially related to fiber gripping and determination of fiber cross-sectional areas.

Reinforcements (individual constituents) improve the mechanical properties such as strength, stiffness, and creep resistance of plastics—and their strength-to-weight and stiffness-to-weight ratios. In a fiber-reinforced polymer, the fibers serve as a reinforcement and, therefore, have to have a high tensile strength and stiffness, whereas the task of the matrix is to hold the fibers together, to transmit the shear forces, and to work as a coating. The material behavior of the matrices usually applied is characterized by a functional relationship of time and temperature, a considerably lower tensile strength, and a comparatively higher elongation. Therefore, the mechanical properties of the fibers determine the stiffness and tensile strength of the composites decisively. Generally, very thin fibers showing a large surface to volume ratio are used for a good adhesion of the fibers and the matrix.

It is evidenced from various literature that the tensile properties of natural fiber–reinforced polymeric composites (both thermoplastics and thermosets) are mainly influenced by the interfacial adhesion between the matrix and the fibers. Several chemical modifications are used to improve the interfacial matrix–fiber bonding resulting in the enhancement of tensile properties of the composites. In general, the tensile strengths of the natural fiber–reinforced polymer composites increase with fiber content, up to a maximum or optimum value, and then the value will drop. However, the Young's modulus of the natural fiber–reinforced polymer composites increase with increasing fiber loading. The results were able to support the previous findings for glass/epoxy composite, therefore corroborating the modulus findings of Moyeenuddin et al. (2011).

HFRUPE composites were subjected to water immersion tests to study the effects of water absorption on the mechanical properties. The tensile and flexural properties of water-immersed specimens subjected to both aging conditions were evaluated and compared alongside dry composite specimens. The percentage of moisture uptake increased as the fiber volume fraction increased due to the high cellulose content. The tensile and flexural properties of HFRUPE specimens were found to decrease with increase in percentage moisture uptake. Moisture-induced degradation of composite samples was significant at elevated temperature (Dhakal et al. 2007a).

As can be seen from Table 15.2, the microfibril or spiral angle (Figure 15.6) in wood fiber is more variable (3°–50°). Wood fibers, therefore, are suitable to study the correlation between fiber tensile properties and MFAs. The correlation between MFA and tensile properties can be realized clearly from Table 15.2. For angles below 10°, the tensile stress is in the range of 359.5–791.5 MPa and the stiffness is in the range of 36.37–60.52 GPa. For angles between 10° and 50°, the tensile stress and modulus are reduced to the range of 343–405 and 10.2–16.5 GPa, respectively. Similar correlation between MFA and tensile properties has been reported by Page et al. (1997). It is shown that at small angles (<5°), stiffness is in the range of 50–80 GPa, and at large angles (40°–50°) the stiffness is reduced to about 20 GPa. Also shown in the report is a theoretical model. The cell wall is modeled by a planar model of a homogenous and orthotropic material. The elastic constants of the wood fiber cell wall are estimated to be about 80 GPa for axial stiffness, 9 GPa for transverse stiffness, 7 GPa for shear stiffness, and 0.3 for Poisson's ratio. Other models, in which cell wall geometry and structure are more correctly modeled, are proposed to predict the tensile properties of plant fibers.

In Gassan et al. (2001) two structural models are presented to predict elastic properties of natural fibers. In model A, three cell wall layers are taken into account by using a laminated plate model where a natural fiber cell is considered as antisymmetrical laminated structure. In model B, the elliptic geometry and the hollow structure of the cross section of the fiber cell (central lumen) as shown in Figure 15.2 (Gassan et al. 2001) are further taken into account by using a thick laminated tube model. The report suggests, based on both models, that the elastic properties are dependent on spiral angle, cellulose content, and fiber cross section. The report concludes that, in general, modulus in fiber axis decreases with increasing spiral angle as well as the degree of anisotropy, whereas shear modulus reaches a maximum value for a spiral angle of 45°, and fiber cell modulus increases with increasing cellulose content.

(a)   (b)

FIGURE 15.6 Flexural tested hemp composite specimen (a) and scanning electron microscopic image of fractured surface intensile testing (b).

The cell wall of a fiber is made up of a number of layers: the so-called primary wall (the first layer deposited during cell development) and the secondary wall (S), which again is made up of three layers (S1, S2, and S3). In all lignocellulosic fibers these layers contain cellulose, hemicellulose, and lignin in varying amounts (Table 15.2). The individual fibers are bonded together by a lignin-rich region known as the middle lamella. Cellulose attains its highest concentration in the S2 layer (about 50%), and lignin is most concentrated in the middle lamella (about 90%), which, in principle, is free of cellulose. The S2 layer is usually, by far, the thickest layer and dominates the properties of the fibers.

15.6.3 FLEXURAL PROPERTIES

With the increased use of engineering polymer for metal substitution, a greater emphasis has been placed on presenting the mechanical properties of composite materials in a format similar to that used for metals. Although tensile strength and modulus are important parameters, design applications frequently involve a bending rather than a tensile mode. As a result, flexural strength and modulus are frequently quoted.

Flexural strength is the ability of the composite materials to withstand bending forces applied perpendicular to its longitudinal axis. The flexural strength is equivalent to the modulus of rupture.

The flexural strength and modulus values are normally calculated from Equations 15.21 and 15.22.

Flexural stress:

$$\sigma_f = \frac{3FL}{2bh^2} \tag{15.21}$$

Flexural modulus:

$$E_f = \frac{L^3}{4bh^3} \frac{\Delta F}{\Delta d} \tag{15.22}$$

In Equations 15.21 and 15.22, σ_f is the flexural stress in megapascals, F is the force applied in newtons, L is the span in millimeters, b is the width of the specimen in millimeters, h is the thickness of the specimen in millimeters, E_f is the modulus in megapascals, ΔF is a chosen difference in force in newtons, and Δd is the difference in deflection corresponding to the difference in force ΔF in millimeters.

Flexural properties of natural fiber–reinforced polymer composites have been widely reported. Hemp fiber–reinforced polyester composites were prepared using the resin transfer molding (RTM) technique and the flexural and impact behavior investigated (Sebe et al. 2000). The study reports that flexural stress at break and flexural modulus showed an increasing trend with the introduction of hemp fiber as reinforcement into polyester matrix. A strong interfacial adhesion between hemp and polyester was reported due to chemical modification of hemp fiber. The modification consisted of introducing reactive vinylic groups at the surface of the fibers, via esterification of hemp hydroxyl groups, using methacrylic anhydride.

Natural fiber–reinforced polyester composites were evaluated for both strength performance and cost factors (d'Almeida 2001). The report showed that the flexural strength of the composite fabricated with untreated chopped natural fibers is comparable only to the performance of low-strength wooden agglomerates and plywood. However, it is reported that on a cost basis the composite fabricated with high-strength natural fibers can even compete with glass fiber/mat polyester composites.

Flexural strength of the composites was found to decrease with increased fiber content; however, flexural modulus increased with increased fiber content. The reason for this decrease in flexural strength could be due to fiber defects that could induce stress concentration points in the composites during flexural test, accordingly decreasing flexural strength. Alkali and silane fiber treatments were found to improve flexural strength and flexural modulus, which could be due to enhanced fiber/matrix adhesion.

The flexural properties investigated by Dhakal et al. (2007a) on nonwoven hemp fiber–reinforced polyester laminates noted approximately for 15, 21, and 26 fiber volume fraction of hemp fiber were 83, 94, and 110 MPa, respectively. Similarly, the flexural moduli recorded for similar fiber volume fraction of hemp fiber were 5.34, 7.30, and 6.49 GPa, respectively. Both the flexural strengths and the modulus values recorded in their study were in excellent agreement with the corresponding values obtained by the work carried out by Sebe et al. (2000) on untreated hemp fiber–reinforced polyester composites produced by the RTM method.

The SEM image of fractured surface in flexural test for hemp polyester samples as shown in Figure 15.6 clarifies that the matrix cracking, delamination, fiber pull-out, and fiber breakage were the main failure modes observed for the specimens.

15.6.4 IMPACT PROPERTIES

Low-velocity impact of fiber-reinforced plastics has been the subject of many experimental and analytical investigations (Bogdanovich and Friedrich 1994; Dhakal et al. 2007b; Naik and Sekher 1998). Susceptibility to low-velocity impact damage of composite materials has been well documented (Richardson and Wisheart 1996). One of the problems with the structural applications of the fiber-reinforced composites, particularly NFRCs, is the response to impact damage. Such damage may be caused by bumps or crashes and falling objects and debris.

One of the important aspects of the behavior of natural plant fiber–reinforced polymeric composites is their response to an impact load and the capacity of the composites to withstand it during their service life. Many natural fiber–reinforced composite materials have been found to be very sensitive to impact loading. In the broader context, assessing impact resistance of a composite material is always difficult since the damage manifests itself in different forms such as delamination at the interface, fiber breakage, matrix cracking, and fiber pull-out. Because of their complexity, many of their characteristics still remain unresolved (Wisheart and Richardson 1999). There are two main types of impact test methods, low-velocity impact test and high-velocity impact test. Normally, with the low-velocity impact test, the composite is not damaged fully and is still capable of performing its primary function, whereas with the high-velocity impact test, the composite is completely ruptured or penetrated by the striker (Thanomsilp and Hogg 2003).

The impact process may involve relatively high contact forces acting over a small area for a short duration. Local strains generated at the point of contact between the two solids result in the absorption of energy.

When a projectile strikes a laminated composite, fracture processes such as delamination, matrix cracking, and fiber fracture frequently occur. Observing the nature of impact damage through the section of the specimen, it appears to be approximately conical. The pattern of the crack growth can be identified as a result of contact stress in the thick and short span plate, or as a result of deformation in the thin and long span plate; the damage has been, therefore, often referred to as pine or reverse pine tree (Cantwell 1985).

Impact properties of unidirectionally aligned continuous natural fiber composite with polyester matrix along with that of randomly oriented short fiber composites are reported by Biswas et al. (2005). They have reported that the impact properties of these fibers, determined by the Charpy impact test with fiber volume fraction of 0.5, are very high (98 KJ/m²) for sisal fiber, which is comparable to that of glass fiber–reinforced composites. Santulli and Cantwell (2001) investigated impact properties of untreated jute fabric–reinforced polyester laminates in comparison with different E-glass fiber–reinforced laminates. The report suggests that jute fiber–reinforced polyester composites show damage at relatively low load during impact tests, providing nevertheless a sufficient damage tolerance.

Dhakal et al. (2007b) investigated the impact properties of untreated HFRUPE laminates in comparison with different E-glass fiber–reinforced laminates (Figure 15.7). The report suggests that hemp fiber–reinforced polyester composites show damage at relatively low load during impact tests, providing nevertheless a sufficient damage tolerance. The impact damage pattern confirmed that the higher the fiber weight fractions, the greater the residual velocity, hence, the higher the impact energy dissipation by the impacted hemp unsaturated polyester specimens (Figure 15.8).

Incorporation of sisal fiber into thermosetting composites has been reported by various researchers. Joseph et al. (1993) and Paramasivam and Abdulkalam (1974) have investigated the feasibility of developing polymer-based composites using sisal fiber. They have reported the impact properties of oriented sisal fiber–polyester

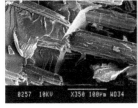  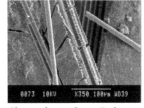

Four-layered hemp composite   Five-layered hemp composite   Chopped strand mat E-glass

FIGURE 15.7 Scanning electron microscopic image of fractured surfaces of impacted hemp and E-glass composites showing matrix cracking, delamination, and fiber breakage. (Adapted from Dhakal H.N., Zhang Z.Y. and Richardson M.O. 2007a. *Composites Science and Technology*, 67(7–8), with permission from Elsevier Ltd. Copyright 2007.)

Front face   Back face
Four-layered hemp   Four-layered hemp

Front face   Back face
five-layered hemp   five-layered hemp

FIGURE 15.8 Impact damage characteristics of four- and five-layered hemp composite specimens. (Adapted from Dhakal H.N., Zhang Z.Y. and Richardson M.O. 2007a. *Composites Science and Technology*, 67(7–8), with permission from Elsevier Ltd. Copyright 2007.)

composites. Unidirectionally aligned sisal fiber–polyester composites containing 0.5 volume fraction of sisal fiber were prepared from unsaturated polyester prepregs. Impact strength was measured by the Charpy test in a pendulum impact-testing machine using a pendulum load of 0.4 kg. They have compared the work of fracture of sisal fiber–polyester composites to those of composites containing other natural fibers. Among sisal, pineapple, and banana fiber–reinforced polymer composites, sisal fiber–reinforced polyester composite is likely to give high work of fracture because of the high toughness of sisal fiber.

Pavithran et al. (1988) have compared the impact properties of unidirectionally oriented sisal fiber–polyester composites with those of composites having ultrahigh-density polyethylene and glass fibers. It was observed that sisal composites show work of fracture identical with that of ultrahigh-modulus polyethylene composites, and the toughness of sisal fiber composites is only 25% less than that of glass fiber–reinforced composites when the density of the latter is taken into account.

Most researchers, nevertheless, report an improvement in toughness, often measured as impact strength, over the unreinforced resin alone with the addition of natural fiber (Karus 2004; Wambua et al. 2003), and furthermore, with some fiber treatments additional improvements can be obtained. For example, studies of flexural strength and flexural modulus of chemically treated random short and aligned long hemp fiber–reinforced polylactide and unsaturated polyester composites were investigated over a range of fiber content (0–50 wt.%).

Although natural fiber–reinforced thermosetting PMCs compare well in terms of stiffness with glass fiber–reinforced materials, their toughness is, in general, significantly poor (Dissanayake et al. 2008; Eichhorn et al. 2001a). The fracture of epoxy was reported to be between 0.1 and 0.3 kJ/m^2, yet, when reinforced with glass fiber to form a composite, fracture energies of between 40 and 100 kJ/m2 can be achieved (Bolton, 1995).

15.6.5 FRACTURE TOUGHNESS BEHAVIOR

Evaluation of fracture behavior, that is, resistance to fracture of NFRCs, is an important area that has not been explored so well in literature.

There are several methods for determining the fracture toughness theoretically, experimentally, and numerically. The stress intensity factor is a very important property in fracture criteria; it depends on the state of applied load and material configuration. The Mode I and II stress intensity factors can be computed (Gamstedt et al. 1999). The axial strength of a continuous fiber composite can be predicted from properties of fiber and matrix when tested in isolation.

Numerous attempts have been made to characterize Mode I fracture behavior of glass/carbon fiber–reinforced polymer composites invented in 1938 by Russel Games; mostly beam type specimens have been investigated. Young's modulus was found experimentally across the fiber orientation (transversally and longitudinally) for the computed tomography specimen to study its fracture toughness. The results revealed that the cracked specimens are tougher along the fiber orientations as compared with across the fiber orientations. The CT specimen is shown to be suitable only to investigate the initiation damage due to the insufficient ligament length for further testing (Wambua et al. 2003).

The work carried out by Pande and Sharma 1984 suggests that the toughness of particulate and fibrous composites has shown a tendency to increase with the increase in weight fractions of reinforcement; in the case of hybrids of particulates and fibers, however, it decreases with the increase in reinforcement. At lower weight fractions of reinforcement, the toughness appears to be a weighted sum of individual toughness of particulate and fibers, but at higher interspersion of weight fractions of reinforcement, a negative synergistic effect was reported.

Other research carried out into the fracture toughness of glass/epoxy composites showed that the fracture toughness of materials can be measured by the critical energy release rate GIC for pure Mode I and GIIC for pure Mode II (Benzeggagh and Kenane 1996). The loading in their test was a simple combination of the double cantilever beam (Mode I) and the end notch flexure (Mode II). The results indicated that the critical Mode I energy release rate for delamination increased monotonically with increasing Mode II loading.

The work done by Benzeggagh and Kenane (1996) and Mathews and Swanson (2007) suggests that the critical Mode I energy release rate for delamination decreases monotonically with increasing Mode II loading. This is in contrast to particular results presented in the literature (Benzeggagh and Kenane 1997).

The delamination crack growth rate was predicted through the Paris power law. The propagation in Mode II was unstable but in Mode I it was stable. The measured crack growth rate data were correlated with the corresponding strain-energy release rate. The GII/GT ratio varied from 0% to 100%; in each investigated mode ratio the results were fitted to the Paris power law. Finally, a semiempirical mixed-mode fatigue criterion was proposed to predict the coefficients of the Paris power law (Pugno et al. 2006).

Natural fibers are characteristically discontinuous and tend to be highly twisted to obtain sufficient strength from the relatively short fibers, as Summerscales et al. (2010) found that the strength of fibers reduced with increasing fiber length. However, highly twisted, this has been noted to be inappropriate for composite reinforcement as enhanced mechanical strength and stiffness arises when the fibers are aligned in the direction of the load. The twisted nature of the yarns also poses problems in impregnating the fibers with viscous thermoplastics such as PP and poly-(lactic acid) (PLA).

Liu and Hughes (2008) investigated epoxy matrix composites reinforced with woven flax fiber textiles having fiber volume fractions greater than 30%, prepared by a vacuum infusion process. Both fracture behavior and K1c were found to be strongly dependent on the linear density of the weft yarn and the direction of crack propagation with respect to the orientation of the textile; however, the fracture toughness was independent of the influence of the weave.

15.6.6 Effect of Manufacturing Methods on the Properties

The quality of polymer composite structures is highly dependent on materials selection and process control (Cullen et al. 2008). The properties of natural fiber–reinforced polymer composites are generally governed by the pretreated process of fiber; the manufacturing process of the composite fiber packing in a composite depends on the method of manufacturing, and fiber packing has been greatly attributed to fracture behavior characteristics. The various composite manufacturing techniques commonly used are hand lay-up, compression molding, resin transfer molding, pultrusion, and so on.

Hand lay-up is the oldest and simplest method of manufacturing composites. The tools required for the process are a mold to accommodate dry manufacturing according to the desired shape and a roller to facilitate uniform distribution of resin. Virtually any size composite can be manufactured using this method. This method is the cheapest method of manufacturing, but it has some disadvantages such as long curing time, low production rate, and, further, the quality of the composite depends on the skill of the worker.

The compression molding method is commonly used for sheet molding or bulk molding. The compression molding machine consists of male and female dies or platens to form the mold. The reinforcement combined with resin is placed in the mold and a hydraulic press is used to apply high pressure by closing the male and female halves of the mold. After the material is cured, the pressure is released and the part is removed from the mold. Exterior body panels for structural members such as automobile bumpers are widely manufactured using this method.

The varying molding methods can produce altered values for the interlaminar fracture energies gained for specimens (Thomason and Adzima 2001). Hand lay-up and RTM can be used in conjunction. Specimens formed by hand lay-up can cause problems during fracture testing; cracks may deviate from the original crack plane invalidating the fracture mechanics approach, a lack of reinforcement symmetry may cause twisting and mixed-mode loading, and the introduction of voids can lead to lower interlaminar fracture energies (Stamboulis et al. 2001). These problems may be resolved by producing specimens via RTM, which tend to show higher interlaminar fracture energies than those produced by hand lay-up. The report by Sumpter suggested that interlaminar fracture energy can be a function of both specimen geometry and manufacturing method. For example, hand lay-up specimens showed a reduction in GIC with increasing thickness while RTM specimens showed the opposite trend. The RTM materials exhibit twice the GIC of the hand lay-up materials ($\gg$ 1200 J m^{-2}, compared to $\gg$ 650 J m^{-2} for hand lay-up).

The hemp fiber composites manufactured with the RTM process were found to have a very homogenous structure with no noticeable defects. The mechanical properties of these materials were found to increase with increasing fiber content (Rouison et al. 2006). However, these properties were much lower than those of a glass fiber composite of comparable fiber content. In addition, the flexural creep measurements showed substantial deformation of the natural fiber composites when they were under high-load fatigue conditions.

The work carried out by Peng et al. (2011) focuses on the mechanical and morphological characterization of the pultruded composite rods made from hemp and wool fiber reinforcements. The results showed that the composite using a polyurethane resin system has higher specific tensile and compressive strengths as well as Young's and compressive moduli compared with the polyester and vinyl ester composites, whereas the polyester composite exhibits better flexural strength. The SEM observation of the cross section and layered polyester and polyurethane composites has found defects on the interface of fibers and resin matrix unlike the RTM hemp specimens (Rouison et al. 2006).

The differences in the value of the stress intensity factor can also be put down to the style of fabrication. Results have found that there is 15% and 24% error in estimation of KI_C. This discrepancy in estimated magnitudes of stress intensity factor was attributed to varied loading condition and method of composite fabrication (Gouda et al. 2011).

Other parameters in composite manufacturing can contribute to the fracture behavior; Saidpour et al. (2003) conducted Mode-II tests on unidirectional carbon/epoxy composites, and reported higher GIIC values of interlaminar fracture energy than Mode I (Joseph et al. 1999b). The GIIC values were well above 1000 Jm^{-2} for composites molded at higher temperature compared with medium- and low-temperature molding composites. They reported that after the initial cure, GIIC values were fairly low for medium-temperature molding composites. However, GIIC values for these materials increased significantly at 200°C after cure, reaching similar values to those obtained for high-temperature molding composites. Saidpour et al. (2003) also reported that postcuring conditions had significant effect on fracture toughness energy giving similar values for high-temperature molding due to better phase separation for medium- and low-temperature molding systems.

The selection criteria are highly dependent on their type, application, and cost. However, there is still uncertainty as to which type of manufacturing process is suitable for producing these composites as their materials and mechanical characteristics are different as compared with traditional carbon and glass fiber composites in general. The original design for some processes was not targeted for natural fibers, whereas their technologies have been well developed for fast and reliable composite production; also fibers may suffer mechanical damage during mechanized harvesting (Summerscales et al. 2010). Therefore, manufacturing processes are vital in ensuring that fiber properties are maintained.

Twin-screw extruders have been widely used for the fabrication of NFRCs. Wang et al. (2001) conducted an experimental study on the influence of processing parameters on the properties of wood flour-filled polyethylene using corotating twin-screw extruders. The study outlines that a suitable combination of processing variables including screw speed, screw configuration, throughput rate, and barrel temperatures were necessary to limit thermal degradation and darkening of the filler. A similar process was used to produce rice hull–reinforced recycled high-density polyethylene (Charlton et al. 2000). In this work, the materials were extruded using a vented screw extruder, and the rheological properties were determined using a parallel plate rheometer. They reported that the addition of rice hulls resulted in a large increase in the dynamic shear properties, and extrudate tearing occurred at all extrusion rates, increasing with increasing throughput and decreasing temperature.

A study on the effect of filler size and the composition on the mechanical properties of wood/PP composites by Albano et al. (1999) was conducted using an injection molding method. The high-fiber weight fraction blends were prepared in an intermeshing co-rotating twin-screw extruder and injection molded. Hemp fiber–reinforced polymer composites were fabricated using injection molding to replace chopped strand mat-reinforced polymer composites to be used in the parcel shelf and high roof for Ford Transit van. Research by automotive industries into plant fiber composites seems to focus on mainly two processing routs: injection molding for nonstructural parts and RTM for semistructural parts.

15.7 OPTIMIZATION OF FIBER AND MATRIX PROPERTIES

15.7.1 FIBER MODIFICATION

From the early development of polymer composites, the optimization of the interface has been of pivotal importance. In conventional man-made fiber composites such as glass and carbon, various methods of modifying the fiber surface, such as silane sizings, electrochemical oxidation, and so on, have been found to be very successful in controlling the interface (Thomason and Adzima 2001).

Because of the heterogenous and inconsistent chemical structure of plant fiber, it is always difficult to control fiber/matrix compatibility of plant fiber composites. However, a whole range of methods have been developed to enhance fiber/matrix compatibility and these methods are useful in plant fiber composites (Zafeiropoulos et al. 2002). The thermal characteristics, crystallinity index, reactivity, and surface morphology of untreated and chemically modified natural fibers were investigated by Mwaikambo and Ansell (2004). In their study, hemp, sisal, jute, and kapok fibers were subjected to alkalization using sodium hydroxide. The result showed that alkalization modifies plant fibers promoting the development of fiber resin adhesion, which then will result in increased interfacial energy and, hence, improvement in the mechanical and thermal stability of NFRCs. The alkalization effects on natural fibers are of particular importance for fiber–matrix adhesion and the creation of high-fiber surface area required for the optimization of fiber-resin reinforcement. The modification of cellulose fibers, therefore, develops into hydroxyl groups. These changes will effectively result in improved surface tension, wetting ability, swelling, adhesion, and compatibility with polymeric materials. The work of Eichhorn et al. (2001a) has revealed that hemp fiber appears to have the highest crystallinity index and thermal stability when alkalized and acetylated with and without acid catalyst followed by jute, sisal, and kopak fibers. In their report, they have outlined that kapok fiber has the highest reaction affinity to chemicals followed by jute, sisal, and hemp fibers. The chemical modification of hemp fiber has been investigated by Sebe et al. (2000). They have found that the interfacial properties of hemp and polyester were enhanced by the chemical modification of fiber.

Investigation on the effect of various anhydride modifications of plant fibers and their effect on the mechanical properties and water absorption of composites have been carried out by Khalil and Ismail (2001). The report suggests that fiber modification has enhanced the mechanical properties such as tensile strength and modulus of the composites. The reason for this has been that the modified fiber is more hydrophobic, exhibiting better fiber–matrix adhesion.

The effect of treatments with other chemicals, for instance, sodium alginate and sodium hydroxide, has been reported for coir, banana, and sisal fibers by Bledzki and Gassan (1999). The treatment resulted in an increase in debonding stress and thus improved the ultimate tensile strength up to 30%. Based on these past works, the proposed research study aims to further investigate the effect of these various fiber surface treatments toward the enhancement of fiber/matrix interface.

15.7.2 MATRIX PROPERTIES ENHANCEMENT

Most of the reported studies on natural fiber composite involved the study of mechanical properties as a function of fiber content, effect of various treatments on the mechanical properties, prediction of modulus and strength using models, and comparison with experimental data. However, there are very few reports dealing with the improvement of matrix and other novel methods of enhancing overall matrix properties. One way of enhancing the properties of matrix is by hybridizing nanoparticles into polymer matrix.

15.7.3 FIBER/POLYMER MATRIX INTERFACE AND INTERPHASE OPTIMIZATION

To optimize the performance of the NFRCs, effective utilization of the fiber reinforcement properties is essential and it is predicted by the rule of mixtures to be a function of the fiber length and orientation with respect to the tensile strength of the composites.

15.8 FUTURE TRENDS

Serious environmental as well as economical problems caused by the use of nonrenewable (synthetic fibers) reinforcements in composites have led to the search for more ecologically friendly and sustainable materials such as natural fibers as reinforcements.

Nanoparticles added to hybrid composites will also be vital in the near future due their property-modifying characteristics. Silva et al. (2012) reported that the addition of silica microparticles into the matrix phase led to improvement in various natural fiber–reinforced polymeric composites. Similar effects of hybrid composites have been reported by Jawaid and Khalil (2011) and Hariharan and Khalil (2005).

Development of biocomposites (both resin and matrix degradable) as substitute materials to glass and carbon fiber–reinforced composites in the context of sustainable materials development for engineering applications is another area that deserves further investigation.

15.9 CONCLUSIONS

Properties and characterization of various natural fiber–reinforced polymeric composites have been addressed in this chapter. The investigation and the review of various literature suggests that NFRCs have great potential for use in composite structures in load-bearing applications. However, it is important that the outstanding issues related to natural fiber composites and biocomposites such as their susceptibility to moisture absorption and other important properties such as mechanical, thermal, and surface properties characterization are well understood.

REFERENCES

Albano, C., Ichazo, M.N., Gonzalez, J., Molina, K. and Espejo, L. 1999. On the properties of injection moulded polypropylene and polypropylene-wood flour composites. In: *Proceedings of the Antec'99 Conference Volume III.* New York.

Andersen, T.L. and Lilholt, H. 1999. Natural fibre composites: Compaction of mats, press consolidation and material quality. In: *Proceedings of the 7th Euro-Japanese Symposium.* Paris, France, 1–12.

Benzeggagh, M.L. and Kenane, M. 1996. Measurement of mixed-mode delamination fracture toughness of unidirectional glass/epoxy composites with mixed-mode bending apparatus. *Composites Science and Technology* 56: 439–449.

Benzeggagh, M.L. and Kenane, M. 1997. Mixed-mode delamination fracture toughness of unidirectional glass/epoxy composites under fatigue loading. *Composites Science and Technology* 57: 597–605.

Berman, H. 1939. A torsion microbalance for the determinations of specific gravities of minerals. *American Mineralogist* 24: 434–440.

Bisanda, E.T.N. and Ansell, M.P. 1991. The effect of silane treatment on the mechanical and physical properties of sisal-epoxy composites. *Composites Science and Technology* 41: 165–178.

Biswas, S., Sriknath, G. and Nangia, S. 2005. Development of natural fibre composites in India. News and Views, Technology Information, Forecasting and Assessment Council (TIFAC), Delhi, India.

Bledzki, A.K. and Gassan, J. 1999. Composites reinforced with cellulose base fibres. *Progress in Polymer Science* 24: 221–274.

Bledzki, A.K., Reihmane, S., and Gassan, J. 1996. Properties and modification methods for vegetable fibres for natural fibre composites. *Journal of Applied Polymer Science* 59: 1329–1336.

Bledzki, A.K. and Gassan, J. 1997. Natural fibre reinforced plastics. In: *Handbook of Engineering Polymeric Materials.* Ed. Cheremisinoff, N.P. New York: Marcel Dekker.

Bledzki, A.K., Sperber, V.E. and Faruk, O. 2002. *Natural and Wood Fibre Reinforcement in Polymer.* U.K. Rapra Review Reports.

Bogdanovich, A.E. and Friedrich, K. 1994. Initial and progressive failure analysis of laminated composites structures under dynamic loading. *Composites Structures* 27: 439–456.

Bolton, J. 1995. The potential of plant fibres as crops for industrial use. *Outlook on Agriculture* 24: 85–89.

Cantwell, W.J. 1985. Impact Damage in Carbon Fibre Composites, PhD thesis, University of London, London.

Charlton, Z., Vlachopoulos J. and Suwanda, D. 2000. Profile extrusion of highly filled recycled HDPE. In: *Proceedings of the Antec 2000 Conference.* Orlando, FL.

Chen, X, Guo, Q. and Mi, Y. 1998. Bamboo fibre-reinforced polypropylene composites: A study of the mechanical properties. *Journal of Applied Polymer Science* 69: 1891–1899.

Chow, C.P.L. and Li, R.K.Y. 2003. A study on the mechanical properties of sisal fibre reinforced polypropylene composites after moisture absorption. In: *Proceedings of the 3rd International Conference on Eco-Composites.* Queen Mary University of London, London, U.K., 1–12.

Cullen, R., Grove, S.M., Vronsky, T., Summerscales, J. and Dhakal, H. N. 2008. Flow convergence and void formation in resin-infused cored sandwich structures. In: *Proceedings of the 9th International Conference on Flow Processing in Composite Materials.* Quebec, Canada.

d'Almeida, J.R.M. 2001. Analysis of cost and flexural strength performance of natural fibre-polyester composites. *Polymer Plastic Technology and Engineering* 40: 205–215.

Davies, G.C. and Bruce, D.M. 1997. A stress analysis model for composite coaxial cylinders. *Journal of Materials Science* 32: 5425–5447.

Deng, Q., Li, S. and Chen, Y. 2012. Mechanical properties and failure mechanism of wood cell wall layers. *Computational Materials Science* 62: 221–226.

De Rosa, I.M., Dhakal, H.N., Santulli, C., Sarasini, F. and Zhang, Z.Y. 2012. Post-impact static and cyclic flexural characterisation of hemp fibre reinforced laminates. *Composites Part B* 43: 1382–1396.

Dhakal, H.N. and Zhang, Z.Y. 2012. Polymer matrix composites: Moisture effects and dimensional stability. In: *Wiley Encyclopedia of Composites,* Vol. Set 5, 2nd edition. Nicolais L. and Borzacchiello A., Eds. 2179–2185. Hoboken, New Jersey: John Wiley & Sons. ISBN (printed set): 978-0-470-12828-2.

Dhakal, H.N., Zhang, Z.Y., Bennett, N. and Reis, P.N.B. 2012. Low-velocity impact response of non-woven hemp fibre reinforced unsaturated polyester composites: Influence of impactor geometry and impact velocity. *Composite Structures* 94: 2756–2763.

Dhakal, H.N., Zhang, Z.Y. and Richardson, M.O.W. 2009. Creep behaviour of hemp fibre reinforced unsaturated polyester composites. *Journal of Biobased Materials and Bioenergy* 3: 232–237.

Dhakal, H.N., Zhang, Z.Y. and Richardson, M.O. 2007a. Effect of water absorption on the mechanical properties of hemp fibre reinforced unsaturated polyester composites. *Composites Science and Technology* 67(7–8): 1674–1683.

Dhakal, H.N., Zhang, Z.Y., Richardson, M.O.W. and Errajhi, O.A.Z. 2007b. The low velocity impact response of non-woven hemp reinforced unsaturated polyester composites. *Composites Structures* 81: 559–567.

Dissanayake, N.P.J., Dhakal, H.N., Grove, S.M., Singh, M.M. and Summerscales, J. 2008 Optimisation of energy use in the production of flax fibre as reinforcement for composites. In: *Proceedings of the International Conference on Flax and Other Bast Plants.* Saskatoon, Canada.

Eichhorn, S.J., Baillie, C.A., Zafeiropoulos, N., Maikambo, L.Y., Ansell, M.P., Dufresne, A., Entwistle, K.M. et al. 2001a. Review. Current international research into cellulosic fibres and composites. *Journal of Materials Science* 36: 2107–2131.

Eichhorn, S.J., Sirichaist, R.J. and Young, R.J. 2001b. Deformation mechanisms in cellulose fibres, paper and wood. *Journal of Materials Science* 36: 3129–3135.

Eichhorn, S.J. and Young, R.J. 2004. Composite micromechanics of hemp fibres and epoxy resin microdroplets. *Composites Science and Technology* 64: 767–772.

Errajhi, O.A.Z., Osborne, J.R.F., Richardson, M.O.W. and Dhakal, H.N. 2005. Water absorption characteristics of aluminised e-glass fibre reinforced unsaturated polyester composites. *Composite Structure* 71: 333–336.

Faruk, O., Bledzki, A.K., Fink, H.P. and Sain, M. 2012. Biocomposites reinforced with natural fibres: 2000–2010. *Progress in Polymer Science* 37: 1552–1596.

Gamstedt, E.K., Berglund, L. A. and Peijs, T. 1999. Fatigue mechanisms in uni-directional glass-fibre reinforced polypropylene. *Composites Science and Technology* 59: 759–768.

Gassan, J. and Bledzki, A.K. 1998. Possibilities for improving the mechanical properties of jute/epoxy composites by alkali treatment of fibres. *Composites Science and Technology* 50: 1303–1309.

Gassan, J., Chate, A. and Bledzki, A.K. 2001. Calculation of elastic properties of natural fibres. *Journal of Materials Science* 36: 3715–3720.

Gayer, U. and Schuh, T.G. 1996. Automotive application of natural fibres composites. In: *Proceedings of the First International Symposium on Lignocellulosic Composites*. UNESP-Sao Paulo State University, Brazil.

George, J., Bhagawan, S.S. and Thomas, S. 1998. Effects of environment on the properties of LDPE composites reinforced with pineapple-leaf fibre. *Composites Science and Technology* 58: 1471–1485.

Gouda, P.S.S., Kudari, S.K., Prabhuswamy, S. and Jawali, D. 2011. Fracture toughness of glass-carbon (0/90)s fibre reinforced polymer composite - an experimental and numerical study. *Journal of Minerals & Materials Characterization & Engineering* 10(8): 671–682.

Hariharan, A.B.A. and Khalil, H.P.S.A. 2005. Lignocellulose-based hybrid bilayer laminate composite: Part I – Studies on tensile and impact behavior of oil palm fiber–glass fiber-reinforced epoxy resin. *Journal of Composite Materials* 39(8): 663–684.

Hull, D. and Clyne, T.W. 1996. *An Introduction to Composite Materials*. 2nd edition. Cambridge, United Kingdom: Cambridge University Press.

Jawaid, M. and Khalil, H.P.S.A. 2011. Cellulosic/synthetic fibre reinforced polymer hybrid Composites: A review. *Carbohydrate Polymer* 86: 1–18.

Joseph, K., Filho, R.D.T., James, B., Thomas, S. and Carvalho, L.H. 1999a. A review on sisal fibre reinforced polymer composites. *Revista Brasileira de Engenharia Agricola e Ambiental* 3: 367–379.

Joseph, K., Thomas, S. and Pavithran, C. 1993. Tensile properties of short sisal fibre reinforced polyethylene composites. *Journal of Applied Polymer Science* 47: 1731.

Joseph, P.V., Joseph, K. and Thomas, S. 1999b. The effect of processing variables on the physical and mechanical properties of short sisal fibre reinforced polypropylene composites. *Composites Science and Technology* 59: 1625–1640.

Joshi, S.V., Drzal, L.T., Mohanty, A.K. and Arora, S. 2004. Are natural fibre composites environmentally superior to glass fibre reinforced composites? *Composites Science and Technology* 35: 371–376.

Karus, M. 2004. *European Hemp Industry 2004: Cultivation, Processing and Product Lines*. European Industrial Hemp Association (EIHA) report.

Khalil, H.P.S.A. and Ismail, H. 2001. Effect of acetylation and coupling agent treatments upon biological degradation of plant fibre composites. *Polymer Testing* 20: 65–75.

Kuruvilla, J., Thomas, S. Pavithran, C., Joseph, K., Thomas, S., Pavithran, C. and Brahamakumar, M. 1993. Tensile properties of short sisal fibre reinforced polyethylene composites. *Journal of Applied Polymer Science* 47: 1731–1739.

Lilholt, H. and Lawther, J.M. 2000. Natural organic fibres. In: *Comprehensive Composites Materials* (6 vols). Vol. 1, chap. 10. Kelly, A. and Zweben, C., Eds. 32–41. New York: Elsevier.

Liu, Q. and Hughes, M. 2008. The fracture behaviour and toughness of woven flax fibre reinforced epoxy composites. *Composites: Part A* 39: 1644–1652.

Lystrup, A. 1998. *Hybrid Yarns for Thermoplastic Fibre Composites.* Riso-R-1034 (EN). Roskilde, Denmark: Riso National Laboratory.

Madsen, B. and Lilholt, H. 2003. Physical and mechanical properties of unidirectional plant fibre composites— an evaluation of the influence of porosity. *Composites Science and Technology* 63: 1265–1272.

Marcovich, N.E., Aranguren, M.I. and Reboredo, M.M. 2001. Modified woodflour as the thermoset fillers. 1. Effect of the chemical modification and percentage of filler on the mechanical properties. *Polymer* 42: 815–825.

Mathews, M.J. and Swanson, S.R. 2007. Characterization of the interlaminar fracture toughness of a laminated carbon/epoxy composite. *Composites Science and Technology* 67: 1489–1498.

Matthews, F.L. and Rawlings, R.D. 1995. *Composite Materials: Engineering and Science.* London: Chapman & Hall.

Moyeenuddin, A. S., Pickering, K.L. and Fernyhough, A. 2011. Effect of fibre treatments on interfacial shear strength of hemp fibre reinforced polylactide and unsaturated polyester composites. *Composites Part A* 42: 310–319.

Mwaikambo, L.Y. and Ansell, M.P. 2004. Chemical modification of hemp. Sisal, jute and kopok fibres by alkalisation. *Journal of Applied Polymer Science* 84: 2222–2234

Naik, N.K. and Sekher Y.C. 1998. Damage in laminated composites due to low velocity impact. *Journal of Reinforced Plastic* 17: 1232–1263.

Page, D.H., EI-hosseiny, F., Winkler, K. and Lancaster, A.P.S. 1977. Elastic modulus of single wood pulp fibres. *Tappi* 60: 114–117.

Pande, S.J. and Sharma, D.K. 1984. Fracture toughness of short glass fibre and glass particulate hybrid composites. *Fibre Science and Technology* 21: 307–317.

Paramasivam, T. and Abdulkalam, A.P.J. 1974. On the study of natural fibre composites. *Fibre Science and Technology* 1: 85–98.

Pavithran, C., Mukherjee, P.S., Brahmakumar, M. and Damodaran, A.D. 1988. Impact performance of sisal-polyester composites. *Journal of Materials Science Letters* 7: 825–826.

Peijs, T. 2000. Composites Turn Green. E-polymers 2002, no. T_002. http://www.e-polymers. org (pp. 1–12).

Peng, X., Fan, M., Hartley, J. and Al-Zubaidy, M. 2011. Properties of natural fibre composites made by pultrusion process. *Journal of Composite Materials* 46(2): 237–246.

Prattern, N.A. 1981. Review. The precise measurement of the density of small samples. *Journal of Materials Science* 16: 1737–1747.

Pugno, N., Ciavarella, M., Cornetti, P. and Carpinteri, A. 2006. A generalised Paris' law for fatigue crack growth. *Journal of the Mechanics and Physics of Solids* 54: 1333–1349.

Richardson, M.O.W., Santana, M.T.J. and Hague, J. 1998. Natural fibre composites-the potential for the Asian markets. *Progress in Rubber and Plastics Technology* 14: 174–188.

Richardson, M.O.W. and Wisheart, M.J. 1996. Review of the low velocity impact properties of composites materials. *Journal of Composites Part A* 27: 1123–1131.

Rouison, D., Sain, M., and Couturier, M 2006. Resin transfer moulding of hemp fibre composites: Optimisation of the process and mechanical properties of the materials. *Composites Science and Technology* 66: 895–906.

Rout, J., Misra, M., Tripahty, S.S., Nayak, S.K. and Mohanty, A.K. 2001. The influence of fibre treatment on the performance of coir/polyester composites. *Composites Science and Technology* 61: 1303–1310.

Saha, A.K., Das, S., Basak, R.K., Bhatta, D. and Mitra, B.C. 2000. Improvement of functional properties of jute-based composite by acrylontrile pre-treatment. *Journal of Applied Polymer Science* 78: 495–506.

Saidpour, H., Barikani, M. and Sezen, M. 2003. Mode-II interlaminar fracture toughness of carbon/epoxy laminates. *Iranian Polymer Journal* 13: 389–400.

Santulli, C. and Cantwell, W.J. 2001. Impact damage classification of jute fibre reinforced composites. *Journal of Materials Science Letters* 20: 477–479.

Sebe, G., Cetin, N.S., Hill, C.A.S. and Hughes, M. 2000. RTM hemp fibre reinforced polyester composites. *Applied Composites Materials* 7: 341–349.

Shen, C.H. and Springer, G. 1999. Moisture absorption and disorption of composite materials. *Journal of Composite Materials* 10: 2–20.

Silva, L.J.P., Christoforo, T.H., Rubio, A.L., Campos, J.C. and Fabrizio, S. 2012. Micromechanical analysis of hybrid composites reinforced with unidirectional natural fibres, silica microparticles and maleic anhydride. *Materials Research* 15: 1003–1012.

Stamboulis, A., Baillie, C.A. and Peijs, T. 2001. Effects of environmental conditions on mechanical and physical properties of flax fibers. *Composites Part A: Applied Science and Manufacturing* 32: 1105–1115.

Summerscales, J., Dissanayake, N.P.J., Virk, A.S. and Hall, W. 2010. A review of bast fibres and their composites. Part 1 – Fibres as reinforcements. *Composites Part A* 41: 1329–1335.

Thanomsilp, C., and Hogg, P.J. 2003. Penetration impact resistance of hybrid composites based on commingled yarn fabrics. *Composites Science and Technology* 63: 467–482.

Thomason, J.L. and Adzima, L.J. 2001. Sizing up the interphase: An insider's guide to the science of sizing. *Composites Part A: Applied Science and Manufacturing* 32: 313–321.

Wambua, P.W., Ivens, J. and Verpoest, I., 2003. Natural fibres: Can they replace glass in fibre reinforced plastics? *Composites Science and Technology* 63: 1259–1264.

Wang, Y., Chan, H.C., Lai, S.M. and Shen, H.F. 2001. Twin screw compounding of PE-HD wood flour composites. *International Polymer Processing* 16: 100–107.

Wisheart, M. and Richardson, M.O.W. 1999. Low velocity response of a complex geometry pultruded glass/polyester composite. *Journal of Materials Science* 34: 1107–1116.

Zafeiropoulos, N.E., Baillie, C.A. and Hodgkinson, J.M. 2002. Engineering and characterisation of the interface in flax fibre/polypropylene composite materials. Part II. The effect of surface treatments on the interface. *Composites Part A. Applied Science and Engineering* 33: 1185–1190.

Zeidman, M. and Sawhney, P.S. 2002. Influence of fibre length distribution on strength efficiency of fibres in yarn. *Textile Research Journal* 72: 216–220.

16 Vegetable Oils for Green Composites

Vijay K. Thakur, Mahendra Thunga, and Michael R. Kessler

CONTENTS

16.1 INTRODUCTION

Rising concerns about the future availability of energy resources and increased environmental awareness have led to the utilization of renewable materials in a number of applications (Alawar et al. 2009; Abdelmouleh et al. 2004; Cai et al. 2012; Demirbas 2009; Duwensee et al. 2010). In fact, the industrial exploitation of valuable feedstocks that can be easily obtained from renewable resources is currently the focus of considerable research efforts (Huber et al. 2006; Zhang et al. 2012). For example, renewable materials such as natural fibers, vegetable oils, starch, and others are already used in applications such as packaging, automotive, biomedical, and as structural composite materials for housing and infrastructure projects (Andjelkovic et al. 2009; Averous 2004; Cheng et al. 2011; Huber et al. 2012). Concerns regarding the disposal of synthetic polymers have initiated another approach to the development of novel materials from renewable resources (Arno 1989; Bunker and Wool 2002; Ouajai and Shanks 2009), and one of the current priorities in green polymer chemistry is the exploration of biodegradable polymers. These emerging bio-based materials have the potential to play a significant role in the next generation of materials applications. Intense research efforts around the globe are ongoing to effectively use biopolymers and to successfully implement them in useful products such as in packaging, structural, and biomedical applications (Yang et al. 2010a; Zain et al.

2011). Current studies show that it is imperative to develop renewable materials from different renewable resources as suitable alternatives to the Earth's limited petroleum derivatives in order to decrease the dependence on oil, to reduce carbon dioxide emissions, and to generate more economic opportunity for the agricultural sector both in the developing and developed countries (Clark et al. 2006; Gandini 2008).

Today petroleum-dependent industries suffer from the decline in feedstock availability, price volatility, and fast changing environmental restrictions (driven by regulations and public perception), bringing agricultural raw materials that have preceded petrochemicals for millennia in nonfood applications back into focus (Williams and Hillmeyer 2008). The development of biopolymers and their successful use in the manufacture of eco-friendly industrial products have become the focal point of recent sustainability efforts (Andjelkovic and Larock 2006; Bozell 2008; Ververis et al. 2004). Among various biopolymeric materials from renewable resources, vegetable oils represent an ideal alternative to chemical feedstock (Baboi et al. 2007; Can et al. 2006a; De Espinosa et al. 2009a,b). Derivatives of vegetable oils, such as epoxidized vegetable oil (EVO), can be used as raw materials for the synthesis of a variety of chemicals, including glycol, polyols, and carbonyl compounds, as well as lubricants and plasticizers for polymers because of the high reactivity of their oxirane rings (De Espinosa et al. 2010). Vegetable oils have been used for lubrication and for coatings and paints for many centuries before the availability of abundant and cheap mineral oils. Figure 16.1 shows the structure of one of the most important natural oils (castor oil).

Because of their inherent eco-friendly nature combined with low cost when compared with petroleum-based products, vegetable oils have become an integral (and growing) part of the oleochemical industry. Currently, different derivatives of vegetable oils are used as polymerizable monomers and a number of polymer resins were synthesized (Badrinarayanan et al. 2009; Bharathi et al. 2009; Bonnaillie and Wool 2007; Berh et al. 2002). The traditional synthetic resins such as epoxy, phenolic, polyurethanes (PUs), polyester, and vinyl ester pose problems regarding biodegradability, initial processing cost, energy consumption, health, and environmental hazards as none of the components of these polymers come from renewable resources. For example, epoxy resin that is used in many industrial areas such as glues, adhesive paint, and composites is prepared by the condensation reaction between an epoxy monomer and a hardener, while phenolic resins are prepared by the polycondensation of a phenol and an aldehyde with elimination of water and formation of a three-dimensional network. PUs are prepared from the chemical reactions of polyisocyanate with groups containing labile hydrogen. For example, hydroxyl, amine, and vinyl ester resins are obtained by polycondensation reactions

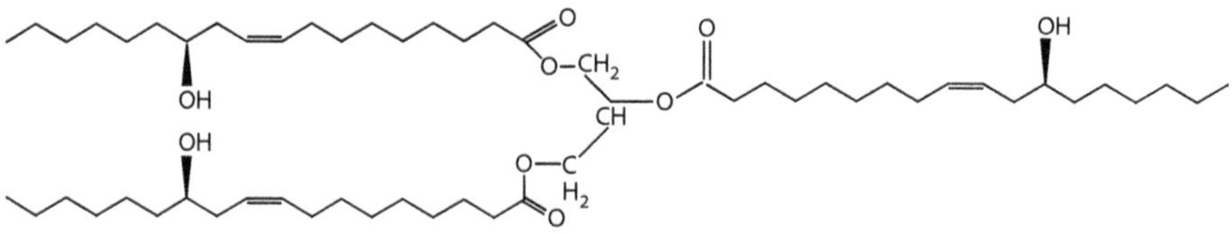

FIGURE 16.1   General chemical structures of castor oil (vegetable oil).

of an unsaturated carboxylic acid and an epoxy resin. On the other hand, vegetable oil-based resins offer specific advantages over these synthetic resins (Dutta et al. 2004). The major advantages of vegetable oil-based polymers are renewability, low environmental impact, and reduction of the dependence on the limited resources of petrochemicals.

Vegetable oils are obtained from plant seeds and consist mainly of mixtures of triglycerides (Cakmakli et al. 2005). In other words, most vegetable oils contain high levels of unsaturated fatty acids (De Espinosa et al. 2009a,b), which can be easily converted into epoxy fatty acid by conventional epoxidation, metal catalyst epoxidation, catalytic acidic ion exchange, or other reactions (Dinda et al. 2008). EVOs enjoy growing interest, because they are easily obtained from sustainable, renewable natural resources and are environmental friendly. While petroleum-based epoxy and polyester resins are brittle resin systems, the long fatty acid chains of vegetable oils impart desirable flexibility and toughness to the resulting polymer resin (Dwan'Isa et al. 2003; Lligadas et al. 2006a). These oils are multicomponent mixtures of various triglycerides (esters of glycerol and fatty acids) and exhibit a number of outstanding properties that can be successfully utilized in new, valuable, eco-friendly polymeric materials such as epoxy, polyester amide, alkyd, and PU (Firdaus 2012; Petrovic 2008). Triglycerides from different plants sources such as palm, soybean, rapeseed, or sun flower can be utilized (Goud et al. 2006). Bio-based thermosetting polymers from vegetable oils such as epoxy from soybean oil, palm oil, canola oil, castor oil, and linseed oil were synthesized via different routes (Guner et al. 2006; Jin and Park 2008). In order to ensure the required reactivity, the triglyceride compound must be isolated and purified carefully, followed by suitable functionalization. A number of chemical modification techniques are used for functionalization of vegetable oils and one of the most commonly used is an epoxidation reaction (Mungroo et al. 2008).

From the literature, it is quite evident that the use of plant oil-based resins, compared with their synthetic counterpart, reduces the emission of some of the synthetic components (e.g., volatile organic compounds), and the health and environmental risks associated with them. Along with this, it will also lead to promoting global sustainability. Figure 16.2 shows the route from biomass to polymer for vegetable oils.

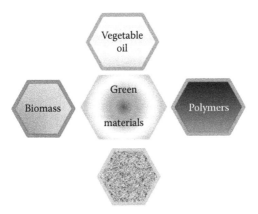

FIGURE 16.2 Route from biomass to polymer for vegetable oils.

The use of vegetable oils for polymer applications has a long tradition. For example, one can differentiate between their use as polymer additives (e.g., epoxidized soybean oil (ESO) as plasticizer), as building blocks for thermoplastic polymers, and as a basis for thermosets. Indeed in the last few years, a number of chemical routes have been developed for using natural or modified vegetable oils as a basis for polymers, adhesives, and composites with specific properties and applications.

16.2 CHEMISTRY OF VEGETABLE OILS

Lipids are one of the largest families of chemical compounds and include vegetable oils and fats. Vegetable oils and fatty acids have been used for a number of polymer applications (Lligadas et al. 2005). These oils are likely candidates for conversion into novel polymers because of their molecular structure. Triglycerides, also known as triacylglycerol, are the predominant molecules in vegetable oils. Triglyceride molecules are composed of three fatty acid chains of unsaturated or saturated fatty acids joined at a glycerol center. The saturated fatty acids have only single bonds between two carbon atoms, while the unsaturated fatty acids contain a number of double or triple bonds between two carbon atoms. Figure 16.3a and b shows the structure and synthesis of triglyceride molecules.

With few exceptions, these fatty acid chains range from 14 to 22 carbon atoms in length. The formation of a triglyceride is shown in Figure 16.3b.

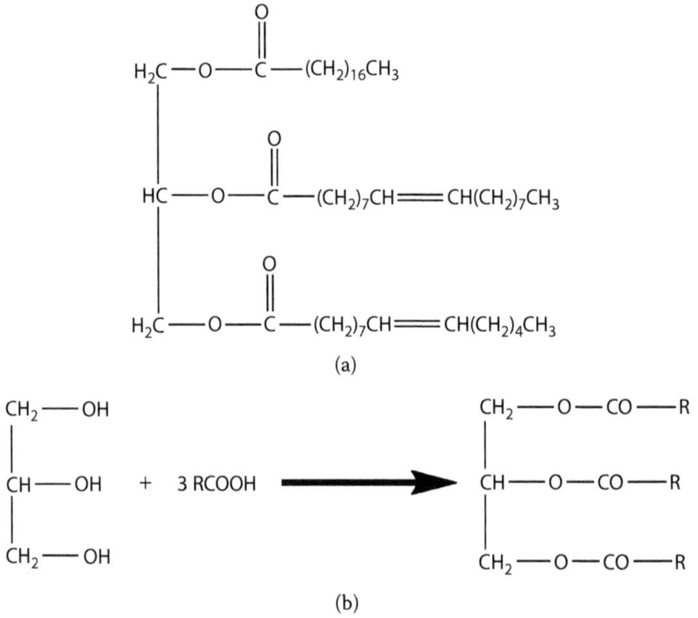

FIGURE 16.3 (a) General structure of triglyceride. (b) Synthesis of triglyceride.

The composition of the major fatty acids in fats and oils determines their bulk physical and chemical properties (Petrovic et al. 2002). Properties such as melting characteristics and oxidative stability depend on the amount and type of fatty acids on the glycerol molecule. The oxidative stability of different fatty acids follow the following order: C18:1 > C18:2 > C18:3 where the first number represents the number of carbon atoms and the second, the number of double bonds in the hydrocarbon chain. Unsaturation is associated with lower melting points, greater solubility, and chemical reactivity. The properties of natural oils vary to a considerable extent, depending on the composition of the different fatty acids. By modifying the fatty acids attached to the glycerol, the behavior of fat can be modified to meet the requirements of a specific application (Goud et al. 2007a). A triglyceride with one stearic and two oleic acids is 15 times more stable with regard to oxidation than a triglyceride with three linoleic acids. Structural investigations of naturally occurring fats show that they contain fatty acids with chain lengths varying between 14 and 22 carbon atoms. Figure 16.4a through q shows the structures of some of the common fatty acids.

Oleic acid, linoleic acid, and linolenic acid are among the predominant unsaturated fatty acids present in most vegetable oils. These acids contain one, two, and three double bonds between two carbon atoms, respectively. Triglycerides contain a number of reactive positions, acting as starting points for diverse reactions: ester groups, C = C double bonds, allylic positions, and the α-position of ester groups. Among these reactive positions, the C = C double bonds are the most easily accessible reaction sites for chemical modification, for example, by epoxidation (Goud et al. 2007a,b,c). Table 16.1 shows the chemical composition of some of the industrially important vegetable oils (Petrovic 2008).

In order to find acceptance and broad application, polymers derived from triglyceride oils have to exhibit the same or better properties than the currently available petrochemical-based polymers (Lligadas et al. 2006b; Sharma and Kundu 2006). The following properties are essential for the successful industrial implementation of polymers derived from triglycerides:

- Thermal stability
- Biocompatibility
- Biodegradability
- Mechanical strength
- Chemical resistance
- Electrical conductivity or resistivity
- Adhesion to metallic substances
- Gas permeability
- Fire resistance

During the last few years, the modification of vegetable oils underwent its most important advances (Biswas et al. 2008; Park et al. 2004; Vlcek and Petrovic 2006). Exhaustive literature surveys show that the unsaturation present in vegetable oils can be used for chemical modification, resulting in "value-added products."

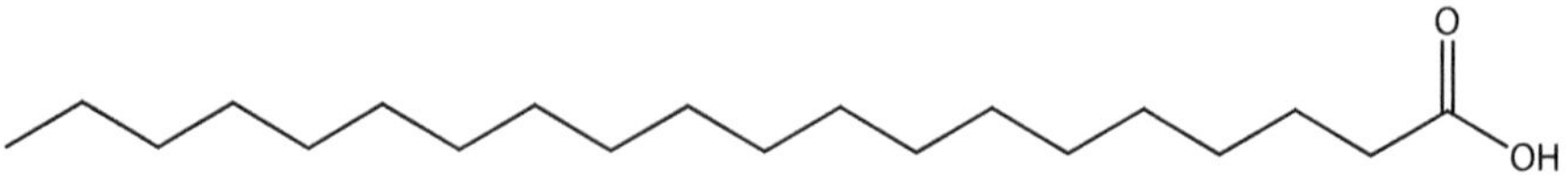

Arachidic acid

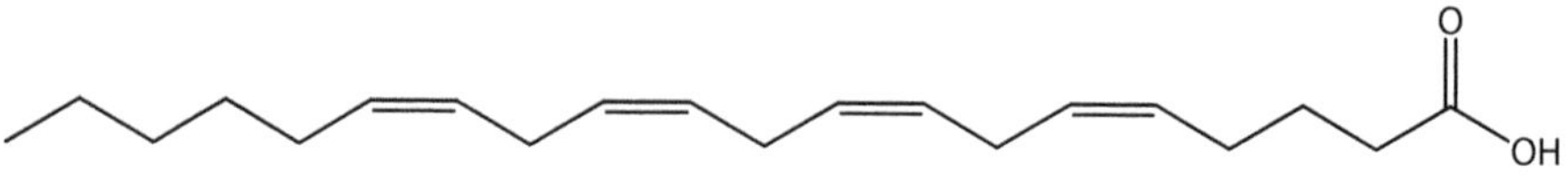

Arachidonic acid

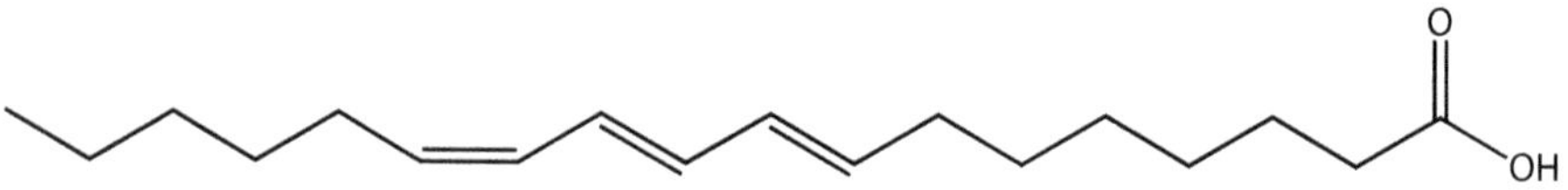

Calendic acid

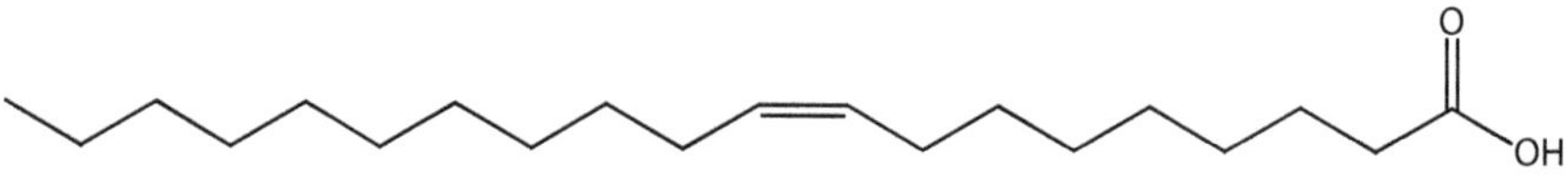

Gadoleic acid

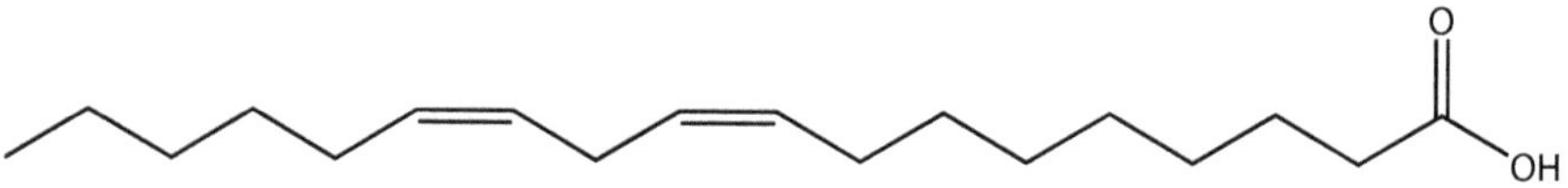

Linoleic acid

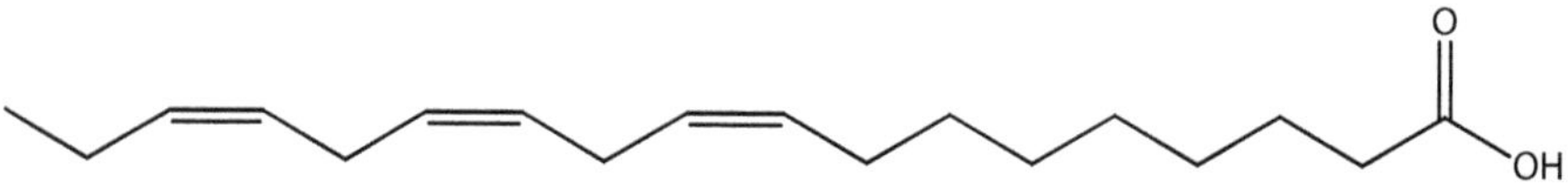

Linolenic acid

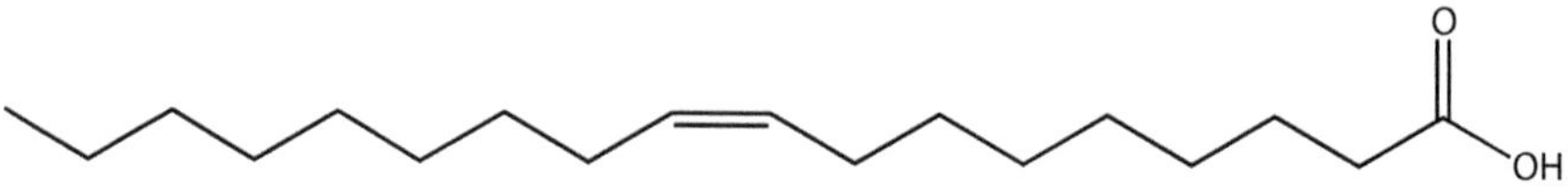

Oleic acid

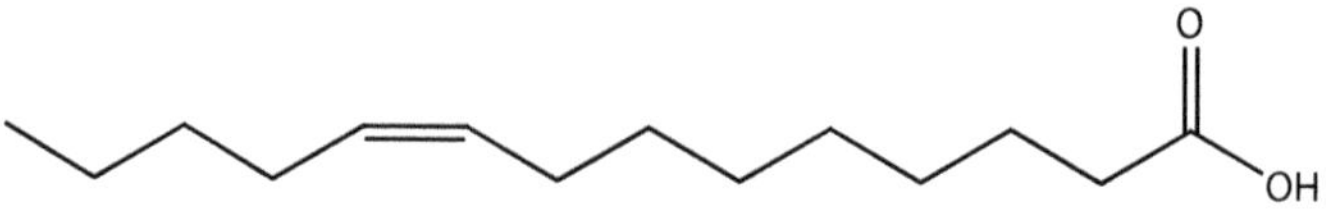

Myristoleic acid

FIGURE 16.4   Structures of some of the most important fatty acids.

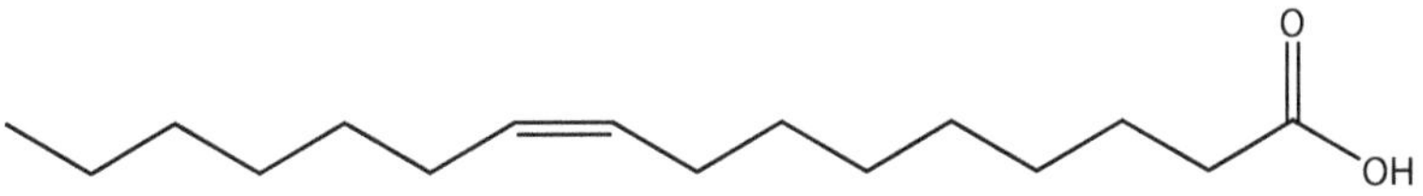

Palmitoleic acid

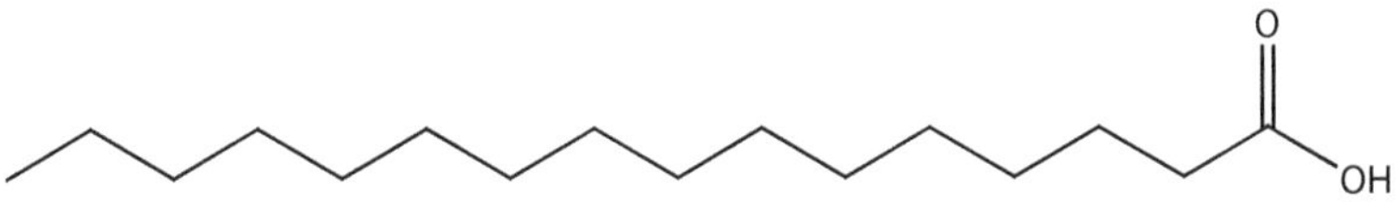

Palmitic acid

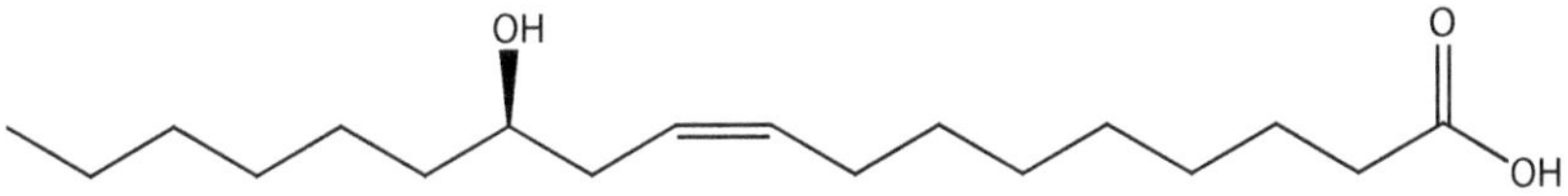

Ricinoleic acid

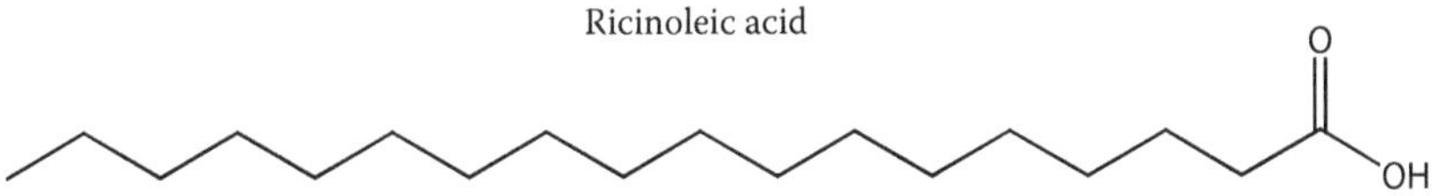

Stearic acid

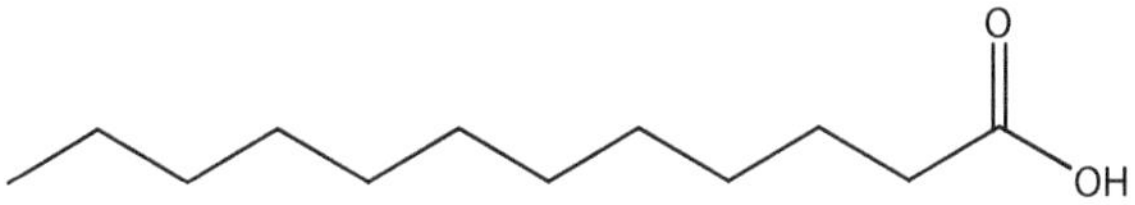

Lauric acid

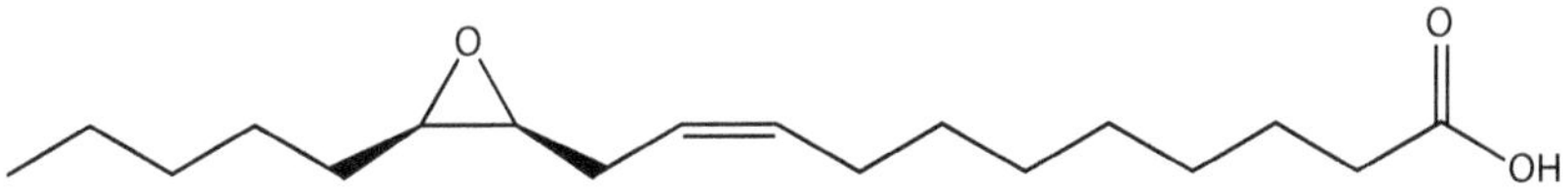

Vernolic acid

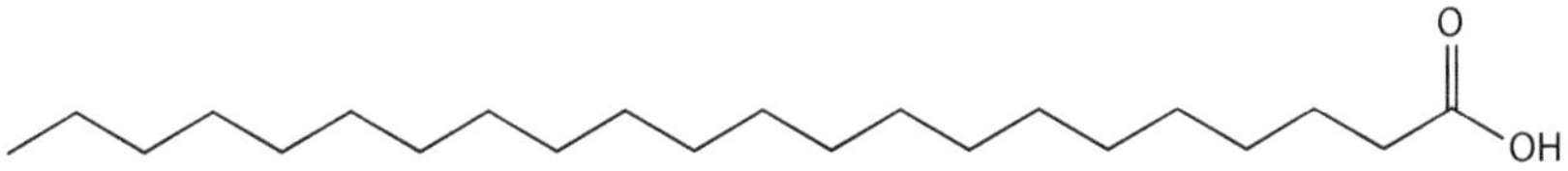

Behenic acid

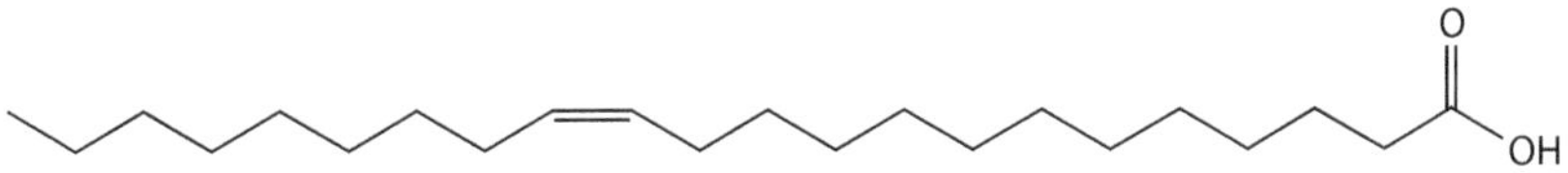

Erucic acid

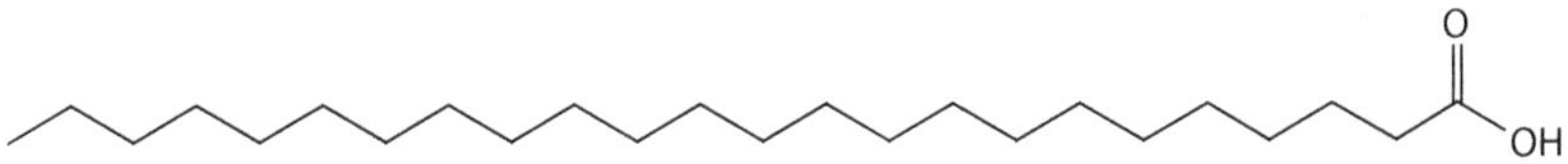

Lignoceric acid

FIGURE 16.4 (*Continued*) Structures of some of the most important fatty acids.

TABLE 16.1

Chemical Composition of Some Industrially Important Vegetable Oils

Carbon Atoms:Double Bonds	8:0	10:0	12:0	14:0	16:0	16:1	18:0	18:1	18:2	18:3	20:0	20:1	22:0	22:1	24:0	Iodine Value
Canola oil				0.1	4.0	0.3	1.8	60.9	21.0	8.8	0.7	1.0	0.3	0.7	0.2	100–115
Castor oil					2.0	1.0	7.0	3.0								81–91
Coconut oil	7.1	6.0	47.1	18.5	9.1		2.8	6.8	1.9							7–12
Corn				0.1	10.9	0.2	2.0	25.4	59.6	1.2						118–128
Cottonseed oil				0.7	21.6	0.6	2.6	18.6	54.4	0.7	0.3					98–118
Linseed oil					6.0		4.0	22.0	16.0	52.0	0.5					>177
Olive oil					9.0	0.6	2.7	80.3	6.3	0.7	0.4					76–88
Palm oil				1.0	44.4	0.2	4.1	39.3	10.0	0.4	0.3	0.1				50–55
Palm kernel oil	3.3	3.4	48.2	16.2	8.4		2.4	15.3	2.3							14–19
Peanut oil				0.1	11.1	0.2	2.4	46.7	32.0	1.3	1.6	2.9	1.5		1.0	84–100
Rapeseed oil				0.1	3.8	0.3	1.2	18.5	14.5	11.0	0.7	6.6	0.5	41.1	1.0	100–115
Safflower oil				0.1	6.8	0.1	2.3	12.0	77.7	0.4	0.3	0.1	0.2			140–150
Safflower oil (high oleic)					3.6	0.1	5.2	81.5	7.3	0.1	0.2	1.2		0.3		82–92
Soybean oil				0.1	10.6	0.1	4.0	23.3	53.7	7.6	0.3					123–139
Sunflower oil				0.1	7.0	0.1	4.5	18.7	67.5	0.8	0.4	0.1	0.7			125–140
Sunflower oil (high oleic)					3.7	0.1	5.4	81.3	9.0	0.1						81–91

Source: Petrovic, Z.S. Polym. Rev., 48, 109–155, 2008. Copyright 2008 Taylor & Francis.

16.3 POLYMERIZATION METHODS FOR VEGETABLE OILS

As discussed in the preceding section, vegetable oils mainly consist of triglycerides that can be either directly or indirectly synthesized into a variety of polymers (Henna et al. 2007; John et al. 2002). Diacids/polyacids, polyaldehydes, diamines/polyamines, hydroxy acids, diols/polyols, amino acids, and vinyl compounds are the most commonly used monomers derived from vegetable oils (Lligadas et al. 2007a; Lu and Larock 2009; Narine et al. 2007a) and are used for the preparation of different kinds of polymers. A number of approaches that have been applied to obtain the polymeric materials from vegetable oils can be grouped into two categories: (1) direct polymerization and (2) chemical modification. A brief introduction to some of the important techniques is given below.

16.3.1 DIRECT POLYMERIZATION

Direct polymerization of oils has been used in coatings for a long time for different applications. During polymerization, triglycerides play a significant role, providing many active sites that are susceptible to chemical reactions. These active sites can be easily used to polymerize triglycerides directly. As evident from the structures of vegetable oils; triglyceride molecules contain $C = C$ double bonds that are capable of being polymerized. These double bonds can be polymerized through a cationic mechanism or a free-radical mechanism depending on the application. Cationic polymerization has been widely studied by a number of researchers. Most of this research has focused on the direct cationic polymerization of the $C = C$ double bonds of natural oils, in the presence of petroleum-derived comonomers. Strong Lewis acids are generally used as initiators in cationic polymerizations (Andjelkovic et al. 2005; Lu and Larock 2009). A three-dimensional cross-linked polymer network is formed when vegetable oils are subjected to cationic polymerization conditions, as the unsaturated fatty acid chain in the triglyceride can participate in the cationic reaction. Different types of thermosetting polymers ranging from rubbers to hard plastics have been prepared by the cationic polymerization of a variety of oils such as soybean oil, fish, oil, and tung oil (TUN). The resulting thermosetting materials possess thermal and mechanical properties comparable to those of industrial plastics. Vegetable oils such as linseed and TUNs have also been classically exploited as drying oils due to their highly unsaturated nature. The reaction with atmospheric oxygen leads to the formation of a cross-linked network. This polymerization is used mostly in paints and coatings, and also in inks and resins. Larock's group at Iowa State University intensively studied the cationic polymerization of the double bonds of some vegetable oils such as soybean, corn, and TUNs without any modification; however, other comonomers such as styrene (ST), divinylbenzene (DVB), and dicyclopentadiene (DCPD) must be added into the mixture to improve the polymer properties. Figure 16.5 shows the cross-linking reaction of these vegetable oils (Andjelkovic et al. 2005).

Li et al. (2001) have also conducted extensive research on the cationic copolymerization of soybean oil and low-saturation soybean oil with various petroleum-derived cross-linking agents. The copolymerization of 50–60 wt% of these vegetable oils with DVB results in the formation of a densely cross-linked polymer network, interpenetrated by

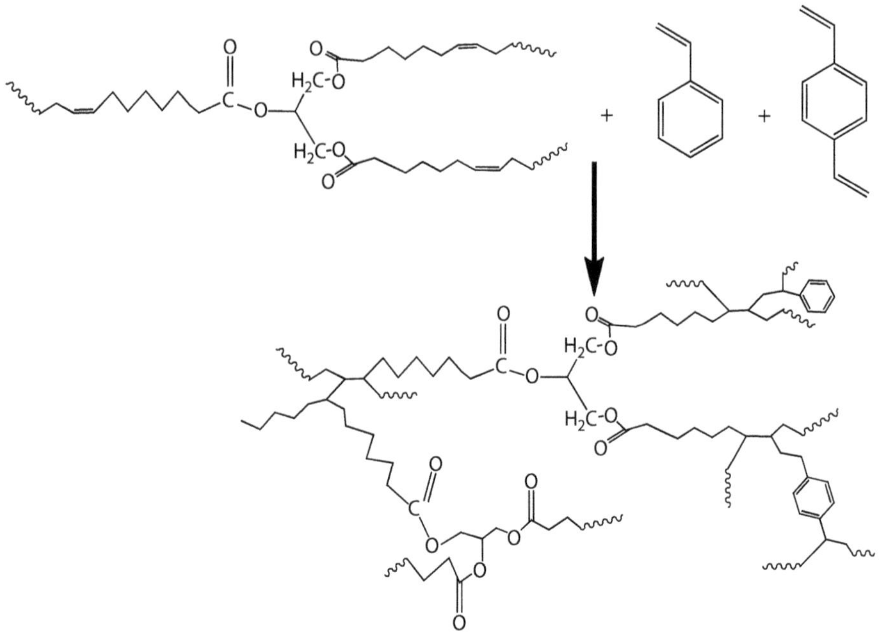

FIGURE 16.5 Direct copolymerization of a triglyceride with styrene and divinylbenzene. (From Andjelkovic, D.D., Valverde, M., Henna, P., Li, F. and Larock, R.C. *Polymer*, 46, 9674–9685, 2005.)

12–31 wt% of unreacted free oil or oligomers (Li et al. 2001). The amount of unreacted free oil left after cure of the resin was found to be directly dependent on the amount and reactivity of the oil initially employed. With respect to the concentration of the comonomers, an increase in room temperature storage modulus of the final copolymer has been observed for higher amounts of DVB in the original composition. Furthermore, the unreacted free oil or oligomers present in the final copolymers significantly affect the thermal stability of the thermosets (Li et al. 2001). In order to overcome the problems faced by the use of DVB especially to increase the structural uniformity of the cross-linked copolymers, Li and Larock (2001) added ST to the original composition using soy oil and low-saturation soybean oil. It has been observed that with the substitution of 25–50 wt% of DVB by ST, the overall properties of the resulting plastics were significantly improved (Li and Larock 2001). The thermophysical properties of the thermosets were also considerably affected by the cross-link density of the bulk copolymers and the yield of cross-linked material is strongly dependent on the concentration and reactivity of the cross-linking agent and the reactivity of the oil used (Li and Larock 2001). A number of thermosets using different vegetable oils such as canola, peanut, olive, sesame, sunflower, safflower, walnut, and linseed oils were cationically copolymerized with DVB and/or ST to form a range of thermosets with properties that can be tailored according to desired applications (Andjelkovic et al. 2005). Overall, the properties of these new materials gradually increase with the number of carbon–carbon double bonds in the oil while the gel times of these copolymers were found to be independent of the reactivity of the used vegetable oil (Andjelkovic et al. 2005).

The free-radical polymerization of triglyceride double bonds has received little attention due to the presence of chain-transfer processes to the many allylic positions in the molecule. However, some vegetable oils such as linseed and TUNs have been found to be susceptible to free-radical polymerization and classically exploited as drying oils because the drying power of these oils is directly related to their highly unsaturated nature (Tallman et al. 2004). These oils are used mostly in paints and coatings, and also in inks and resins since drying oils will auto-oxidize in the presence of oxygen from air. It has been reported that auto-oxidation leads to the self-polymerization of the carbon–carbon double bonds in the oil to form a cross-linked network (Cakmakli et al. 2005). Figure 16.6 shows the free-radical mechanism for polymerization of the vegetable oil.

Free-radical macroinitiators from vegetable oils such as linseed and soybean oil were prepared and used to initiate the polymerization of methyl methacrylate (MMA) and n-butyl methacrylate (BMA) (Cakmakli et al. 2004, 2005). The final grafted poly-MMA and -BMA copolymers had been found to be biocompatible and partially biodegradable due to the presence of sufficient content of the vegetable oils in the system, suggesting the potential applications of these polymers in tissue engineering (Cakmakli et al. 2004, 2005). Translucent vegetable oil-based thermosets were also prepared through free-radical copolymerization of reactive unsaturated triglycerides and vinyl comonomers (Henna et al. 2007; Valverde et al. 2008). The free-radical copolymerization of vegetable oils with petroleum-based comonomers requires that the carbon–carbon double bonds in the oil be sufficiently reactive in order to form a homogeneous material, which typically requires the use of conjugated oils. Soybean oil-based thermosets were prepared by copolymerization of mixtures containing 40–85 wt% of conjugated low-saturation soybean oil with various amounts of acrylonitrile, DVB, and/or DCPD in the presence of azobisisobutyronitrile (AIBN) (Valverde et al. 2008). A wide range of thermal and mechanical properties was obtained by simply changing the stoichiometry of the resin components. Indeed conjugated vegetable oils have been found to be more reactive than their non-conjugated counter parts and constitute better starting materials for the preparation of bio-based free-radical thermosets (Andjelkovic and Larock 2006; Li and Larock 2001).

Vegetable oils can also be thermally polymerized by heating to high temperatures. Among various oils, TUN has been found to be very reactive and readily polymerizable as nearly 84% of its fatty acid chains are α-eleostearic acid, which facilitates its polymerization. It has been reported that copolymers containing 0.1%–2.0% of TUN and ST

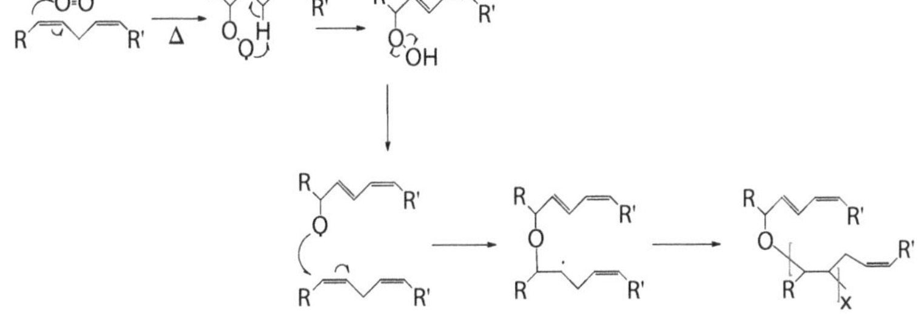

FIGURE 16.6 Free-radical reaction mechanism for the auto-oxidation of vegetable oils in the presence of oxygen from air. (From Cakmakli, B., Hazer, B., Tekin, I.O. and Comert, F.B. *Biomacromolecules*, 6, 1750–1758, 2005.)

can be easily obtained by heating the materials to 125°C for 3 days (Stoesser and Gabel 1940). A series of copolymers of TUN, ST, and DVB, in which the amount of TUN was varied from 30 to 70 wt%, was prepared by thermal polymerization in the temperatures ranging from 85°C to 160°C (Li and Larock 2003). The material properties of these materials have been found to range from elastomeric to tough and rigid. Furthermore, the addition of catalytic amounts of Co, Ca, and Zr salts accelerates the thermal copolymerization and improves the properties of the resulting copolymers (Li and Larock 2003). Linseed oil was also thermally polymerized using 30–70 wt% of a commercially available conjugated linseed oil (CLO). In this oil, 87% of the carbon–carbon double bonds were found to be conjugated and the resulting polymer possess properties ranging from rubbery to hard depending on the stoichiometry of the comonomers (Kundu and Larock 2005). The scope, limitations, and possibility of utilizing such methods with triglycerides for various applications have recently been highlighted (Guner et al. 2006).

16.3.2 CHEMICAL MODIFICATION

Direct polymerization (especially cationic polymerization) of vegetable oils leads to some good products; however, it is not universally applicable to all the vegetable oils (especially nonconjugated oils) as the internal double bonds in the triglyceride molecules are not sufficiently reactive for any viable polymerization process. Direct polymerization of such oils normally yields materials with poor properties due to the low reactivity of both double bonds and secondary hydroxyl groups in oils such as castor oil. So in order to overcome the shortcoming of direct polymerization, considerable efforts are being devoted to modifications that could facilitate a subsequent polymerization of the triglycerides to produce polymers with appropriate properties. These modifications consist of (1) introduction of different reactive groups into the aliphatic chains, which would exhibit a higher aptitude to polymerize and (2) the reduction of the triglycerides to monoglycerides through glycerolysis or chemical transformation to produce platform chemicals for subsequent polymer synthesis.

The most important modification of the vegetable oils is the epoxidation. Epoxidation can be carried out in several ways in a very efficient manner. The methods for converting vegetable oils into epoxidized polymer resins are well documented in the literature (Goud et al. 2006). Currently, there are four generally accepted methods of producing epoxides from olefinic molecules (unsaturated chemical compounds containing at least one carbon–carbon double bond) (Goud et al. 2006). These are as follows: (1) Epoxidation in situ with peroxyacetic or peroxyformic acid, with peroxyacetic acid showing a higher conversion to oxirane (Dinda et al. 2008). (2) Epoxidation with organic and inorganic peroxides catalyzed by a transition metal catalyst. The most common of these is nitrile hydrogen peroxide, which is an inorganic chemical (Goud et al. 2006). (3) Epoxidation with halohydrines, using hypohalous acids (HOX) and their salts as reagents for the epoxidation of olefins with electron-deficient double bonds. (4) Epoxidation with molecular oxygen (Goud et al. 2006).

EVOs have the potential to be used directly in a wide range of polymer applications, such as stabilizers of polyvinylchloride or the direct photochemically initiated cationic polymerization and cross-linking with polyols and diamines. One of the most interesting applications of EVOs is as intermediate in the synthesis of

macromonomers for different applications. Different synthesis techniques allow for
the introduction of new functional groups, enhancing the properties of the vegetable
oils in their utilization as macromonomers. Figure 16.7 shows some of the commercially important derivatives obtained by ring-opening polymerization.

The ring opening of the epoxide group yields an alcohol group and, depending
on the procedure, other functionality can also be obtained. Common uses of these
derivatives are the synthesis of PUs by reaction of the polyols with an isocyanate or
the radical polymerization in the case of the ring opening with acrylic acid. Using
some standard techniques such as hydrolysis or methanolysis fatty acids and their
esters as well as glycerol can be isolated from triglycerides in good purities. Thiol–
ene coupling has been applied to the synthesis of monomers derived from vegetable
oils, and these monomers can be used for the synthesis of linear polyesters and PUs,
as well as monomers for branched structures. Figure 16.8 shows some of the reactions used in the synthesis of monomers via thiol–ene coupling for polyester, polyamide, and PU synthesis (Lligadas et al. 2010; Turunc and Meier 2010, 2013).

One of the most important reactions recently used for the transformation of fatty
acids into monomers and polymers is the metathesis reaction (Henna and Larock
2007). It was first used to prepare polyolefins via an exchange of alkylidene groups.
Metathesis reactions have been known for a long time and applied in organic chemistry for catalytic C–C bond formation reactions. The metathesis reaction has become
one of the most useful reactions applied to polymer chemistry and especially useful
for oleochemicals for the following reasons: (1) functional group introduction via
cross metathesis; (2) direct use of inherently present double bonds on fatty acids; and
(3) breakthrough fatty acid derivatization via ruthenium-based catalysts.

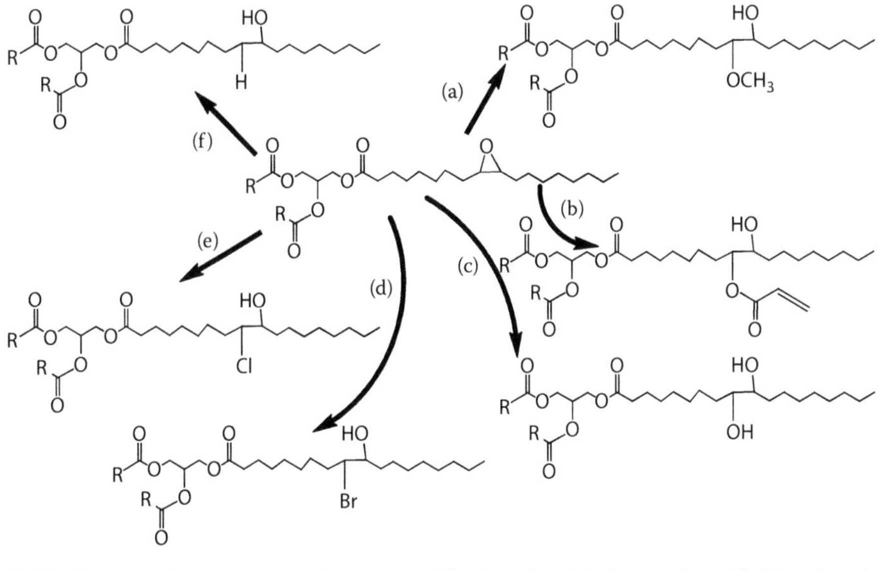

FIGURE 16.7   Ring-opening products of epoxidized triglycerides by reaction with (a) methanol,
(b) acrylic acid, (c) H$_2$O, (d) HBr, (e) HCl, and (f) catalytic hydrogenation.

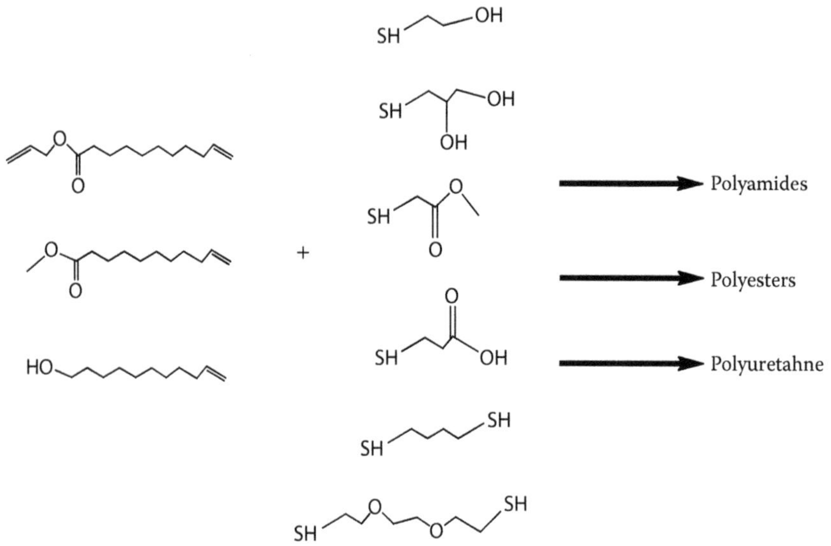

FIGURE 16.8 Some reactions used in the synthesis of monomers via thiol–ene coupling for polyester, polyamide, and polyurethane synthesis. (From Lligadas, G., Ronda, J.C., Galià, M. and Cádiz, V. *Biomacromolecules,* 11, 2825–2835, 2010; Turunc, O. and Meier, M.A.R. *Macromol. Rapid Commun.,* 31, 1822–1826, 2010; Turunc, O. and Meier, M.A.R. *Eur. J. Lipid Sci. Technol.,* 115, 41–54, 2013.)

Self-metathesis of fatty acid derivatives results in the formation of diacids and diols. In the same way, cross metathesis of fatty acid derivatives with methyl acrylate yields monomers for polymerization.

In contrast to the use of fatty acids, triglycerides can also be directly utilized in metathesis reactions, resulting in branched structures. Different kinds of polymers such as polyamides, polyethers, and polyesters from triglycerides have been prepared using this reaction. Figure 16.9 shows the general schematic for the synthesis of different kinds of polymers using metathesis reactions (Guner et al. 2006).

Bierman et al. (2010) polymerized high oleic sunflower oil via acyclic triene metathesis (ATMET) reactions. Metathesis reactions of vegetable oils are generally classified into ring-opening metathesis polymerization (ROMP) and acyclic diene metathesis (ADMET) polymerization.

It has been observed that during the ROMP process, strained, unsaturated, cyclic molecules are opened at the carbon–carbon double bond by interaction with a ruthenium carbene catalyst and subsequent coordination of a new molecule of the strained ring, and metathesis of the carbon–carbon double bonds result in an unsaturated polymer. This polymer can be additionally cross linked through reaction of the remaining carbon–carbon double bonds (Leitgeb et al. 2010). Figure 16.10 shows the illustration of the ROMP process, with cyclopropene as an example.

If more than one type of strained, unsaturated ring is present in the reaction medium, a copolymer is formed.

Castor oil had been modified with bicyclo [2,2,1] hept-5-ene-2,3-dicarboxylic anhydride for use in ROMP reactions (Henna and Larock 2007). Varying amounts

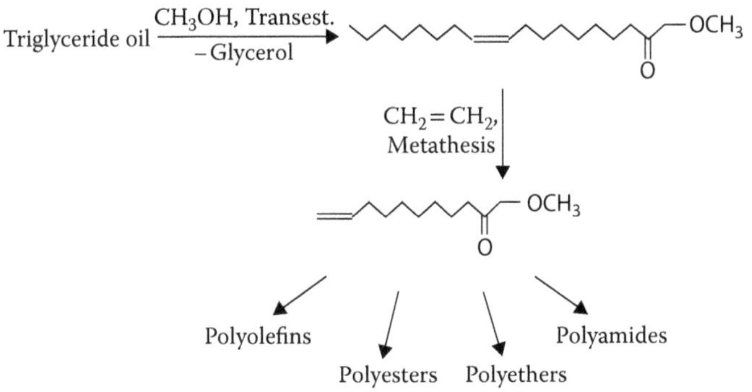

FIGURE 16.9 Preparation of various polymers via metathesis reaction. (From Guner, F.S., Yagci, Y. and Erciyes, A.T. *Prog. Polym. Sci.*, 31, 633–670, 2006. With permission. Copyright 2006 Elsevier.)

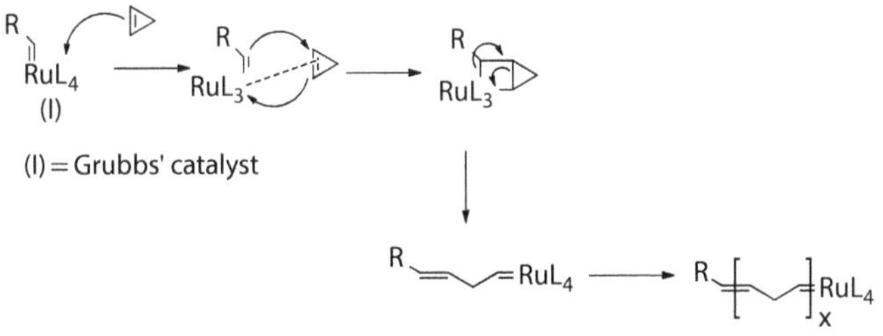

FIGURE 16.10 Mechanism of the ring-opening metathesis polymerization of cyclopropene.

of the resulting norbornenyl-functionalized triglyceride were copolymerized with cyclooctene in the presence of 0.5 wt% of second-generation Grubbs catalyst to afford transparent, rubbery thermosets (Henna and Larock 2007). Thermal stability of thermosets thus prepared was found to increase with the increase in the content of modified oil and is closely related to the presence of unreacted triglycerides in the final material (Henna and Larock 2007). Bio-based ROMP thermosets were also prepared from the vegetable oil–derived, functionalized fatty alcohols (Xia and Larock 2010b). Norbornenyl-functionalized fatty alcohols from soybean oil (NMSA), Dilulin (NMDA), ML189 (NMMA), and castor oil (NMCA) were prepared and polymerized in the absence of other comonomers in order to avoid problems, such as homogeneous mixture of the catalyst, associated with the high viscosity of castor oil during ROMP (Xia and Larock 2010b). In comparison with ROMP, ADMET polymerization is based on the coordination of aliphatic dienes to a ruthenium carbene catalyst, followed by metathesis of the carbon–carbon double bonds. These dienes can be polymerized under acyclic diene metathesis condition. Figure 16.11 shows the synthesis of monomers by metathesis reaction, and synthesis of fatty acid-derived monomers for ADMET polymerization.

ADMET polymerization of soybean oil using first-generation Grubbs catalyst produced materials ranging from sticky oils to rubbers (Refvik et al. 1999; Tian and Larock 2002). High-molecular-weight polymers and block copolymers were synthesized by using long-chain aliphatic α,ω-dienes from plant oil derivatives, such as

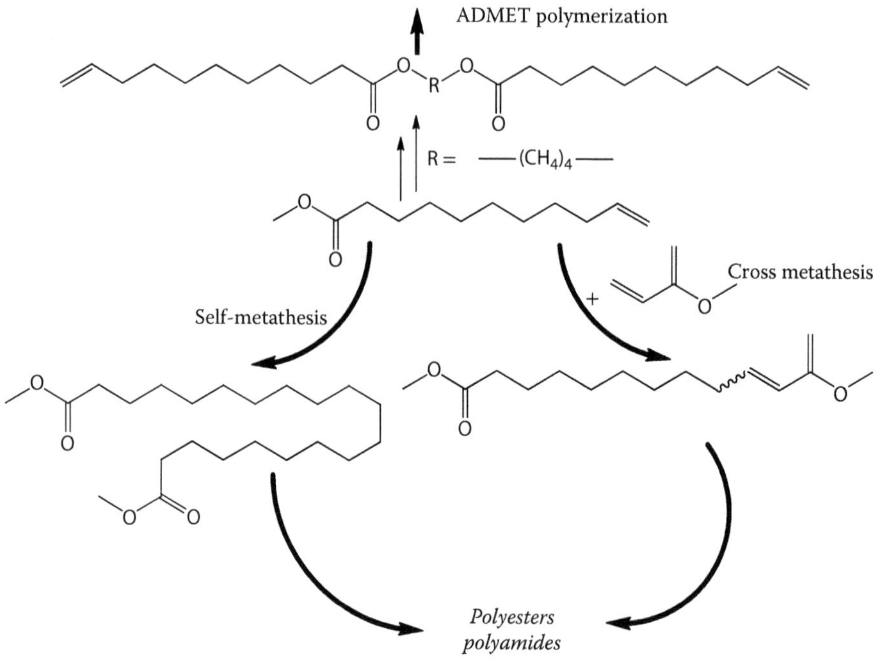

FIGURE 16.11 Synthesis of monomers by metathesis reaction, and synthesis of fatty acid derived monomers for ADMET polymerization.

undec-10-enyl undec-10-enoate, using ADMET polymerization (Mutlu and Meier 2009; Rybak and Meier 2008). Recently, it has been reported by De Espinosa et al. (2010) that ADMET polymerization of phosphorous-containing vegetable oil–based α,ω-dienes affords polymers with relatively good flame retardancy.

Petrovic (2008) reported that ozonolysis of oils and fatty acids produced a number of difunctional monomers and chemicals as shown in Figure 16.12.

Different polyols have been prepared from vegetable oils by different methods (Narine et al. 2007b). Modification of the polyols has also been carried out to increase their functionality. Polyols can be easily prepared by oxidation and epoxidation of vegetable oil route. It has been mentioned that hydroxyl groups can be introduced at the position of double bonds by direct oxidation. Sometimes during polyol synthesis, mitigation of negative properties such as low functionality, variability in composition, odor, higher acid value, and dark color may require posttreatment depending on the applications. Another issue related to the preparation of polyols is the presence of high unsaturation, which makes the products susceptible to oxidation and color and property change with time (Petrovic 2008). In order to optimize synthesis results, oxidation is controlled by epoxidation, that is, epoxy groups are inserted exactly at the position of double bonds. Ring opening of epoxy groups can also be achieved using alcohols or water in the presence of acid catalysts, with organic/inorganic acids, and by hydrogenation (Petrovic 2008).

Figure 16.13 schematically shows reactions of ring opening with alcohols, inorganic acids, and hydrogenation.

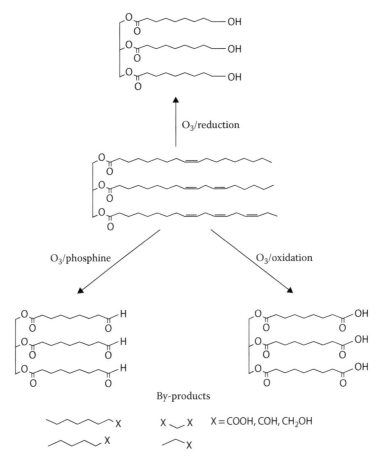

FIGURE 16.12 Reaction products of triglycerides with ozone followed by oxidation, reduction, or reductive amination. Epoxidized soybean oil, soy polyol, where Y = OCH$_3$, Cl, Br, or H. (From Petrovic, Z.S. *Polym. Rev.*, 48, 109–155, 2008. With permission. Copyright 2008 Taylor & Francis.)

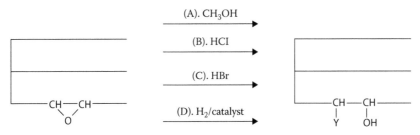

FIGURE 16.13 Formation of different kinds of polyols form epoxidized soybean oil. (From Petrovic, Z.S. *Polym. Rev.*, 48, 109–155, 2008. With permission. Copyright 2008 Taylor & Francis.)

The polyols obtained by ring opening using alcohols are liquid, while those obtained using HCl, HBr, or by hydrogenation are solid at room temperature. Except alcohols, all ring openers reacted quantitatively with epoxy groups.

16.4 BIO-BASED POLYURETHANES AND POLYESTERS

Vegetable oils are excellent raw materials for the development of new monomers and polymers, because they are economical and offer different degrees of unsaturation, which facilitates their easy conversion into polymers (Lligadas et al. 2007b; Sharma and Kundu 2008). Some of the most common examples of polymers produced from triglyceride oils include polyesters as well as PUs and are described in the following sections.

16.4.1 POLYURETHANES

During the last few years, PUs have become one of the most used polymers due to their wide range of applications, properties, and versatility. The versatility of PUs has allowed the synthesis of different materials such as foams, adhesives, coatings, sealants, and elastomers. These materials can be used in a number of fields such as in automobiles, footwear, in construction as insulators, and most recently in medical devices. However, due to environmental concerns, intense research efforts are now being focused on the development of PUs derived from renewable resources. These PUs also possess the potential to be used in a number of applications analogous to their synthetic counterparts. Palmeri oil, castor oil, vernonia oil, cardanol oil, and others have been frequently used to synthesize PU polyols with multiple functionality to replace the petrochemical-based polyols (Pourjavadi et al. 1998; Bhunia et al. 1998). In addition to this, epoxidized plant oils such as canola, midoleic sunflower, soybean, linseed, sunflower, and corn are also used to prepare polyols for PU synthesis. Figure 16.14 shows the schematic for synthesis of different types of polyols for PUs.

Thermoplastic PUs have also been prepared from the reaction of polyricinoleate diols with diphenylmethane diisocyanate (MDI) and butanediol. The polyricinoleate diols for this reaction were obtained from the polycondensation (transesterification) of methyl ricinoleate using diethylene glycol as an initiator (Xu et al. 2008).

Castor oil has long been used in the PU industry. Relatively, it is stable to hydrolysis due to its long fatty acid chain but sensitive to oxidation due to the presence of unsaturated fatty acid. Commercially, it can only be used in the coating and adhesive industries. Castor oil (see Figure 16.1), which is rich in ricinoleic acid, has been designated to be the first and almost the only vegetable oil used in the synthesis of PUs and in interpenetrating polymer networks. The rest of the vegetable oils have to be modified, introducing hydroxyl groups into their structure. These urethanes are generally prepared by the reaction of diisocyanates with hydroxyl-containing materials. The monomers play a significant role in determining the overall properties of the resulting PU. Depending on the requirements, both organic solvent-soluble PUs and water-soluble PUs have been prepared. Figure 16.15 shows the schematic synthesis of water-soluble PU prepared by the

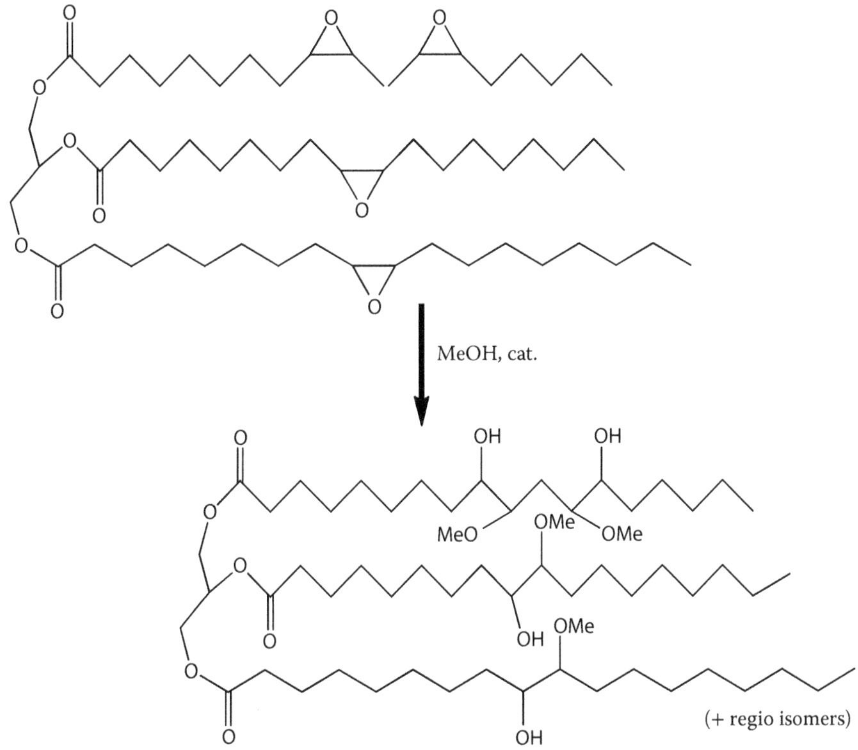

FIGURE 16.14 Polyols prepared for polyurethane synthesis.

reaction of glycerol, methylethyl ketone, and fatty acid, and addition of maleic anhydride (Guner et al. 2006).

PUs obtained using this method exhibited good physical and mechanical properties. Vegetable oil–based polyols with a range of hydroxyl numbers have also been prepared by hydroformylation, reduction, and partial esterification of the hydroxyl groups with formic acid. These polyols have been reacted with MDI to give PUs with different cross-link densities (Petrovic et al. 2008). Bio-based PUs containing tertiary amines have been prepared from castor and methoxylated and acrylated ESO-based polyols by the reaction of reacted N-methyl diethanol amine and MDI (Xia and Larock 2010a). It was observed that the treatment of the PU with acetic acid results in protonation of amine groups that gives rise to cationic hydrophilic sites, and subsequent dispersion into water results in the desired waterborne cationic PU (Xia and Larock 2010a).

Different vegetable oils such as sunflower, soybean, corn, linseed oils, and others have been epoxidized and reacted with boiling methanol with a catalyst to obtain polyols that were further used in the synthesis of PU. The differences in the properties of these PU networks have been found to be resulted primarily from different cross-link densities and less from the position of the hydroxyl functionalities. Linseed oil–derived PU has been found to exhibit the highest glass transition and tensile strength, and the lowest elongation at break, which can be related with the

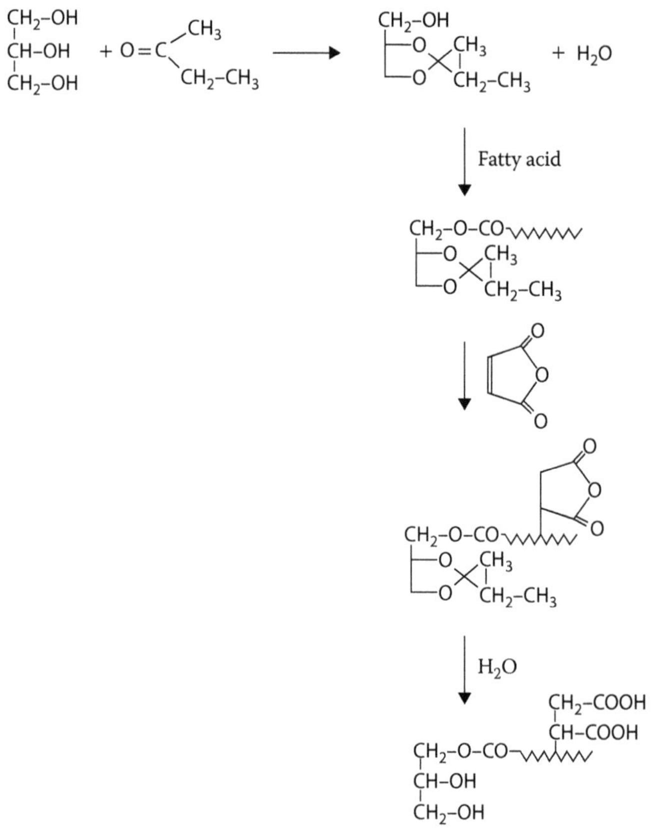

$$CH_2-OH$$
$$CH-OH \quad + \; O=C \overset{CH_3}{\underset{CH_2-CH_3}{}}$$
$$CH_2-OH$$
$$\longrightarrow$$
$$CH_2-OH$$
$$\begin{matrix} -O & CH_3 \\ -O & CH_2-CH_3 \end{matrix} \quad + \; H_2O$$

Fatty acid

$$CH_2-O-CO-\!\!\!\sim\!\!\!\sim\!\!\!\sim$$
$$\begin{matrix} -O & CH_3 \\ -O & CH_2-CH_3 \end{matrix}$$

$$CH_2-O-CO-\!\!\!\sim\!\!\!\sim\!\!\!\sim\!\!\!O$$
$$\begin{matrix} -O & CH_3 \\ -O & CH_2-CH_3 \end{matrix}$$

$$H_2O$$

$$CH_2-COOH$$
$$CH-COOH$$
$$CH_2-O-CO-\!\!\!\sim\!\!\!\sim\!\!\!\sim$$
$$CH-OH$$
$$CH_2-OH$$

FIGURE 16.15 Maleinization of triglyceride oil. (From Guner, F.S., Yagci, Y. and Erciyes, A.T. *Prog. Polym. Sci.*, 31, 633–670, 2006. With permission. Copyright 2006 Elsevier.)

high functionality of the obtained polyols (Zlatanic et al. 2004). The thermal stability of the PUs obtained using this method has been found to be better than that for polypropylene oxide–based PU both in nitrogen and air atmospheres (Javni et al. 2000). Regular and modified castor oil and ricinoleic acid were also used in the preparation of bio-based PUs (Petrovic 2008; Yeganeh and Mehdizadeh 2004). PUs were also prepared from soy-polyol by ring opening with methanol (Javni et al. 2003). The effect of isocyanate on the final properties of the PU was studied. The properties of the PU were strongly dependent on cross-linking density and the structure of isocyanate. PU prepared using aromatic triisocyanate demonstrated the highest density, glass transition, modulus, and tensile strength, whereas aliphatic isocyanates gave rubbery materials with the highest elongation at break and swelling atmospheres (Javni et al. 2003). Different types of nitrogen-containing monomers such as fatty amides, fatty amines, fatty imidazolines, and polymers (PUs) made from soybean oil–based triglyceride oils, fatty acid esters, and methyl esters of triglyceride oils have recently been reviewed (Biswas et al. 2008). Among these, fatty amines and amides have been found to be industrially significant and widely used.

16.4.2 POLYESTERS

Vegetable oil–based polyesters can also be successfully synthesized from different kinds of vegetable oils depending on their availability in a particular region. The prime routes used for the synthesis of vegetable oil–based polyesters include polycondensation of dicarboxylic acids (or anhydrides) and diols, condensation of hydroxyl acids, or ring-opening polymerization of lactones (Guner et al. 2006). Figure 16.16 shows the general schemes for polyester synthesis.

Soybean oil has been modified via anhydride functionalization of the double bonds and polymerized with the resulting monomers by polycondensation with low-molecular-weight polyols and long diols. The polyesters formed were resilient and soft rubbers at room temperature that could find application as adhesives, film formers, textile and paper sizes, and tackifiers (Eren et al. 2003). In the family of polyesters, alkyd resin is one of the oldest polymers prepared from triglyceride oils. The resin is prepared by the esterification of polyhydroxy alcohols with polybasic acids and fatty acids. Alkyd resins are often used because they offer several advantages, including economic viability, ease of application, and biological degradability (caused by the presence of oil and glycerol parts).

Figure 16.17 shows the synthesis of liquid crystalline (LC) polymers, which are extensively used in the plastic and fiber industries (Guner et al. 2006).

LC resins are prepared by three different grafting methods: (1) p-hydroxybenzoic acid (PHBA) to hydroxy-terminated alkyd resin; (2) PHBA to carboxy-terminated alkyd resin; and (3) PHBA to an excess SA-modified alkyd resin. The reaction is performed at room temperature in the presence of p-toluene sulfonic acid as catalyst and the resins exhibit low viscosity and good film properties. Vegetable oil–based polyesters have been recently obtained from ω-carboxyl fatty acids (Duwensee et al. 2010; Mifune et al. 2009) using a biocatalytic process that involves the use of immobilized *Candida antarctica* Lipase B (N435) as the biocatalyst. Unsaturated polyesters were prepared by the reaction of fatty acid with diols. These polyesters exhibit much lower melting points (23–40°C) than analogous saturated polyesters (88°C), together with high thermal stabilities (Yang et al. 2010b). Moreover, unlike polyesters from saturated diacids and diols, unsaturated polyesters can be modified or cross linked to develop curable coatings for medical applications.

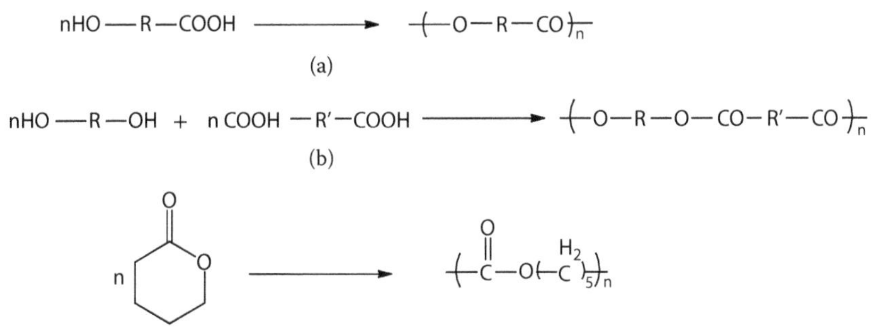

FIGURE 16.16 Polyester synthesis: (a) polycondensation of hydroxyl acid, (b) polycondensation of diacid, and (c) diol ring-opening polymerization of lactones.

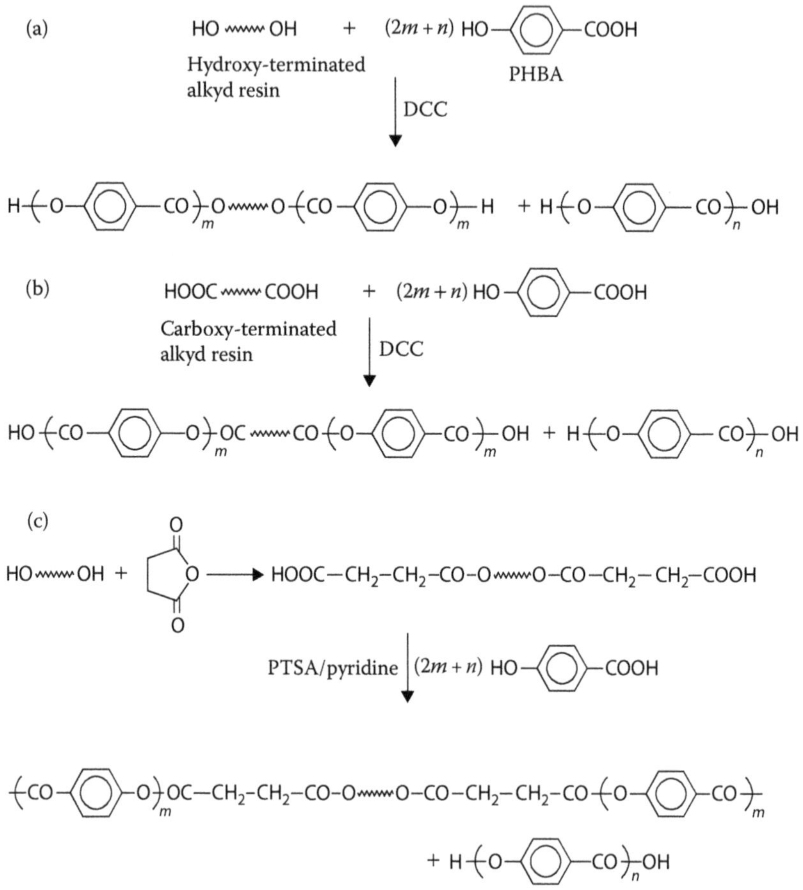

FIGURE 16.17 Preparation of liquid crystalline alkyd resin by three methods (a)(b)(c). (From Guner, F.S., Yagci, Y. and Erciyes, A.T. *Prog. Polym. Sci.*, 31, 633–670, 2006. With permission. Copyright 2006 Elsevier.)

16.5 GREEN COMPOSITES FROM VEGETABLE OILS

Polymer composites play an integral role in present day civilization and are the most important materials used in diverse applications such as in construction, aerospace, military, electronics, and medicine (Casado et al. 2009; Can et al. 2006b; Cipriani et al. 2010; Henna et al. 2008; Ragauskas et al. 2006; Raquez et al. 2010). However, their widespread use, often followed by immediate disposal, increases the amount of solid waste and ultimately results in environmental pollution, because most polymers produced from fossil-fuel feedstock are non-biodegradable. Recent recycling efforts are limited by economic constraints (cost effectiveness of collection, separation, and reprocessing), while energetic reclaim (combustion) raises ecological concerns. Vegetable oil–based polymers and composites exhibit a number of advantages compared with polymers from petroleum-based monomers (Casado et al. 2009; Cipriani et al. 2010; Henna et al. 2008; Meier et al. 2007; Ragauskas et al. 2006). They are biodegradable and often cheaper than synthetic polymers (Goud et al. 2007b). With the recent emergence of vegetable oil–based polymers, a number of composite materials were prepared based on materials from renewable resources using both synthetic and natural polymers as reinforcements (Haq et al. 2008; O'Donnell et al. 2004). Bio-based composites have been prepared from glass fiber–reinforced vegetable oil–based resins, whereas the polymer matrix was prepared by ROMP. Composite test specimens were prepared by cutting the short glass fibers chopped strand mat. The resins and the corresponding composites were subsequently characterized thermophysically and thermomechanically. The toughness of the composite samples was found to increase with increasing glass fiber content due to the ability of the fiber to inhibit crack propagation and fracture. It was also observed that glass fiber content significantly improved the tensile modulus of the resin from 28.7 to 168 MPa. Higher DCPD content yields had been found to yield materials with higher glass transition temperatures. Thermogravimetric analysis (TGA) of these resins revealed that the presence of glass fiber did not alter the degradation mechanism of the polymeric matrix. These bio-based composites utilized only a limited amount of a petroleum-based monomer, while employing substantial amounts of a renewable resource (Henna et al. 2007).

Polymer composites using organomodified, montmorillonite clay as reinforcement and corn oil-based cationic resins as the polymer matrix were also prepared (Lu and Larock 2007). In order to get the best results, the amount of montmorillonite clay was varied in the matrix and evaluated for different mechanical properties. It was observed that clay loadings between 2 and 3 wt% yield better properties as a result of an increase in cross-link density of the matrix (Lu and Larock 2007).

The effect of surface modification of glass fibers using silane coupling agent was also assessed and the functionalized glass fibers were used as reinforcement in vegetable oil–based composites (Cui and Kessler 2012). Vegetable oil–based polymer resin was prepared by the ROMP of a modified linseed oil-based monomer and DCPD. In this study, the effect of two types of silane coupling agents, norbornenylethyldimethylchlorosilane (MCS) and norbornenylethyltrichlorosilane, on glass fiber–reinforced composites was examined. Figure 16.18 shows the reactions between silane coupling agents and glass fiber.

The temperature dependence of storage modulus (E') for glass fiber/bio-based polymer composites with varied surface treatments is shown in Figure 16.19a and the loss factor, tan δ, is shown in Figure 16.19b as a function of temperature.

Dynamic mechanical analysis of composite panels made with untreated and silane-treated fibers reveals that significant improvements in interfacial adhesion had been achieved with silane functionalized fibers that were consistent with microbond testing results, indicating that covalent bonding between fiber and matrix can enhance the load transfer at the interphase and increase stiffness. Short beam shear tests and scanning electron microscopy (SEM) of fracture surfaces were also used to assess the interfacial adhesion of the composites (Figure 16.20).

These measurements also confirmed that coupling agents improve interfacial adhesion between the fibers and the matrix. The fracture surface morphology was also investigated by SEM to examine the microstructure and adhesion between fibers and matrix (Figure 16.21).

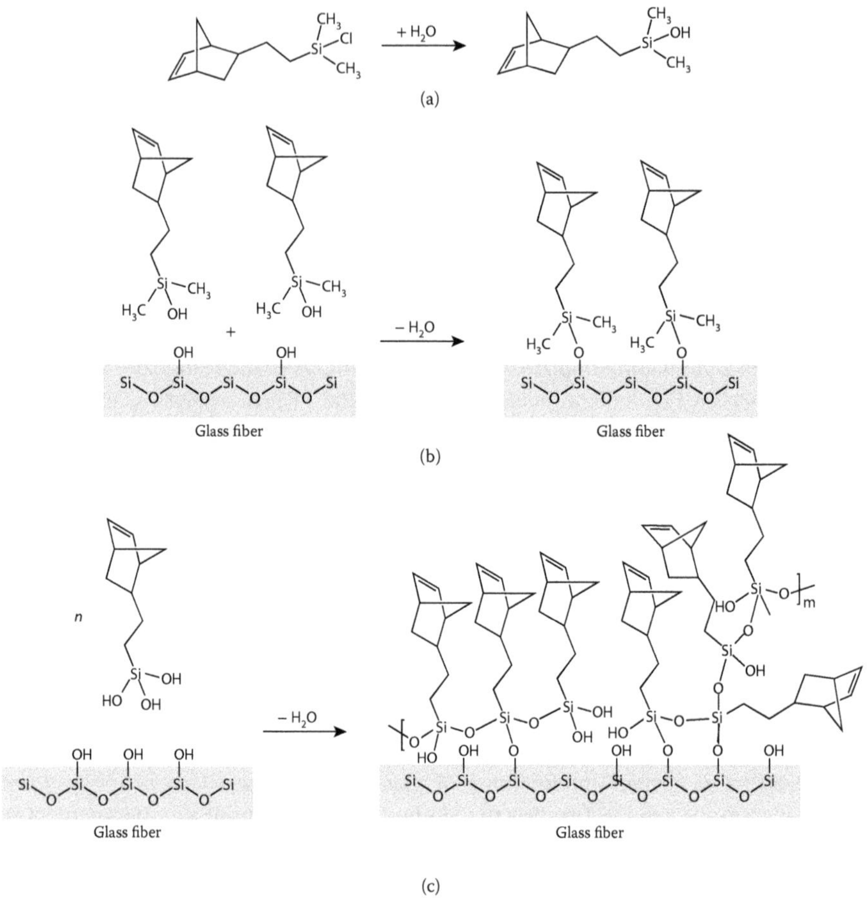

FIGURE 16.18 Silane reactions: (a) hydrolysis of MCS, (b) surface condensation reaction of monosilanol, and (c) surface condensation reaction of trisilanol. (From Cui, H. and Kessler, M.R. *Compos. Sci. Technol.*, 72, 1264–1272, 2012. With permission. Copyright 2012 Elsevier.)

In the case of composites made with functionalized fibers, a high number of short fiber fragments and significant amount of adhered resin on the fiber surfaces were observed, which demonstrated signs of good adhesion between fibers and matrix (Cui and Kessler 2012).

Mendesa et al. (2002) reported their studies on the synthesis of bio-based PU composites with graphite reinforcement. The polymer resin was based on castor oil, and composites with different graphite content were prepared. The composites with 60% (graphite, w/w) loading exhibited good mechanical and appropriated electric resistance and easy preparation when compared with other loading levels (Mendesa et al. 2002). Polymer composites were also prepared using spent germ, co-product of wet mill ethanol production, as the reinforcement and TUN-based resin as the matrix (Pfister et al. 2008). The resin consisted of a copolymer of TUN (50 wt%), BMA, and DVB and was cured in the presence of *t*-butyl peroxide. Spent germ reinforcement was used in particle sizes of different dimension. It was observed that the spent germ-reinforced

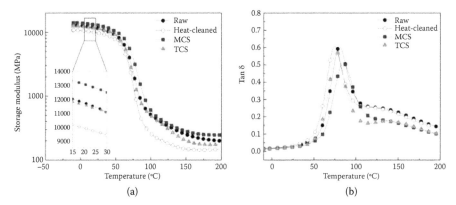

FIGURE 16.19   Storage modulus curves (a) and tan δ curves (b) for composites made with different type of fibers. (From Cui, H. and Kessler, M.R. *Compos. Sci. Technol.*, 72, 1264–1272, 2012. With permission. Copyright 2012 Elsevier.)

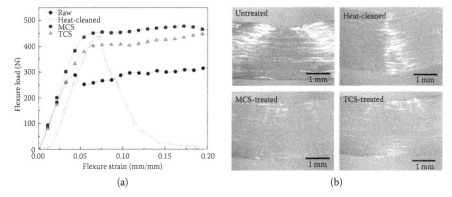

FIGURE 16.20   (a) Load–strain curves for short beam shear test. (b) Oleyl Methacrylate images for laminates after short beam shear test. (From Cui, H. and Kessler, M.R. *Compos. Sci. Technol.*, 72, 1264–1272, 2012. With permission. Copyright 2012 Elsevier.)

thermosets exhibited higher storage moduli than the pure resin and incorporation of shorter particles and higher DVB content resulted in even better properties. However, the filler loading beyond 40 wt% leads to agglomeration and formation of microvoids in the composite and decreases the overall mechanical properties (Pfister et al. 2008).

Among various oils used to prepare composites, soybean oil is the most commonly used oil, creating durable and strong composite materials (Khot et al. 2001). Soybean oils are biodegradable and available in bulk quantities. Natural fiber–reinforced composites using structural foam from ESO and flax mats, chicken feathers, and recycled paper were prepared (Dweib et al. 2004). In this case, a modified vacuum-assisted resin transfer molding process was used for the fabrication of the composites. The same research groups also developed green composites using recycled paper, sisal, jute twine, and jute fabric as reinforcement and ESO as a matrix for mainstream civil construction materials such as building materials and appliance housing (Hu et al. 2007). Biocomposites were also prepared using an unsaturated thermosetting resin (a mixture of acrylated ESO and ST) as the matrix and natural fibers such as flax as the potential

(a)   (b)

(c)   (d)

FIGURE 16.21 Scanning electron microscopic images of fracture surfaces for composites made with different type of fibers: (a) untreated, (b) heat cleaned, (c) MCS-treated, (d) TCS-treated. (From Cui, H. and Kessler, M.R. *Compos. Sci. Technol.,* 72, 1264–1272, 2012. With permission. Copyright 2012 Elsevier.)

reinforcement (Thielemans and Wool 2004). The effect of butyrated kraft lignin as renewable compatibilizer in the composites was assessed instead of traditional compatibilizer. The composites were prepared by the vacuum-assisted transfer molding process with varying amounts of butyrated kraft lignin dissolved in the unsaturated resin system. It was observed that the butyrated kraft lignin improved the interface between the resin and reinforcing flax fibers, which ultimately enhances the mechanical properties of the composites. For example, the flexural strength increased by 40% for a 5 wt% butyrated lignin addition (Thielemans and Wool 2004).

Different mechanical properties, such as flexural and impact strength, of composites prepared with modified soybean oil as the matrix and glass/flax as reinforcement were also studied (Morye and Wool 2005). The mechanical properties of symmetric and asymmetric composites depend on the glass/flax ratio and the arrangement of fibers in the composite. With appropriate fiber arrangement in the composites, glass fibers and flax fibers acted synergistically, providing improved flexural and impact performance. Other biocomposites were prepared using ESO as the matrix and Wollastonite mineral fimine as the reinforcing materials (Xu et al. 2002). These biocomposites exhibited strong viscoelastic solid properties, similar to those of composites based on synthetic rubbers. Quirino and Larock (2009) reported that the presence of bio-based fillers minimizes shrinkage of the polymer resin prepared from soybean oil. The biocomposites were prepared by using soybean resin as the matrix (prepared by the free-radical polymerization of conjugated soybean oil) and soy hull as reinforcement. The tensile and flexural properties of the thermosetting biocomposites were determined for various resin compositions. The effects of filler/resin ratio, particle size, and pressure during the cure process were evaluated. The properties of the composites were found to decrease when higher filler/resin ratios or larger particle sizes were used. This behavior was closely related to impregnation of the filler by the resin; whenever good dispersion of the filler in the matrix was compromised, the mechanical properties of the composites were negatively affected. Morphological investigations also demonstrated that there were minimal microcracks in the soybean hull composites with smaller soybean hull size (Quirino and Larock 2009). Green composites from CLO-based resin and wheat straw as the reinforcement were also prepared (Pfister and Larock 2010). The effect of different parameters, such as loading level of wheat straw, matrix cross-link density, size of the wheat straw fibers, incorporation of a compatibilizer (maleic anhydride), and molding pressure on the structure, water absorption, and thermal and mechanical properties of the composites was studied. It was observed that increasing loading levels of wheat straw, cross-link density of the matrix, and molding pressure led to improved thermal and mechanical properties of the respective composites. The addition of maleic anhydride as a compatibilizer significantly improved the mechanical properties of the composites.

Dried, ground sugarcane bagasse fillers were used for composites prepared by free-radical polymerization of regular or modified vegetable oils with DVB and *n*-butyl methacrylate (Quirino and Larock 2012). The effect of different parameters, such as cure times and temperatures, was investigated to determine the optimum cure sequence for the composite materials while the effects of varying filler loadings and resin compositions was assessed by tensile tests, dynamic mechanical analysis, TGA, and Soxhlet extraction. The initial washing and drying of the filler influenced the filler's interaction and impact on the final properties of the composites (Quirino and Larock 2012). PU matrix-based composites were prepared using natural sisal fibers as reinforcement and castor oil as source materials for polymer synthesis (Milanese et al. 2012). The polymer matrix was prepared by the reaction of the hydroxyl groups of castor oil triglycerides with 4,4′-methylene diphenyl diisocyanate (MDI). The composites were prepared by compression molding at room temperature using woven sisal fiber as reinforcement. The woven sisal was used both with and without thermal treatment (at 60°C for 72 hours) prior to the molding process. The moisture content in the sisal fibers influenced the flexural behavior of the composites. The effect of moisture content on the properties of polymer composites prepared using TUN-based PU as the matrix and wood floor and microcellulose as reinforcement was studied (Mosiewicki et al. 2012). It was reported that the mechanical properties of the composites such as tensile strength were strongly affected by moisture.

The biodegradation of PU matrix-based composites was systematically studied (Aranguren et al. 2012). The PU polymer was synthesized from TUN, and sawdust was used as reinforcement. To evaluate the biodegradability of the composite, soil and vermiculite aerobic degradation media were used for a test duration of 383 days. Figures 16.22 and 16.23 show the thickness and surface micrographs of polymer composite samples before and after degradation.

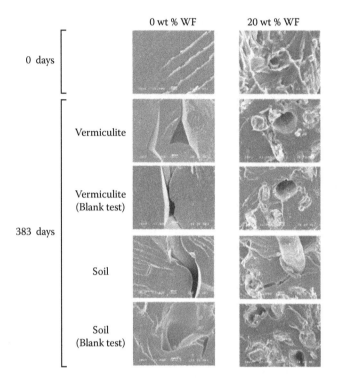

FIGURE 16.22 Thickness scanning electron microscopy micrographs of unfilled polyurethane and the composite with 20 wt% wood flour (WF) before and after exposure to soil and vermiculite media. (From Aranguren, M.I., Gonzalez, J.F. and Mosiewicki, M.A. *Polym. Test.*, 31, 7–15, 2012. With permission. Copyright 2012 Elsevier.)

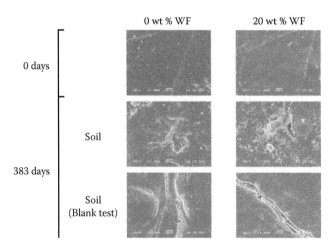

FIGURE 16.23 Surface scanning electron microscopy micrographs of unfilled polyurethane and the composite with 20 wt% wood flour (WF) before and after exposure to soil and vermiculite media. (From Aranguren, M.I., Gonzalez, J.F. and Mosiewicki, M.A. *Polym. Test.*, 31, 7–15, 2012. With permission. Copyright 2012 Elsevier.)

The degraded samples (thickness and surface) exposed morphological changes in the composite samples compared with their unexposed counterparts. Cracks appeared in the parent PU matrix as a consequence of degradation in soil and vermiculite. With increase in exposure time, more cracks, small holes, and fragmentation of films occurred, while the cracks also became deeper with time. The authors assumed that hydrolytic degradation was the most important mechanism of deterioration in all cases. In addition, the contact angle of water decreased with exposure to soil and vermiculite media, indicating changes in the surface of the material that increased its polarity.

16.6 CONCLUSION

This chapter provided a brief overview of the preparation of different polymers and composites from vegetable oils. The discussed examples highlight the potential of vegetable oils as novel renewable polymeric materials. Increasing implementation of vegetable oils in a variety of industrial applications alleviates environmental and energy concerns and promotes the principles of green chemistry. Potential applications for these vegetable oil–based materials include paneling, coatings, exterior body panels for automobiles, and furniture parts as well as noise and vibration damping materials.

REFERENCES

Abdelmouleh, M., Boufi, S., Belgacem, M.N., Duarte, A.P., Ben Salah, A. and Gandini, A. 2004. Modification of cellulosic fibers with functionalized silanes: Development of surface properties. *International Journal of Adhesion and Adhesives* 24(1): 43–54.
Alawar, A., Hamed, A.M. and Al-Kaabi, K. 2009. Characterization of treated date palm tree fiber as composite reinforcement. *Composites Part B* 40(7): 601–606.
Andjelkovic, D.D. and Larock, R.C. 2006. Novel rubbers from cationic copolymerization of soybean oils and dicyclopentadiene. 1. Synthesis and characterization. *Biomacromolecules* 7: 927–936.
Andjelkovic, D.D., Lu, Y., Kessler, M.R. and Larock, R.C. 2009. Novel rubbers from the cationic copolymerization of soybean oils and dicyclopentadiene. 2. Mechanical and damping properties. *Macromolecular Materials and Engineering* 294: 472–483.
Andjelkovic, D.D., Valverde, M., Henna, P., Li, F. and Larock, R.C. 2005. Novel thermosets prepared by cationic copolymerization of various vegetable oils—Synthesis and their structure-property relationships. *Polymer* 46: 9674–9685.
Aranguren, M.I., Gonzalez, J.F. and Mosiewicki, M.A. 2012. Biodegradation of a vegetable oil based polyurethane and wood flour composites. *Polymer Testing* 31: 7–15.
Arno, P. 1989. *Concise Encyclopedia of Wood and Wood-Based Materials*, 1st ed. Elmsford, NY: Pergamon Press, 271–273.
Averous, L. 2004. Biodegradable multiphase systems based on plasticized starch: A review. *Journal of Macromolecular Science Part C Polymer Reviews* 44(3): 231–274.
Baboi, M., Srinivasan, G., Jane, J.L. and Grewell, D. 2007. Improvement of the mechanical properties of soy protein isolate based plastics through formulation and processing. *International Polymer Processing* 5: 489–496.
Badrinarayanan, P., Lu, Y., Larock, R.C. and Kessler, M.R. 2009. Cure characterization of soybean oil–styrene–divinylbenzene thermosetting copolymers. *Journal of Applied Polymer Science* 113: 1042–1049.

Berh, A., Naendrup, F. and Obst, D. 2002. The synthesis of silicon oleochemicals by hydro-silylation of unsaturated fatty acid derivatives. *European Journal of Lipid Science and Technology* 104: 161–166.

Bharathi, N.P., Khan, N.U., Shreaz, S. and Hashmi, A.A. 2009. Seed oil based zinc bioactive polymers: synthesis, characterization, and biological studies. *Journal of Inorganic and Organometallic Polymers and Materials* 19: 558–565.

Bhunia, H.P., Jana, R.N., Basak, A., Lenka, S. and Nando, G.B. 1998. Synthesis of polyure-thane from Cashew Nut Shell Liquid (CNSL), a renewable resource. *Journal of Applied Polymer Science* 36: 391–400.

Biermann, U., Metzger, J.O. and Meier, M.A.R. 2010. Acyclic triene metathesis oligo- and polymerization of high oleic sun flower oil. *Macromolecular Chemistry and Physics* 211: 854–862.

Biswas, A., Sharma, B.K., Willett, J.L., Erhan, S.Z. and Cheng, H.N. 2008. Soybean oil as a renewable feedstock for nitrogen-containing derivatives. *Energy and Environmental Science* 1: 639–644.

Bonnaillie, L.M. and Wool, R.P. 2007. Thermosetting foam with a high bio-based content from acrylated epoxidized soybean oil and carbon dioxide. *Journal of Applied Polymer Science* 105: 1042–1052.

Bozell, J.J. 2008. Feedstocks for the future—Biorefinery production of chemicals from renew-able carbon. *Clean-Soil Air Water* 36: 641–647.

Bunker, S.P. and Wool, R.P. 2002. Synthesis and characterization of monomers and poly-mers for adhesives from methyl oleate. *Journal of Polymer Science, Part A: Polymer Chemistry* 40: 451–458.

Cai, T., Yang, G., Zhang, H., Shao, H. and Hu, X. 2012. A new process for dissolution of cel-lulose in ionic liquids. *Polymer Engineering and Science* 52: 1708–1714.

Cakmakli, B., Hazer, B., Tekin, I.O. and Comert, F.B. 2005. Synthesis and characterization of polymeric soybean oil-*g*-methyl methacrylate (and *n*-butyl methacrylate) graft copoly-mers: Biocompatibility and bacterial adhesion. *Biomacromolecules* 6: 1750–1758.

Cakmakli, B., Hazer, B., Tekin, I.O., Kizgut, S., Koksal, M. and Menceloglu, Y. 2004. Synthesis and characterization of polymeric linseed oil grafted methyl methacrylate or styrene. *Macromolecular Bioscience* 4: 649–655.

Can, E., Wool, R.P. and Kusefoglu, S. 2006a. Soybean- and castor-oil-based thermosetting polymers: Mechanical properties. *Journal of Applied Polymer Science* 102: 1497–1504.

Can, E., Wool, R.P. and Kusefoglu, S. 2006b. Soybean and castor oil based monomers: Synthesis and copolymerization with styrene. *Journal of Applied Polymer Science* 102: 2433–2447.

Casado, U., Marcovich, N.E., Anguren, M.I. and Mosiewicki, M.A. 2009. High strength com-posite based on tung oil polyurethane and wood flour: Effect of filler concentration on the mechanical properties. *Journal of Polymer Engineering and Science* 49: 713–721.

Cheng, G., Varanasi, P., Li, C., Liu, H., Melnichenko, Y.B., Simmons, B.A., Kent, M.S. and Singh, S. 2011. Transition of cellulose crystalline structure and surface morphology of biomass as a function of ionic liquid pretreatment and its relation to enzymatic hydroly-sis. *Biomacromolecules* 12: 933–941.

Cipriani, G., Salvini, A., Baglioni, P. and Bucciarelli, E. 2010. Cellulose as a renewable resource for the synthesis of wood consolidants. *Journal of Applied Polymer Science* 118: 2939–2950.

Clark, J.H., Budarin, V.F. and Deswarte, F.I. 2006. Green chemistry and the biorefinery: A partnership for a sustainable future. *Green Chemistry* 8: 853–860.

Cui, H. and Kessler, M.R. 2012. Glass fiber reinforced ROMP-based bio-renewable poly-mers: Enhancement of the interface with silane coupling agents. *Composites Science and Technology* 72: 1264–1272.

De Espinosa, L.M., Meier, M.A.R., Ronda, J.C., Galia, M. and Cadiz, V. 2010. Phosphorous-containing renewable polyester-polyols via ADMET polymerization: Synthesis, functionalization, and radical crosslinking. *Journal of Polymer Science, Part A: Polymer Chemistry* 48: 1649–1660.

De Espinosa, L.M., Ronda, J.C., Galia, M. and Cadiz, V. 2009a. A new route to acrylate oil: Crosslinking and properties of acrylate triglycerides from high oleic sunflower oil. *Journal of Polymer Science, Part A: Polymer Chemistry* 47: 1159–1167.

De Espinosa, L.M., Ronda, J.C., Galia, M. and Cadiz, V. 2009b. A straightforward strategy for the efficient synthesis of acrylate and phosphine oxide-containing vegetable oils and their crosslinked materials. *Journal of Polymer Science, Part A: Polymer Chemistry* 47: 4051–4063.

Demirbas, A. 2009. Biofuels securing the planet's future energy needs. *Energy Conversion and Management* 50: 2239–2249.

Dinda, S., Patwardhan, A.V., Goud, V.V. and Pradhan, N.C. 2008. Epoxidation of cottonseed oil by aqueous hydrogen peroxide catalysed by liquid inorganic acids. *Bioresource Technology* 99: 3737–3744.

Dutta, N., Karak, N. and Dolui, S.K. 2004. Synthesis and characterization of polyester resins based on nahar seed oil. *Progress in Organic Coatings* 49: 146–152.

Duwensee, J., Wenda, S., Ruth, W. and Kragl, U. 2010. Lipase-catalyzed polycondensation in water: A new approach for polyester synthesis. *Organic Process Research and Development* 14: 48–57.

Dwan'Isa, J.P.L., Mohanty, A.K., Misra, M., Drzal, L.T. and Kazemizadeh, M. 2003. Novel biobased polyurethanes synthesized from soybean phosphate ester polyols: Thermomechanical properties evaluations. *Journal of Polymers and the Environment* 11: 161–168.

Dweib, M.A., Hu, B., O'Donnell, A., Shenton, H.W. and Wool, R.P. 2004. All natural composite sandwich beams for structural applications. *Composite Structures* 63: 147–57.

Eren, T., Kusefoglu, S.H. and Wool, R. 2003. Polymerization of maleic anhydride-modified plant oils with polyols. *Journal of Applied Polymer Science* 90: 197–202.

Firdaus, F.E. 2012. Role of silicone on molded flexible polyurethane from soy oil. *International Journal of Chemistry* 4: 45–49.

Gandini, A. 2008. Polymers from renewable resources: A challenge for the future of macromolecular materials. *Macromolecules* 41: 9491–9504.

Goud, V.V., Patwardhan, A.V. and Pradhan, N.C. 2006. Epoxidation of karanja (*Pongamia glabra*) oil by H_2O_2. *Journal of the American Oil Chemists' Society* 83: 635–640.

Goud, V.V., Patwardhan, A.V. and Pradhan, N.C. 2007a. Studies on the epoxidation of mahua oil (*Madhumica indica*) by hydrogen peroxide. *Bioresource Technology* 97: 1365–1371.

Goud, V.V., Patwardhan, A.V. and Pradhan, N.C. 2007b. Kinetics of epoxidation of jatropha oil with peroxyacetic and peroxyformic acid catalysed by acidic ion exchange resin. *Chemical Engineering Science* 62: 4065–4076.

Goud, V.V., Patwardhan, A.V., Dinda, S. and Pradhan, N.C. 2007c. Epoxidation of karanja oil with peroxyacetic and peroxyformic acid catalysed by acidic ion exchange resin. *European Journal of Lipid Science and Technology* 109: 575–584.

Guner, F.S., Yagci, Y. and Erciyes, A.T. 2006. Polymers from triglyceride oils. *Progress in Polymer Science* 31: 633–670.

Haq, M., Burgueno, R., Mohanty, A.K. and Misra, M. 2008. Hybrid bio-based composites from blends of unsaturated polyester and soybean oil reinforced with nanoclay and natural fibers. *Composite Science and Technology* 68: 3344–3351.

Henna, P.H. and Larock, R.C. 2007. Rubbery thermosets by ring-opening metathesis polymerization of a functionalized castor oil and cyclooctene. *Macromolecular Materials and Engineering* 292: 1201–1209.

Henna, P.H., Andjelkovic, D.D., Kundu, P.P. and Larock, R.C. 2007. Biobased thermosets from the free-radical copolymerization of conjugated linseed oil. *Journal of Applied Polymer Science* 104: 979–985.

Henna, P.H., Kessler, M.R. and Larock, R.C. 2008. Fabrication and properties of vegetable-oil-based glass fiber composites by ring-opening metathesis polymerization. *Macromolecular Materials and Engineering* 293: 979–990.

Hu, B., Dweib, M.A., Wool, R.P. and Shenton, H.W. 2007. Bio-based composite roof for residential construction. *Journal of Architectural Engineering* 13: 136–143.

Huber, G.W., Iborra, S. and Corma, A. 2006. Synthesis of transportation fuels from biomass: chemistry, catalysis, and engineering. *Chemical Reviews* 106: 4044–4098.

Huber, T., Pang, S. and Staiger, M.P. 2012. All-cellulose composite laminates. *Composites Part A: Applied Science and Manufacturing* 43: 1738–1745.

Javni, I., Petrovic, Z.S., Guo, A. and Fuller, R. 2000. Thermal stability of polyurethanes based on vegetable oils. *Journal of Applied Polymer Science* 77: 1723–1734.

Javni, I., Zhang, W. and Petrovic, Z.S. 2003. Effect of different isocyanates on the properties of soy-based polyurethanes. *Journal of Applied Polymer Science* 88: 2912–2916.

Jin, F.L. and Park, S.J. 2008. Thermomechanical behavior of epoxy resins modified with epoxidized vegetable oils. *Polymer International* 57: 577–583.

John, J., Bhattacharya, M. and Turner, R.B. 2002. Characterization of polyurethane foams from soybean oil. *Journal of Applied Polymer Science* 86: 3097–3107.

Khot, S.N., Lascala, J.J., Can, E., Morye, S.S., Williams, G.I., Palmese, G.R., Kusefoglu, S.H. and Wool, R.P. 2001. Development and application of triglyceride-based polymers and composites. *Journal of Applied Polymer Science* 82: 703–723.

Kundu, P.P. and Larock, R.C. 2005. Novel conjugated linseed oil–styrene–divinylbenzene copolymers prepared by thermal polymerization. 1. Effect of monomer concentration on the structure and properties. *Biomacromolecules* 6: 797–806.

Leitgeb, A., Wappel, J. and Slugovc, C. 2010. The ROMP toolbox upgraded. *Polymer* 51: 2927–2946.

Li, F., Hanson, M.V. and Larock, R.C. 2001. Soybean oil–divinylbenzene thermosetting polymers: Synthesis, structure, properties and their relationships. *Polymer* 42: 1567–1579.

Li, F.K. and Larock, R.C. 2001. New soybean oil–styrene–divinylbenzene thermosetting copolymers. I. Synthesis and characterization. *Journal of Applied Polymer Science* 80: 658–670.

Li, F.K. and Larock, R.C. 2003. Synthesis, structure and properties of new tung oil–styrene–divinylbenzene copolymers prepared by thermal polymerization. *Biomacromolecules* 4: 1018–1025.

Lligadas, G., Callau, L., Ronda, J.C., Galià, M. and Cádiz, V. 2005. Novel organic-inorganic hybrid materials from renewable resources: Hydrosilation of fatty acid derivatives. *Journal of Polymer Science, Part A: Polymer Chemistry* 43: 6259–6307.

Lligadas, G., Ronda, J.C., Galia, M. and Cadiz, V. 2006a. Development of novel phosphorus-containing epoxy resins from renewable resources. *Journal of Polymer Science, Part A: Polymer Chemistry* 44: 6717–6727.

Lligadas, G., Ronda, J.C., Galià, M. and Cádiz, V. 2006b. Novel silicon containing polyurethanes from vegetable oils as renewable resources. Synthesis and properties. *Biomacromolecules* 7: 2420–2426.

Lligadas, G., Ronda, J.C., Galià, M. and Cádiz, V. 2007a. Polyurethane networks from fatty-acid-based aromatic triols: Synthesis and characterization. *Biomacromolecules* 8: 1858–1864.

Lligadas, G., Ronda, J.C., Galià, M. and Cádiz, V. 2007b. Poly (ether urethane) networks from renewable resources as candidate biomaterials: synthesis and characterization. *Biomacromolecules* 8: 686–692.

Lligadas, G., Ronda, J.C., Galià, M. and Cádiz, V. 2010. Plant oils as platform chemicals for polyurethane synthesis: current state-of-the-art. *Biomacromolecules* 11: 2825–2835.

Lu, Y. and Larock, R.C. 2007. Bio-based nanocomposites from corn oil and functionalized organoclay prepared by cationic polymerization. *Macromolecular Materials and Engineering* 292: 863–872.

Lu, Y. and Larock, R.C. 2009. Novel polymeric materials from vegetable oils and vinyl monomers: preparation, properties and applications. *ChemSusChem* 2: 136–147.

Meier, M.A.R., Metzger, J.O. and Schubert, U.S. 2007. Plant oil renewable resources as green alternatives in polymer science. *Chemical Society Reviews* 36: 1788–1802.

Mendesa, R.K., Claro-Netob, S. and Cavalheiroa, E.T.G. 2002. Evaluation of a new rigid carbon–castor oil polyurethane composite as an electrode material. *Talanta* 57: 909–917.

Mifune, J., Grage, K. and Rehm, B.H.A. 2009. Production of functionalized biopolyester granules by recombinant lactococcus lactis. *Applied and Environmental Microbiology* 75: 4668–4675.

Milanese, A.C., Cioffi, M.O.H. and Voorwald, H.J.C. 2012. Flexural behavior of sisal/castor oil-based polyurethane and sisal/phenolic composites. *Materials Research-IBERO-American Journal of Materials* 15: 191–197.

Morye, S.S. and Wool, R.P. 2005. Mechanical properties of glass/flax hybrid composites based on a novel modified soybean oil matrix material. *Polymer Composites* 26: 407–416.

Mosiewicki, M.A., Casado, U., Marcovich, N.E. and Aranguren, M.I. 2012. Moisture dependence of the properties of composites made from tung oil based polyurethane and wood flour. *Journal of Polymer Research* 19: Article Number: 9776.

Mungroo, R., Pradhan, N.C., Goud, V.V. and Dalai, A.K. 2008. Epoxidation of canola oil with hydrogen peroxide catalyzed by acidic ion exchange resin. *Journal of the American Chemical Society* 85: 887–896.

Mutlu, H. and Meier, M.A.R. 2009. Unsaturated PA X,20 from renewable resources via metathesis and catalytic amidation. *Macromolecular Chemistry and Physics* 210: 1019–1025.

Narine, S.S., Kong, X., Bouzidi, L. and Sporns, P. 2007a. Physical properties of polyurethanes produced from polyols from seed oils. 1. Elastomers. *Journal of the American Oil Chemists' Society* 84: 55–63.

Narine, S.S., Kong, X., Bouzidi, L. and Sporns, P. 2007b. Physical properties of polyurethanes produced from polyols from seed oils. 2. Foams. *Journal of the American Oil Chemists' Society* 84: 65–72.

O'Donnell, A., Dweib, M.A. and Wool, R.P. 2004. Natural fiber composites with plant oil-based resin. *Composites Science and Technology* 64: 1135–1145.

Ouajai, S. and Shanks, R.A. 2009. Preparation, structure and mechanical properties of all-hemp cellulose biocomposites. *Composites Science and Technology* 69: 2119–2126.

Park, S.J., Jin, F.L. and Lee, J.R. 2004. Synthesis and thermal properties of epoxidized vegetable oil. *Macromolecular Rapid Communications* 25: 724–727.

Petrovic, A.S., Zlatanic, A., Lava, C.C. and Sinadinovic-Fiser, S. 2002. Epoxidation of soybean oil in toluene with peroxoacetic acid peroxoformic acids—Kinetics and side reactions. *European Journal of Lipid Science and Technology* 4: 293–299.

Petrovic, Z.S. 2008. Polyurethanes from vegetable oils. *Polymer Reviews* 48: 109–155.

Petrovic, Z.S., Guo, A., Javni, L., Cvetkovic, I. and Hong, D.P. 2008. Polyurethane networks from polyols obtained by hydroformylation of soybean oil. *Polymer International* 57: 275–281.

Pfister, D.P., Baker, J.R., Henna, P.H., Lu, Y. and Larock, R.C. 2008. Preparation and properties of tung oil-based composites using spent germ as a natural filler. *Journal of Applied Polymer Science* 108: 3618–3625.

Pfister, D.P. and Larock, R.C. 2010. Green composites from a conjugated linseed oil-based resin and wheat straw. *Composites Part A: Applied Science and Manufacturing* 41: 1279–1288.

Pourjavadi, A., Rezai, N. and Zohuriaan-M, M.J. 1998. A renewable polyurethane: Synthesis and characterization of the Interpenetrating Networks (IPNs) from cardanol oil. *Journal of Applied Polymer Science* 68: 173–183.

Quirino, R.L. and Larock, R.C. 2009. Synthesis and properties of soy hull-reinforced bio-composites from conjugated soybean oil. *Journal of Applied Polymer Science* 112: 2033–2043.

Quirino, R.L. and Larock, R.C. 2012. Sugarcane bagasse composites from composites vegetable oils. *Journal of Applied Polymer Science* 126: 860–869.

Ragauskas, A.J., Williams, C.K., Davison, B.H., Britovsek, G., Cairney, J., Eckert, C.A., Frederick, W.J. Jr. et al. 2006. The path forward for biofuels and biomaterials. *Science* 27: 484–489.

Raquez, J-.M., Deléglise, M., Lacrampe, M-.F. and Krawczak, P. 2010. Thermosetting (bio) materials derived from renewable resources: A critical review. *Progress in Polymer Science* 35: 487–509.

Refvik, M.D., Larock, R.C. and Tian, Q. 1999. Ruthenium-catalyzed metathesis of vegetable oils. *Journal of the American Oil Chemists' Society* 76: 93–98.

Rybak, A. and Meier, M.A.R. 2008. Acyclic diene metathesis with a monomer from renewable resources: Control of molecular weight and one-step preparation of block copolymers. *ChemSusChem* 1: 542–547.

Sharma, V. and Kundu, P.P. 2006. Addition polymers from natural oils. A review. *Progress in Polymer Science* 31: 983–1008.

Sharma, V. and Kundu, P.P. 2008. Condensation polymers from natural oils. *Progress in Polymer Science* 33: 1199–1215.

Stoesser, S.M. and Gabel, A.R. 1940. US Patent No. 2,190,906.

Tallman, K.A., Roschek, B. and Porter, N.A. 2004. Factors influencing the autoxidation of fatty acids: Eeffect of olefin geometry of the nonconjugated diene. *Journal of the American Chemical Society* 126: 9240–9247.

Thielemans, W. and Wool, R.P. 2004. Butyrated kraft lignin as compatibilizing agent for natural fiber reinforced thermoset composites. *Composites: Part A* 35: 327–338.

Tian, Q.P. and Larock, R.C. 2002. Model studies and the ADMET polymerization of soybean oil. *Journal of the American Oil Chemists' Society* 79: 479–488.

Turunc, O. and Meier, M.A.R. 2010. Fatty acid derived monomers and related polymers via thiol–ene(click) additions. *Macromolecular Rapid Communications* 31: 1822–1826.

Turunc, O. and Meier, M.A.R. 2013. The thiol–ene(click) reaction for the synthesis of plant oil derived polymers. *European Journal of Lipid Science and Technology* 115: 41–54.

Valverde, M., Andjelkovic, D., Kundu, P.P. and Larock, R.C. 2008. Conjugated low-saturation soybean oil thermosets: Free-radical copolymerization with dicyclopentadiene and divinylbenzene. *Journal of Applied Polymer Science* 107: 423–430.

Ververis, C., Georghiou, K., Christodoulakis, N., Santas, P. and Santas, R. 2004. Fiber dimensions, lignin and cellulose content of various plant materials and their suitability for paper production. *Industrial Crops and Products* 19: 245–254.

Vlcek, T. and Petrovic, Z.S. 2006. Optimization of the chemenzymatic epoxidation of soybean oil'. *Journal of the American Oil Chemists' Society* 83: 247–252.

Williams, C.K. and Hillmeyer, M.A. 2008. Polymers from renewable resources: A perspective for a special issue of polymer reviews. *Polymer Reviews* 48: 1–10.

Xia, Y. and Larock, R.C. 2010a. Castor oil-based thermosets with varied crosslink densities prepared by ring-opening metathesis polymerization (ROMP). *Polymer* 51: 2508–2514.

Xia, Y. and Larock, R.C. 2010b. Ring-opening metathesis polymerization (ROMP) of norbornenyl-functionalized fatty alcohols. *Polymer* 51: 53–61.

Xu, J., Liu, Z., Erhan, S.Z. and Carriere, C.J. 2002. A potential biodegradable rubber-viscoelastic properties of a soybean oil-based composite. *Journal of the American Oil Chemists' Society* 79: 593–596.

Xu, Y.J., Petrovic, Z., Das, S. and Wilkes, G.L. 2008. Morphology and properties of thermoplastic polyurethanes with dangling chains in ricinoleate-based soft segments. *Polymer*
49: 4248–4258.

Yang, Q., Lue, A. and Zhang, L. 2010a. Reinforcement of ramie fibers on regenerated cellulose films. *Composites Science and Technology* 70: 2319–2324.

Yang, Y.X., Lu, W.H., Zhang, X.Y., Xie, W.C., Cai, M.M. and Gross, R.A. 2010b. Two-step
biocatalytic route to biobased functional polyesters from omega-carboxy fatty acids and
diols. *Biomacromolecules* 11: 259–268.

Yeganeh, H. and Mehdizadeh, M.R. 2004. Synthesis and properties of isocyanate curable
millable polyurethane elastomers based on castor oil as a renewable resource polyol.
European Polymer Journal 40: 1233–1238.

Zain, M.F.M., Islam, M.N., Mahmud, F. and Jamil, M. 2011. Production of rice husk ash for
use in concrete as a supplementary cementious material. *Construction and Building
Materials* 25: 798–805.

Zhang, H., Pang, H., Shi, J., Fu, T. and Liao, B. 2012. Investigation of liquefied wood residues
based on cellulose, hemicellulose, and lignin. *Journal of Applied Polymer Science* 123:
850–856.

Zlatanic, A., Lava, C., Zhang, W. and Petrovic, Z.S. 2004. Effect of structure on properties of
polyols and polyurethanes based on different vegetable oils. *Journal of Polymer Science
Part B-Polymer Physics* 42: 809–819.

Index

For Product Safety Concerns and Information please contact our EU
representative GPSR@taylorandfrancis.com
Taylor & Francis Verlag GmbH, Kaufingerstraße 24, 80331 München, Germany

www.ingramcontent.com/pod-product-compliance
Ingram Content Group UK Ltd.
Pitfield, Milton Keynes, MK11 3LW, UK
UKHW052031210425
457613UK00032BA/1030